BIOCHEMISTRY OF LIPIDS

BIOCHEMISTRY OF LIPIDS

By

Dr. S.K. Prasad

School of Studies of Zoology & Biotechnology

Vikram University

Ujjain

DISCOVERY PUBLISHING HOUSE PVT. LTD.

NEW DELHI-110 002

Published by:
Tilak Wasan
DISCOVERY PUBLISHING HOUSE PVT. LTD.
4383/4B, Ansari Road, Darya Ganj
New Delhi-110 002 (India)
Phone : +91-11-23279245, 43596064-65
Fax : +91-11-23253475
E-mail : discoverypublishinghouse@gmail.com
namitwasan9@gmail.com
sales@discoverypublishinggroup.com
web : www.discoverypublishinggroup.com

***Reprinted:* 2019**

***First Edition:* 2010**

ISBN: 978-81-8356-612-4

Biochemistry of Lipids

Printed at:
Infinity Imaging Systems
Delhi

Preface

Biochemistry today has made spectacular progress in unraveling the mysteries of animate nature. This progress has allowed us to gain deeper insight into the principles of vital activity and has to a very significant extent stimulated the development of applied disciplines, especially medicine. The present title *"Biochemistry of Lipids"* is intended for those who wish to understand living organisms, especially man. Biochemistry is essential for this purpose, but it would be almost impossible for a student to survey on his own the massive body of existing knowledge, constantly augmented by a remarkable torrent of brilliat discoveries. The purpose of the book, then is to organize our knowledge into something that can be comprehended in a relatively short time and still convey a reasonable complete picture of the chemical structure and function of man. Readability without sacrifice of coverage has been a prime goal. An important device in gaining that goal is to keep attention constantly focused on function, with repeated use of rationalization to show that the chemical facts are not isolated, but part of a whole.

Biochemistry has two major goals as a fundamental science, it treats the vital functions from the standpoint of physical chemistry, and as an applied discipline, it points out practical applications for the wealth of scientific knowledge it has acquired. The dual purpose has been, as far as possible, take into account in the writing of this book. We have also tried to summarize our pedagogical experience in teaching biochemistry to students specializing in medicine and pharmacy. The material of this book has been organized according to the principle of functionally in order to trace the close relationship between the functions of the living organism and the structures of its constituents molecules as well as the chemical and physico-chemical process in which they are involved.

The aim of this book is to present a core of biochemical knowledge that is desirable for undergraduate and postgraduate students and also those involved in the field of medical, microbiology, biotechnology and pharmaceutical. Every attempt has been made to keep abrest of the advances in the subject and at the same time to include the fundamentals.

To make the work more comprehensive and informative, the author has consulted many authoritative books, research journals, abstracts, monographs etc. He is grateful to all those great scholars whose work are cited or substantially reproduced.

There can be no claim to originality except in the manner of treatment and much of the information has been obtained from the books and scientific journals available in the different libraries.

The author expresses his thanks to his friends and colleagues whose continue inspirations have initiated him to bring out this book.

The author expresses his gratitude to Mr. Wasan and Staff of M/s Discovery Publishing House Pvt. Ltd., for their whole hearted co-operation in the publication of this book.

In the mean time, the author will remain sincerely responsible for any shortcomings of the book and be grateful to the readers for their suggestions and constructive criticism for the continuous betterment of the book. He takes this opportunity to appeal to the readers to send their suggestions straightway to his publisher.

Author

CONTENTS

CHAPTER 1

Introduction

Lipids are organic compounds of biological nature, insoluble in water but soluble in nonpolar solvents such as chloroform, ether, or benzene.

This definition encompasses a large group of compounds whose classification appears to be rather a puzzling problem. The main features enabling assignment of a compound to the lipid class are: (1) biological origin; (2) hydrophobicity and, consequently, solubility in nonpolar liquids and insolubility in water; and (3) occurrence of higher alkyl radicals or carbocycles.

CLASSIFICATION OF LIPIDS

A number of classifications have been proposed for lipids on the basis of their structural, physico-chemical, biological, or physiological properties. The most sophisticated classification appears to be the structural one, which takes into account structural features of lipid compounds:

I. Lipid Monomers

1. Higher hydrocarbons
2. Higher aliphatic alcohols, aldehydes, and ketones
3. Isoprenoids and their derivatives
4. Higher amino alcohols (sphingosines)
5. Higher polyols
6. Fatty acids

II. Multicomponent Lipids

1. Simple lipids (esters composed of lipid monomers)
   - 1.1 Waxes (esters of higher monobasic alcohols)
   - 1.2 Simple diol lipids, or acyl diols (esters of dibasic alcohols)
   - 1.3 Glycerides, or acylglycerides (esters of fatty acids with tribasic alcohol glycerol)
   - 1.4 Esters of sterols

2. Heterolipids, or complex lipids
   - 2.1 Phospholipids (phosphoesters of lipids)
     - 2.1.1 Phosphoglycerides (phosphoesters of glycerides)
     - 2.1.2 Diol phosphatides (phosphoesters of diol lipids)
     - 2.1.3 Sphingophosphatides, or sphingophospholipids (phosphoesters of N-acylsphingosines)
   - 2.2 Glycolipids

Simple molecules, *i.e.* lipid monomers, are found in the cells in a free state, although in small amounts. The major part of cellular lipids occur as esterified derivatives of various alcohols.

The differentiation of lipids according to their **physicochemical properties** takes into account the extent of polarity exhibited by the lipid molecules. By this property, the lipids are classified into *neutral* (or nonpolar) and *polar* species. Lipids possessing no charge belong to the former type. The latter type includes charged lipids exhibiting distinct polar properties (for example, phospholipids and fatty acid).

According to their **physiological importance**, the lipids are divided into *reserve* lipids and *structural* lipids. The reserve lipids are stored in large amounts and then consumed to supply the energetic requirements of the organism. Acylglycerides rank among reserve lipids. The rest of lipids may be assigned to structural lipids. The energetic value of structural lipids is inferior to that of reserve lipids. The former are primarily used in the buildup of biological membranes, covering (protective) layers in plants, insects, and skin in vertebrates.

Lipids account for about 10-20% of the total mass of the human organism. The body of an adult human contains on tho average 10 to 12 kg of lipids, of which the structural lipids (major constituents of biomembranes) account for 2 or 3 kg; the rest are reserve lipids. About 98% of the latter are deposited in adipose tissues.

Among the tissues of the organism, the structural lipids are distributed non-uniformly. The nerve tissue is the most rich one in structural lipids (to 20-25%). In biomembranes of the cell, the lipids make up about 40% (by dry weight).

ONE-COMPONENT LIPIDS, OR LIPID MONOMERS

Higher Hydrocarbons

This group of compounds includes lipids of the simplest type. Mostly, higher hydrocarbons occurring in nature possess normal, branched, or unsaturated carbon chains. They are of minor significance for higher organisms.

Higher Aliphatic Alcohols, Aldehydes, and Ketones

These compounds occur in a free state, but mostly as structural constituents of multicomponent lipids. They are oxygen-containing derivatives of hydrocarbons. Higher aliphatic alcohols are compositional constituents of waxes. Unsaturated aliphatic aldehydes participate in the formation of acetalphosphatides. Higher ketones occur mainly in a free state in bacteria, while branched unsaturated ketones are structural components of insect pheromones.

Isoprenoids and their Derivatives

These constitute a vast group of biologically important lipids. They owe the name "isoprenoid" to the occurrence in them of a structural isoprene moiety, $CH_2{=}C{-}CH = CH_2$, which may also be the starting material in their biosynthesis.

$$\begin{array}{l} CH_2{=}C{-}CH = CH_2 \\ \quad\;\;\; | \\ \quad\; CH_3 \end{array}$$

Among isoprenoids, terpenes and steroids should be distinguished.

Tcrpenes. Terpenes are differentiated by the number of structural isoprene moieties. The terpenes possessing two isoprene moieties are called *monoterpenes,* three isoprene moieties, *sesquiterpenes,* and four, six, or eight isoprene moieties, *piterpenes, triterpenes,* or *tetraterpenes,* respectively.

Triterpenes *squalene* and *lanosterol* are precursors in the synthesis of cholesterol in tissues. *Carotinoids,* which are structurally related to tetraterpenes, play an important role in the vital processes. The most widespread among them are phytoene, lycopene, and α–, β–, γ-carotenes. β-*Carotene* is the provitamin A.

Of special interest is a group of oxygen-containing terpenes with a subgroup made up of numerous odorous isoprenoid alcohols which impart a specific smell to plants. For example, the sesquiterpene alcohol farnesol smells like a lily of the valley, while the monoterpenes geraniol, nerol, linalool (coriandrol) smell like a geranium or a rose. They are found in large amounts in the attar of roses and are used in the fabrication of perfumes. The monoterpene menthol finds application in the confectionery industry.

Phytol and *retionol* belong to the diterpene alcohols. The former is involved in the buildup of the photosynthetic pigment chlorophyll and phylloquinone (vitamin K_1); the latter is a fat-soluble vitamin (vitamin A). The monoterpene ketone *camphor* is widely used as a drug. *Abscisic acid,* a monocyclic derivative of sesquiterpenes, functions as a plant growth regulator.

Steroids. Steroids may be assigned to triterpene cyclic derivatives whose biosyntheses proceed with the involvement of isoprene moieties. Commonly, steroids are referred to as compounds containing a hydrocarbon skeletal framework as that of cyclopentanoperhydrophenanthrene, or gonane (earlier name, sterane):

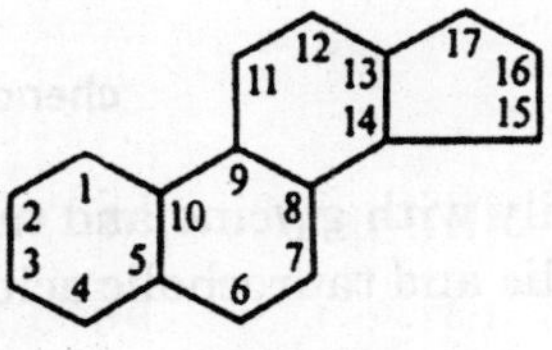

gonane

Most steroids are alcohols, conventionally named *sterols.*

Sterols of animal origin are called *zoosterols* and those of plant origin, *phytosterols.* Sterols never occur in bacteria.

Cholesterol may be regarded as an ancestor of a vast group of biologically important compounds:

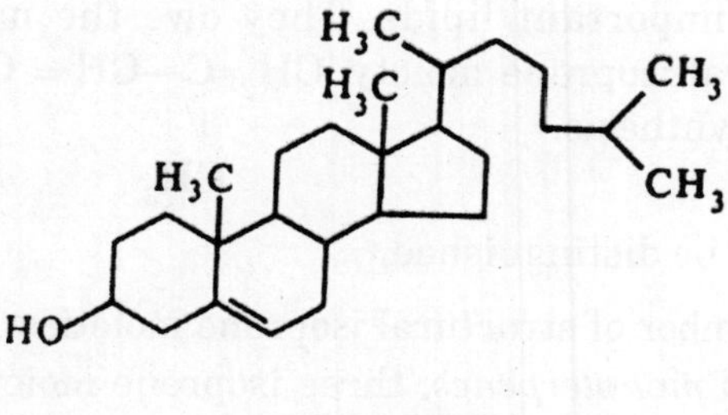

cholesterol

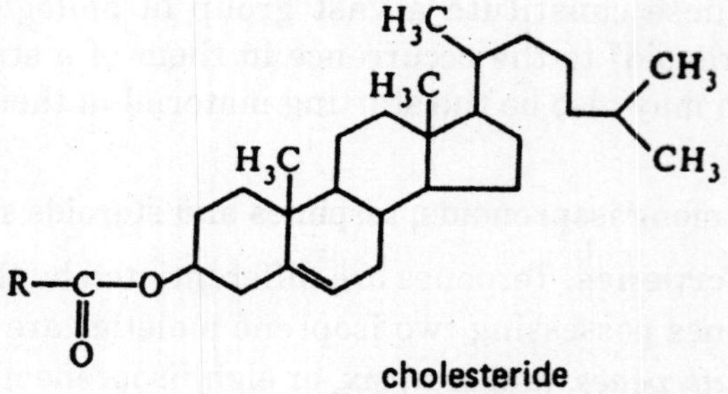

cholesteride

In tissues, chloresterol is found in a free state, or in an esterified form. Animal tissues are rich in cholesterol; it is contained in large amounts in nervous tissues, adrenal glands, and liver. Cholesterol has been classified as a structural lipid. It forms part of biological membranes of the cells with a predominant content in the plasmatic (cellular) membrane rather than in other membranes (those of mitochondria, microsomes, nuclei, etc.).

Among the steroid compounds of animal and plant origin, the following biologically active cholesterol derivatives should be mentioned: bile alcohols, bile acids, hormones, vitamins, steroid glycosides, and steroid alkaloids.

Oxydation of the cholesterol side chain in the liver results in the formation of *bile alcohols* and *bile acids,* which are excreted into the intestine, together with the bile. Bile alcohols are formed in amphibia, while in humans, bile acids only. Formation of bile acids is species-specific. The most widespread bile acids in the human bile are *cholic* and *chenodeoxycholic acids:*

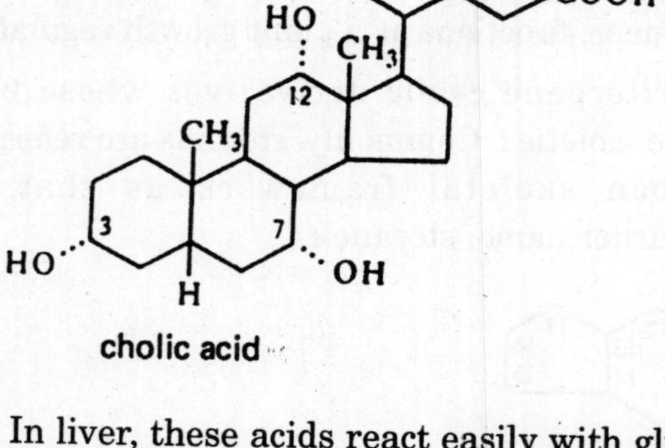

cholic acid

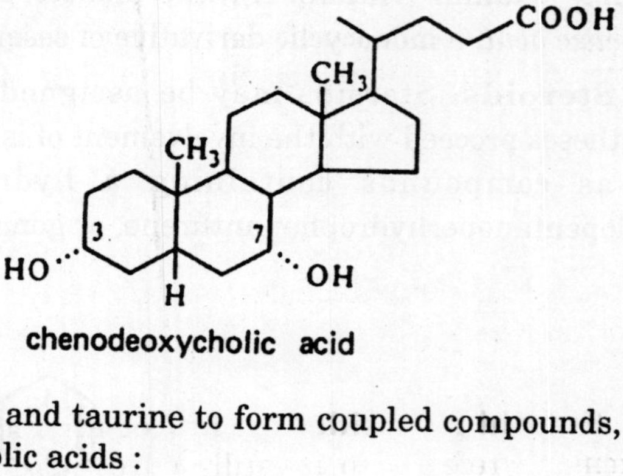

chenodeoxycholic acid

In liver, these acids react easily with glycine and taurine to form coupled compounds, or conjugates, for example, glycocholic and taurocholic acids :

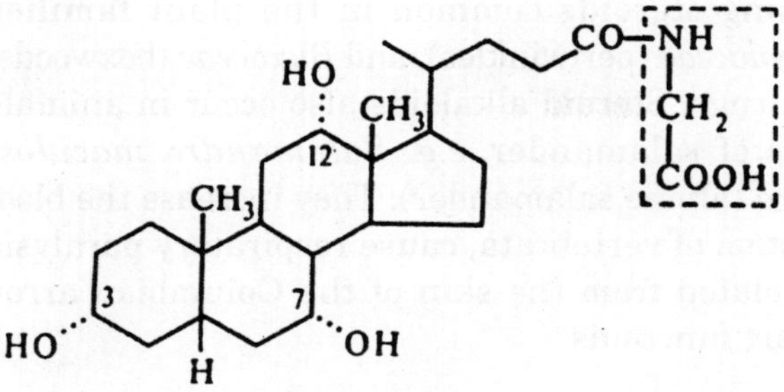

glycocholic acid

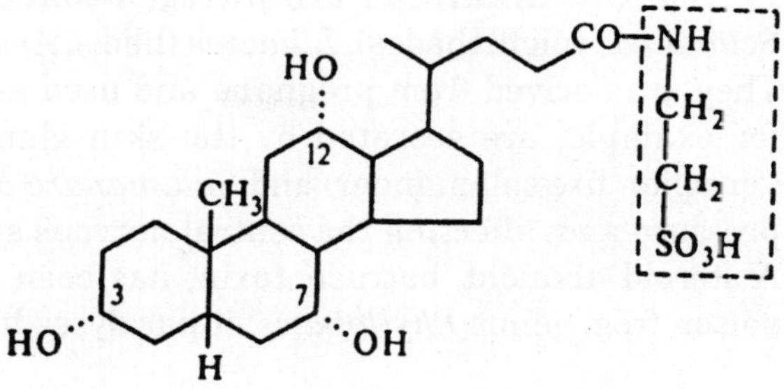

taurocholic acid

In the intestine and bile, the bile acids occur as anions (bile salts). Viewed structurally, the bile acids are amphipathic molecules, since they possess a hydrophobic ring and hydrophilic groups, viz. hydroxyl and carboxyl groups. Owing to this, they exhibit distinct surface-active properties and favour thereby the emulgation of lipids in the intestine. The bile acids are essential for normal digestion and absorption of lipids in the intestine. They also assist in the absorption of fat-soluble vitamins in the digestive tract. Sparing solubility of bile acids is the reason for the formation of gallstones (in humans, the bile acids account for about 1% of the gallstone composition).

Steroid hormones constitute the largest group among all the hormones involved in the cellular metabolism regulation and in the control of growth and development of the cells. Steroid hormones are derived from cholesterol via preoxidation of the cholesterol side chain. Consequently, compounds such as *pregnenolone* and *progesterone* are formed that serve as starting materials for the synthesis of all steroid hormones. The occurrence of steroid hormones has been evidenced in higher plants, insects and animals. In plants, conversion of cholesterol to pregnenolone and progesterone is effected by numerous tissues, while in animals mainly by special organs, viz. endocrine glands (adrenal glands, testes, ovary, and placenta).

In insects and other arthropods, molt-regulating hormones of steroid structure have been found. Although they are incapable of synthetizing the steroid ring, they use cholesterol from plants or other sources.

In plants, gonadal hormones (estrogens and progesterone) have also been found, their metabolic products isolated and shown to be similar to those contained in human urine. These hormones regulate blossoming in higher plants and are essential for sexual reproduction in certain species of fungi. It is probable that steroid compounds, otherwise referred to as products of secondary synthesis, are regulators of growth and reproduction for the plant cells. The formation of steroid hormones emphasizes the unity of the plant and animal kingdoms.

Steroid vitamins. These include vitamins D (calciferols). Plant precursors of calciferols, as well as 7-dehydrocholesterol contained in the skin of animals and humans, are called *provitamins* D. All of them possess a steroid structure. On irradiation with ultraviolet light, they convert into vitamins D.

Steroid glycosides are formed as products of the secondary synthesis in plants and are commonly referred to as cardiac glycosides. They are widely used in therapy as effective heart medicaments. In higher doses, steroid glycosides are toxic.

Steroid alkaloids are nitrogen-containing steroids common in the plant families, *Solanaceae* (nightshades), *Liliaceae* (lilies), *Apocynaceae* (periwinkles), and *Buxaceae* (boxwoods). They are derived from pregnane and used as drugs. Steroid alkaloids also occur in animals, for example, are secreted by the skin glands of salamander, *e.g. Salamandra maculosa* (europian fire salamander) and *Salamandra atra* (alpine salamander). They increase the blood pressure and, affecting the central nervous system of vertebrata, cause respiratory paralysis. A steroid alkaloid, batrachotoxin, has been isolated from the skin of the Columbian arrow poison frog, genus *Phyllobates;* it paralyses heart functions.

Fat-Soluble Vitamins

Vitamins are essential dietary factors indispensable for the normal functioning of the organism. In respect to the action by solvents, they are divided into *water-soluble* and *fat-soluble* vitamins. The latter may be referred to as lipids.

Vitamins A (retinol), D (calciferols), E (tocopherols), K (naphthoquinones), ubiquinone (coenzyme Q), and vitamin F (essential unsaturated fatty acids) belong to fat-soluble vitamins and vitaminoids.

Retinol, tocopherol, and calciferols are easily esterified, for example, with palmitic acid. Probably, the esters are a latent form of fat-soluble vitamins enabling them to be stored in tissues and, at need, released by ester hydrolysis.

Table 1.1. Widely Occurring Fatty Acids of Natural Lipids

Acid	*Structural formula*	*Number of carbon atoms*
Laiuric	$CH_3—(CH_2)_{10}—COOH$	12
Palmitic	$CH_3—(CH_2)_{14}—COOH$	16
Stearic	$CH_3—(CH_2)_{16}—COOH$	18
Lignoceric	$CH_3—(CH_2)_{22}—COOH$	24
Cerebronic	$CH_3—(CH_2)_{21}—CH(OH)—COOH$	24
Oleic	$CH_3—(CH_2)_7—CH=CH(CH_2)_7—COOH$	18
Linolic	$CH_3—(CH_2)_4—(CH=CHCH_2)_2—(CH_2)_6—COOH$	18
Arachidonic	$CH_3—(CH_2)_4—(CH=CHCH_2)_4—(CH_2)_2—COOH$	20
Linolenic	$CH_3—CH_2—(CH=CHCH_2)_3—(CH_2)_6—COOH$	18
Nervonic	$CH_3—(CH_2)_7—CH=CH(CH_2)_{13}—COOH$	24
Hydroxynervonic	$CH_3—(CH_2)_7—CH=CH—(CH_2)_{12}—CH(OH)—COOH$	24

Higher Amino Alcohols

Sphingosine derivatives represent this group of compounds. Commonly, they form part of multicomponent lipids, *sphingolipids*. Sphingolipids contain a higher unsaturated amino alcohol, sphingosine, and a higher saturated amino alcohol, dihydrosphingosine. These two compounds possess a nonpolar portion (hydrocarbon chain) and a polar portion (the rest of the molecule):

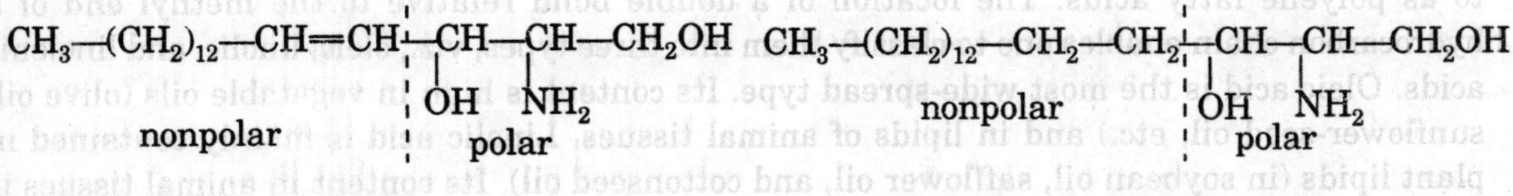

$$CH_3—(CH_2)_{12}—CH{=}CH—\underset{OH}{CH}—\underset{NH_2}{CH}—CH_2OH$$

nonpolar | polar

sphingosine

$$CH_3—(CH_2)_{12}—CH_2—CH_2—\underset{OH}{CH}—\underset{NH_2}{CH}—CH_2OH$$

nonpolar | polar

dihydrosphingosine

Higher Polyols

These constitute a relatively small group of lipid monomers. They occur in microorganisms and participate in the formation of simple and complex diol lipids of animal tissues.

Fatty Acids

Fatty acids are derivatives of aliphatic hydrocarbons that contain a carboxyl group. Higher fatty acids are chief hydrophobic components for simple and complex lipids. Over 200 fatty acids have been isolated from various lipids. They differ among themselves in chain length, number and position of double bonds, as well as in the nature of substituents (oxy-, keto-, epoxygroups, cyclic structures). Depending on the absence, or presence, of double bonds they are classified into saturated and unsaturated species.

In *saturated fatty acids,* the hydrocarbon chain has a zigzag configuration. Fatty acids with an even number of carbon atoms (from 12 to 18) are more prevalent in animal tissues. Especially widespread are palmitic and stearic acids possessing 16 and 18 carbon atoms, respectively. Fatty acids with an odd number of carbon atoms are occasionally found in small amounts in the animal tissues. In bacterial and plant lipids, fatty acids are structurally more diversified.

Natural source	*Acid concentration,%*
Lipids of milk	0.1—1.7
Lipids of all animal tissues	21—25
Lipids of all animal tissues	2—8.5
Lipids of brain	0.1
Lipids of brain	—
Lipids of tissues and natural oils	39—47
Phospholipids of tissues and natural oils	0.05—0.4
Phospholipids of tissues	0.2—22
Phospholipids of tissues	4—24
Cerebrosides of spinal cord	—
Lipids of brain	

Unsaturated fatty acids are classified into monoenic, dienic, trienic, tetraenic, etc. specie depending on the number of double bonds present in them. In general terms, they are referre to as polyene fatty acids. The location of a double bond relative to the methyl end of a hydrocarbon chain enables one to classify them into three types, viz., oleic, linolic, and linolenic acids. Oleic acid is the most wide-spread type. Its content is high in vegetable oils (olive oil, sunflower-seed oil, etc.) and in lipids of animal tissues. Linolic acid is mainly contained in plant lipids (in soybean oil, safflower oil, and cottonseed oil). Its content in animal tissues is small. In trace amounts, linolic acid has been found in human and animal lipids, but it is abundant in certain vegetable oils (linseed oil and hempseed oil). Arachidonic acid, although not synthesizable by the organism, frequently occurs in the animal lipids. Its synthetic source is linolic acid supplied in vegetable food. Structural formulas and brief characteristics of the most widespread saturated and unsaturated fatty acid are listed.

In the organism, higher fatty acids in a free state occur in quite small amounts.

All of the fatty acids are amphypatic compounds, since they carry a nonpolar "tail" (hydrocarbon chain), and a polar "head" (carboxyl group) Higher fatty acids, as any compounds of the lipid class, are practically insoluble in water and soluble in nonpolar solvents. Short-chained fatty acids (for example, butyric or capronic acid), although available in lipids in an esterified form, cannot be ascribed to lipids, since they are soluble in water.

MULTICOMPONENT LIPIDS

Simple Lipids

Among multicomponent lipids, a large group of simple lipids is distinguished. It is composed of lipid monomers which are linked through ester or amide bonds to alcohols.

Waxes are mixtures of ethers and esters derived from higher monobasic alcohols and higher fatty acids. High hydrophobicity of waxes is attributable to their structural organization. Consequently, waxes form a water-repellent protective layer (grease) on the leaves and fruit of plants, skin and hair of animals, feathers of birds, and the external skeleton of insects. They constitute part of honey wax.

Simple diol lipids make up a relatively recent group of lipids which structurally are esters of fatty acids with dibasic alcohols (ethyleneglycol or G_3-alkane diols). There are known a variety of diol lipids, such as mono-and diacyldiols in which a fatty acid radical may be linked via an ether (I) or ester (II) bond with alcohol groups:

$$\begin{array}{l} CH_2{-}O{-}CH_2{-}R \\ | \\ CH_2{-}O{-}CH_2{-}R' \end{array} \qquad \begin{array}{l} CH_2{-}O{-}\overset{\displaystyle O}{\overset{\|}{C}}{-}R \\ | \\ CH_2{-}O{-}\overset{\displaystyle O}{\overset{\|}{C}}{-}R' \end{array}$$

$$\text{I} \qquad\qquad\qquad \text{II}$$

Mixtures of diol lipids, although in small amounts, have been isolated from the animal tissues and plant seeds. The lipid contents have been found to increase with functional activity of the organism, for example, at the stage of cell regeneration or during the maturation of plant seeds.

Glycerides, or acylglycerides, constitute the most widespread group of simple lipids. By their chemical structure, they are esters of fatty acids with a tribasic alcohol, glycerol. Glycerides (acylglycerides) are also called *neutral lipids* because of their rather inert character. Glycerides are divided into mono-, di-, and tri-acylglycerides containing one, two and three acyl groups (RGO—), respectively:

glycerol	monoacylglyceride	diacylglyceride	triacylglycerde
CH_2—OH	CH_2—O—C(=O)—R	CH_2—O—C(=O)—R	CH_2—O—C(=O)—R
CH—OH	CH—OH	CH—O—C(=O)—R′	CH—O—C(=O)—R′
CH_2—OH	CH_2—OH	CH_2—OH	CH_2—O—C(=O)—R′

The *name* for a neutral lipid is derived from the name of the constituent fatty acid and the ending "glyceride". For example, palmitoylglyceride is a monoacylglyceride containing a palmitic acid moiety; tripalmitoylglyceride is a triacylglyceride containing three identical acyl residues. Differentiated are simple acylglycerides containing residues of the same fatty acid (as typified by the above case), and complex acylglycerides composed of residues of two or three different fatty acids. Natural neutral lipids are mixtures of simple and complex glycerides with a prevalence of unsaturated triglycerides.

Physico-chemical properties of neutral lipids are largely determined by their composition. The melting points are higher in acylglycerides containing higher saturated acyl groups. Saturated fatty acids (for example, tristearin) are found, as a rule, in solid fats, and in liquid fats, unsaturated acids (for example, triolein). All acylglycerides are less dense than water. They are soluble in nonpolar solvents (chloroform, benzene, ether, and hot ethanol). Only mono- and diacylglycerides containing free polar hydroxyl groups are water-soluble. They form micelles in water. Triacylglycerides are insoluble in water and are incapable of micellation. When subjected to basic hydrolysis or saponification, acylglycerides regenerate glycerol and free fatty acids. In the organism, acylglycerides are hydrolyzed by special enzymes, *upases*.

Sterids are also known as esters of sterols with fatty acids. *Cholesterol esters* are of more frequent occurrence. They are contained in products of animal origin (butter and egg yolk). In human and animal organisms, most of the tissue cholesterol (about 60-70%) occurs as cholesterol esters. In blood, cholesterol esters as constituents of transport lipoproteins make up the major content of the overall cholesterol. It is quite probable that cholesterol esters provide for a specific form of storing cholesterol in tissues. *Lanolin* (wool fat) is also of sterid origin (a mixture of fatty acid esters of lanosterol and agnosterol) and is widely used in the pharmaceutical industry as an ointment base.

Sterids such as fatty acid esters of stigmasterol, ergosterol, and β-sitosterol constitute a significant part of overall sterols in plants.

Complex Lipids, or Heterolipids

Complex lipids (heterolipids), as distinct from simple lipids, contain a nonlipid component, which may be a phosphate, carbohydrate, etc. According to their chemical structure, the representatives of complex lipids, *i.e.*, phospholipids and glycolipids, belong to molecular species with a higher level of structural organization, and, strictly speaking, it would have been more appropriate to discuss them in the subsequent chapter "Heteromacromolecules". However, with the purpose of obtaining a more comprehensive view on the whole variety of lipids, it appears expedient to take a closer look at complex lipids.

Phospholipids are complex lipids representative of phosphate-substituted esters of diverse organic alcohols (glycerol, sphingosines, and diols). All of the phospholipids are polar lipids and are predominantly contained in the cell membranes. In the fat depots, their occurrence is insignificant.

Phosphoglycerides. In phosphoglycerides, one of the hydroxyl groups forms an ester bond with a phosphate, rather than with a fatty acid. *Phosphatidic acid* is the simplest representative of naturally occurring phosphoglycerides:

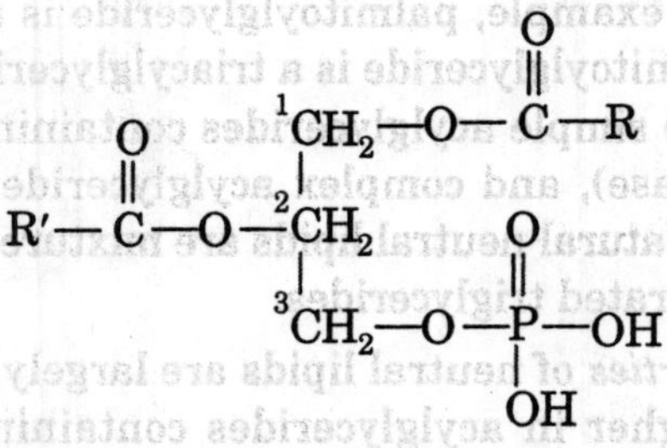

```
                          O
                          ‖
               ¹CH₂—O— C—R
       O        |
       ‖       ²|
R′— C—O—CH     O
                |       ‖
               ³CH₂—O—P—OH
                        |
                        OH
```

The fatty acid radicals adopt a *trans* configuration. As a rule, the glycerol hydroxyl at the 1 position is esterified with a saturated fatty acid; the one at the 2 position, with an unsaturated fatty acid; while the hydroxyl at the 3 position forms a phosphoester bond with phosphoric acid. All phosphoglycerides contain the residue of phosphatidic acid (phosphatidyl) linked to an alcohol residue. The most widespread phosphoglycerides are summarized.

The diversity of phospholipids is determined not only by .the nature of the alcohol residues, but also by the properties of the fatty acid residues.

If one of the fatty acid residues is cleaved off, phosphoglyceride derivatives, called *lysophosphatides,* are formed. Lysophosphatides do not occur under physiological conditions. They are very toxic and cause damage and even destruction of cell membranes.

Physico-chemical properties of phosphoglycerides are determined by the amphipathic character of their molecules. These possess a nonpolar moiety (*i.e.* diacylglyceride residue) and a polar moiety represented by phosphate and alcohol residues. Owing to the polar hydrophilic end, phosphoglycerides are not only easily soluble in nonpolar solvents and their alcoholic mixtures but also soluble, to a certain extent, in water, forming micelles in water milieu. In micelles, the hydrophobic radicals of fatty acids are rearranged to form an inner hydrophobic zone of the micelle. The hydrophilic moieties are located on the outer surface of the micelle facing the aqueous phase.

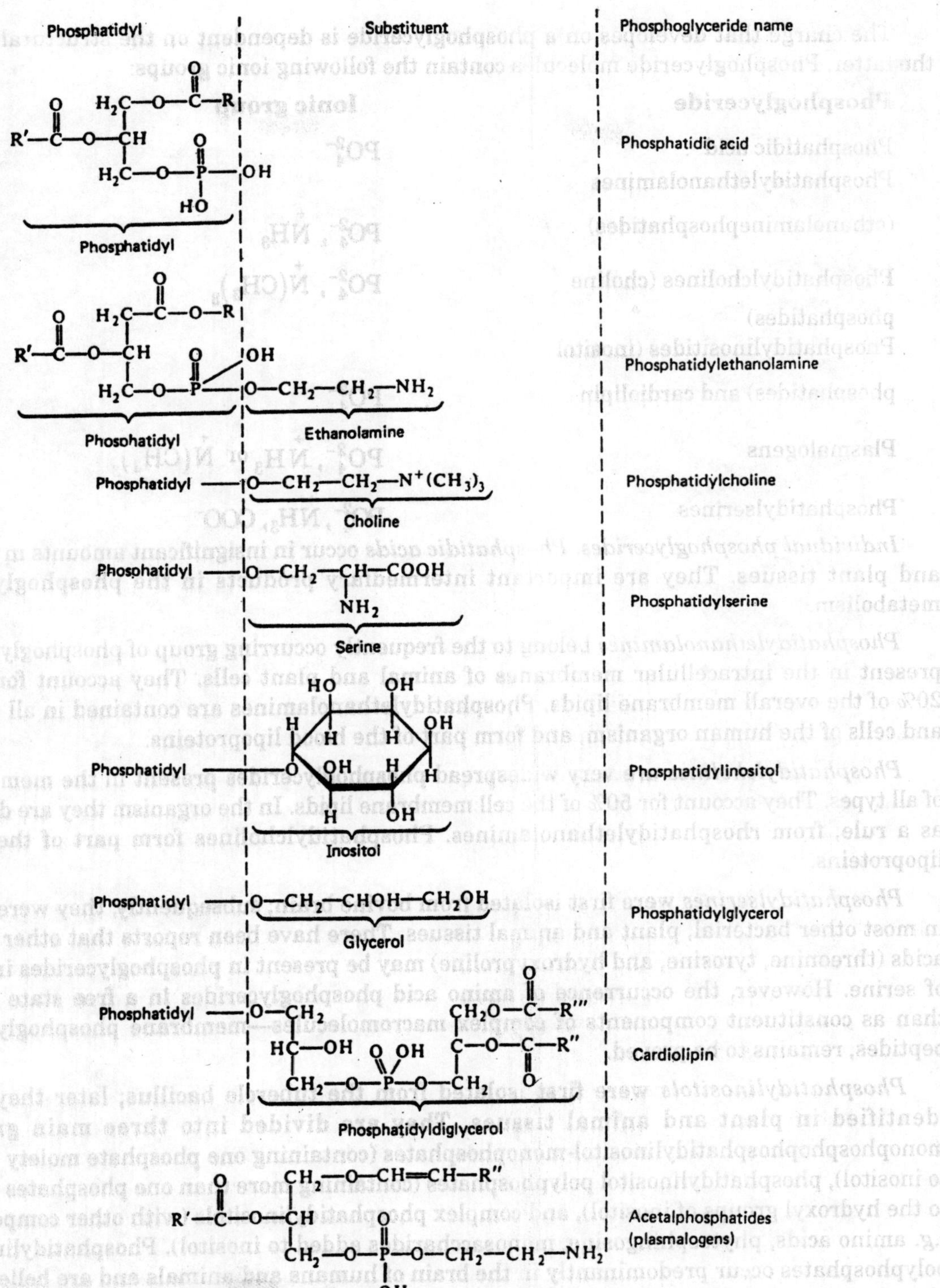

Fig. 1.1. Structures of phosphoglycerides.

The charge that developes on a phosphoglyceride is dependent on the structural type of the latter. Phosphoglyceride molecules contain the following ionic groups:

Phosphoglyceride	**Ionic group**
Phosphatidic acid	PO_4^{2-}
Phosphatidylethanolamines (ethanolaminephosphatides)	PO_4^{2-}, $\overset{+}{N}H_3$
Phosphatidylcholines (choline phosphatides)	PO_4^{2-}, $\overset{+}{N}(CH_3)_3$
Phosphatidylinositides (inositol phosphatides) and cardiolipin	PO_4^{2-}
Plasmalogens	PO_4^{2-}, $\overset{+}{N}H_3$ or $\overset{+}{N}(CH_3)_3$
Phosphatidylserines	PO_4^{2-}, $\overset{+}{N}H_3$, COO^-

Individual phosphoglycerides. Phosphatidic acids occur in insignificant amounts in animal and plant tissues. They are important intermediary products in the phosphoglyceride metabolism.

Phosphatidylethanolamines belong to the frequently occurring group of phosphoglycerides present in the intracellular membranes of animal and plant cells. They account for about 20% of the overall membrane lipids. Phosphatidylethanolamines are contained in all tissues and cells of the human organism, and form part of the blood lipoproteins.

Phosphatidylcholines are very widespread phosphoglycerides present in the membranes of all types. They account for 50% of the cell membrane lipids. In the organism they are derived, as a rule, from rhosphatidylethanolamines. Phosphatidylcholines form part of the blood lipoproteins.

Phosphatidylserines were first isolated from bovine brain; subsequently, they were found in most other bacterial, plant and animal tissues. There have been reports that other amino acids (threonine, tyrosine, and hydroxyproline) may be present in phosphoglycerides in place of serine. However, the occurrence of amino acid phosphoglycerides in a free state rather than as constituent components of complex macromolecules—membrane phosphoglyceride peptides, remains to be proved.

Phosphatidylinositols were first isolated from the tubercle bacillus; later they were identified in plant and animal tissues. They are divided into three main groups: monophosphophosphatidylinositol-monophosphates (containing one phosphate moiety linked to inositol), phosphatidylinositol polyphosphates (containing more than one phosphates linked to the hydroxyl groups of inositol), and complex phosphatidylinositols (with other compounds, *e.g.* amino acids, phytosphingosine, monosaccharides added to inositol). Phosphatidylinositol polyphosphates occur predominantly in the brain of humans and animals and are believed to play an important part in nervous activity. In the plant kingdom, the soya is rich in these valuable materials.

Complex phosphatidylinositols are widely spread in bacteria. They have also been isolated from the seeds of sunflower, groundnut, soya, and other cultivated plants.

Cardiolipin was first isolated from bovine heart; hence its name, from the Greek *kardia,* heart, and *lipos,* fat. Later, cardiolipin was found in numerous animal and human tissues, in the green leaves of higher plants and in yeast. Its content in the cells amounts to 2-5% of the lipid mass. However, in the mitochondrial membranes, cardiolipin is the major phospholipid component.

Acetalphosphatides, or *plasmalogens* (also called phosphatidals), are contained in all tissues of the human organism and account for about 20% of the overall phospholipid amount. Plasmogens are especially abundant in the brain and spinal marrow, constituting 50-90% of the whole of lipids. A specific feature of plasmogen structure is the formation of an ether bond with a higher aliphatic aldehyde.

Diol phosphatides (phospholipids) constitute a new, recently discovered group of compounds. They are derivatives of dibasic alcohols in which one hydroxyl group is esterified to a fatty acid residue and the other one is bound with a phosphate group and with a residual alcohol (for example, choline). In the organism, diol phospholipids can bind to the cell membranes and cause alterations in the properties thereof. They exhibit distinct surface-active behaviour. When present in high concentrations, diol phospholipids cause the hemolysis of erythrocytes. They affect immune reactions and neutralize the mediatorial action of acetylcholine on the cells (this action is referred to as cholinolytic).

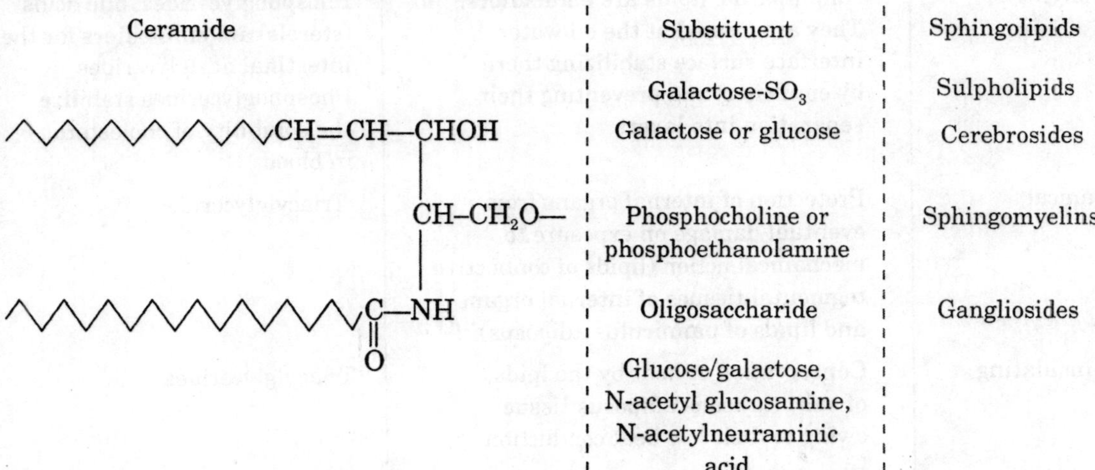

Fig 1.2. Structural schemes for certain sphingolipids—derivatives of cedramide (N-acetylsphingosine)

Sphingophosphatides (sphingolipids). In the group of sphingolipids, sphingosine is the alcohol entity. Its N-acylated derivative, *ceramide,* is the ancestor for all of the sphingolipids. Sphingolipids are divided into *sphingomyelins,* or *sphingophosphatides,* and *glycosphingolipids.* Their structure is shown. They differ in that the sphingomyelins, like all phospholipids, have the hydroxyl group substituted by a phosphate or a residual alcohol (mostly choline), while glycolipids bear, in place of the phosphate group, a monosaccharide (in cerebrosides) or an oligosaccharide (in gangliosides) containing galactose and a number of aminosaccharides.

Table 1.2. Main Biological Functions of Lipids

Function	*Characterization of the function*	*Lipids exercising the function*
Substrate-energetic	The oxidation of lipids in the organism attended by a release of enerygy greater than that from other energetic substrates (proteins and carbohydrates). Energy generated by combustion of 1g of lipids amounts to 39.1 kJ	Acylglycerides, free fatty acids
Structural	Lipids are components of the cell membranes and form a lipid basis thereof	Phospholipids (phosphoglycerides, sphingomyelins), cholesterol and its esters
Transporting	Participation in the transport of materials (for example, cations) across the lipid layer of biomembranes	Phospholipids
Electric insulating	Lipids serve as a specific electro -insulating material in the myelin sheath of nerve tissues	Sphingomyelins, glycosphingolipids
Emulsifying	Amphipathic lipids are emulsifiers. They are located at the oil/water interface surface stabilizing there by emulsions and preventing their separation into layers	Phosphoglycerides, bile acids (sterols) are emulsifiers for the intestinal acylglycerides Phosphoglycerides stabilize the solubility of cholesterol in blood
Mechanical	Protection of internal organs from eventual damage on exposure to mechanical action (lipids of connective tegmental tissues of internal organs and lipids of panniculus adiposus)	Triacylglycerides
Heat-insulating	Conservation of heat by the lpids of subcutaneous adiposus tissue owing to their low heat conduction	Triacylglycerides
Dissolving	Dissolution of certain lipid materials in other lipids under physiological conditions	Bile acids (sterols) are solvents for fat-soluble vitamins in the intestine
Hormonal	Regulation of a large variety of physiological functions with the involvement of steroid hormones (lipids) and prostaglandins (hormone like lipids)	Steroids (sex hormones, corticosteroids).unsaturated fatty acids
Vitaminogenic	Special function with the participation of fat-soluble vitamins (lipids)	Isoprenoids, unsaturated fatty acids

Sphingomyelins are contained in the nerve tissue in large amounts and form part of myelin which serves as a material for the nerve fibre sheath (hence the name coined for these compounds). However, sphingomyelins occur not only in the elements of nervous system. They have been found in the lung, liver, kidney, spleen, blood, and other organs. Characteristically, the appearance and accumulation of sphingomyelin in the nerve tissue is related to the evolution of the brain. For example, the choline-containing sphingomyelin typically occurring in the vertebrates is absent in lower representatives of the animal kingdom (flies, mussels). In these, it is replaced by ethanolamine. Sphingomyelins are absent in the nervous system of gastropods and earthworms; their contents in lower vertebrates do not exceed 4% of the total of lipids, while in higher vertebrates, they amount to 10-12%. The same trend is traced back at the stages of the individual development of the organism. A special period, the so-called *myelinization period,* is singled out during the human and animal brain development. In man, the myelinization period, *i.e.*, the formation of sphingomyelin in the nerve sheath and nervous system, starts immediately after the birth and continues for about four months. Further myelinization and accumulation of sphingomyelin and other sphingolipids occur with the brain growth and the development of higher nervous activity.

In large amounts sphingomyelins are found a in the blood plasma and the erythrocyte envelope: 8-15 and 30-40% of the total lipid mass, respectively. This appears to be associated with the function of erythrocytic membranes.

Glycolipids are complex lipids containing a carbohydrate component. They include the simplest glycolipids, *glycosyldiacylglycerols,* in which one hydroxyl group of the glycerol moiety is replaced by a monosacharide. Glycosphingolipids occur in the animal tissues in large amounts. They are especially abundant in the nerve cells where they appear to be essential for the normal electric activity and transmittance of nervous impulses. Cerebrosides, gangliosides, and sulpholipids belong to these lipids.

Cerebrosides were first isolated from the brain (cerebrum), to which they owe their name. Their carbohydrate moiety is most commonly represented by galactose or, very rarely, by glucose. Fatty acids forming part of Cerebrosides are quite numerous and include lignoceric (It-tetracosanoic), cerebronic (2-hydroxytetracosanoic), nervonic ($\Delta 1^5$-tetracosanoic) and hydroxynervonic fatty acids.

Sulpholipids are sulphated derivatives of cerebrosides. The sulphate moiety is added to the third hydroxyl group of galactose. Sulpholipids exhibit pronounced acidic properties and are capable of easily binding cations. Presumably, they partcipate in the cation transport across the membranes of nervous cells and fibres. Therefore, sulpholipids are necessary for the normal electrical activity of the nervous system.

Gangliosides, in contrast to other gylcosphinglipids, contain an oligosaccharide composed of different monosaccharides. Ganglioside components are diversified and their masses vary over a wide range. The cerebral cortex cells are rich in gangliosides.

CHAPTER 2

Lipids

The word lipid is used loosely to describe the biological material that can be extracted from living tissues by organic solvents. This definition encompasses such chemically heterogeneous substances as fatty acids and their various derivatives, steroids, prostaglandins as well as the 'fat-soluble' vitamins A, D, E and K. However, a more restricted definition of lipids, which is generally more acceptable, is that of naturally occurring esters of long chain fatty acids that are soluble in organic solvents such as chloroform, diethyl ether or hydro-carbons but are insoluble in water. Under such a definition, the free forms of the fat-soluble vitamins are excluded, as are unesterified sterols including cholesterol, bile acids and steroid hormones.

However, because in many cases they often occur naturally as esters of fatty acids and because there are similarities in the approach to their analysis, those of biological importance will be briefly considered. Lipids has been classified in a variety of ways by different workers but generally two categories, simple and complex, are evident. In this text, substances that contain only a long chain fatty acid and an alcohol (which may be either glycerol, a long chain alcohol, a sterol or one of the fat-soluble vitamins) are considered under the general heading of simple lipids. Complex lipids are acyl esters of either glycerol or the long chain amino alcohol, sphingosine, which also contain a hydrophilic group such as a phosphate ester of an organic alcohol, a carbohydrate or a sulphate group.

Lipids rarely exist in an organism in the free state but are usually associated with proteins or combined with polysaccharides. They are important dietary constituents providing energy, vitamins and essential fatty acids and often give flavour and palatability to food. Lipids act as lubricants and insulators and the fat stores in the adipose tissue of the body are a rich source of energy. Combinations of lipid and protein are of particular cellular importance especially in membrane structures and also as a means of transporting lipids in the blood. The steroid hormones are derived from cholesterol and very small amounts of these exert potent physiological effects.

FATTY ACIDS

The major structural units of naturally occurring forms of both simple and complex lipids are the aliphatic, long chain monocarboxylic acids. These are of great importance because of their contribution to the physical and chemical characteristics of the compounds of which they are constituents. In the investigation of the structure of a lipid it may be necessary to determine not only the identity and amount of any fatty acids present but also their position within the molecule.

The most common fatty acids of plant and animal origin are linear chains with even numbers of carbon atoms (4 to 24 but especially 16 and 18) with one terminal carboxyl group. They may be fully saturated or contain one or more (up to six) double bonds. Branched chain and more complex structures may also be found, particularly in fatty acids derived from plants or microorganisms.

Table 2.1. Examples of Straight Chain Saturated Fatty Acids – C_nH_{2n+1}. COOH

Number of carbon atoms	*Systematic name*	*Trivial name*	*Major occurrence*
12	*n*-Dodecanoic	Lauric	Widely distributed. Major component of some seed oils
14	*n*-Tetradecanoic	Mystiric	Seed and nut oils
16	n-Hexadecanoic	Palmitic	Commonest saturated fatty acid in animals and plants
18	*n*-Octadecanoic	Stearic	Very common, particularly in complex lipids
20	*n*-Eicosanoic	Arachidic	Peanut oil
22	*n*-Docosanoic	Behenic	Seed triacylglycerols
24	*n*-Tetracosanoic	Lignoceric	Minor component of seed fats Foud in cerebrosides
26	*n*-Hexacosanoic	Cerotic	Widespread as components of plant and insect waxes
28	n-Octacosanoic	Montanic	Plant waxes

Odd-numbered fatty acids do occur naturally with carbon numbers between 3 and 19.
Those with carbon numbers 15 to 19 are present in large amounts in certain species of fish and bacteria.
Even-numbered fatty acids, 4 to 10, are mainly found in milk and butter fats.

Nomenclature

Fatty acids have been referred to by their trivial names for many years and these are still often used. The systematic nomenclature is based on naming the fatty acid after the saturated hydrocarbon with the same number of carbon atoms, the final -e being changed in saturated acids to -anoic and in the unsaturated acids to -enoic, the carbon atoms being numbered from the carboxyl carbon. By an older system, carbon 2 is referred to as the α-carbon, carbon 3 as the β-carbon with the other carbon atoms lettered accordingly with the end methyl carbon being called ω-carbon. The convention for defining the position of any double bonds may use the symbol Δwith a superscript numeral. Other shorthand versions are also widely used, which may give rise to confusion because the same compound can be written in a variety of ways.

Table 2.2. Some naturally occurring mono-unsaturated fatty acids

Number of carbon atoms	*Systematic name*	*Trivial name*	*Major occurrence*
12	*cis*-5-Dodecenoic	Denticetic	Bacteria
	cis-9-Dodecenoic	Lauroleic	
14	*cis*-5-Tetradecenoic	Physeteric	Bacteria
	cis-9-Tetradecenoic	Myristoleic	Bacteria and plants
16	*cis*-5-Hexadecenoic	—	Plants
	cis-7-Hexadecenoic	—	Plants, algae, bacteria
	cis-9-Hexadecenoic	Palmitoleic	Widespread
	cis-11-Hexadecenoic	Palmitvaccenic	Seed oils
	trans-3-Hexadecenoic	—	Plants
18	*cis*-6-Octadecenoic	Petroselenic	Plants, seed oils
	cis-9-Octadecenoic	Oleic	Very widespread
	cis-11-Octadecenoic	*cis*-Vaccenic	Bacteria
	trans-9-Octadecenoic	Elaidic	Ruminant animal fats
	trans-11-Octadecenoic	*trans*-Vaccenic	Ruminant animal fats
20	*cis*-9-Eicosenoic	Gadoleic	Seed and fish oils
	cis-11-Eicosenoic	—	Seed and fish oils
22	*cis*-13-Docosenoic	Erucic	Seed and fish oils
	trans-13-Docosenoic	Brassidic	Seed and fish oils
24	*cis*-15-Tetracosenoic	Nervonic	Brain cerebrosides

Trans-isomers are much rarer than *cis*-isomers. Many different positional isomers of monoenoic acids may be present in a single, natural lipid and this is not a comprehensive list. Palmitoleic and oleic acids are quantitatively the commonest unsaturated fatty acids in most organisms. Odd-chain monoenoic acids are minor components of animal lipids but are more significant in some fish and bacterial lipids.

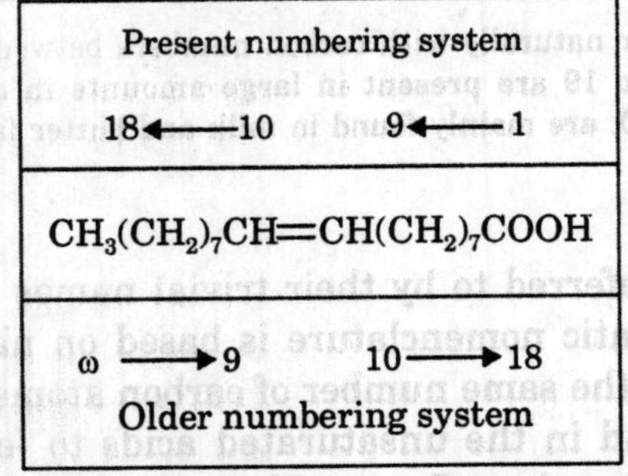

Fig. 2.1. Fatty acid carbon numbering system. The carbon atoms are numbered in sequence with the carboxyl carbon atom as number 1. By this system the terminal methyl carbon of oleic (octadecenoic) acid is number 18. An older system names this methyl carbon as the ω-carbon and the adjacent carbon atom as number 2. Numbering then continues towards the carboxyl carbon and values are given for *n* to denote the number of carbon atoms from the ω-carbon to any double bonds.

Table 2.3. Fatty acid nomenclature

Trivial name	*Oleic acid*	*Linoleic acid*
Number of carbon atoms	18	18
Structure	$CH_3(CH_2)_7CH = CH(CH_2)_7COOH$	$CH_3(CH_2)_4CH = CH.CH_2$ $CH = CH(CH_2)_7COOH$
Systematic names	*cis*-9-Octadecenoic acid (Δ-)9-Octadecenoic acid Δ^9-Octadecenoic acid *ω–9-Octadecenoic acid *(*n*-9)-Octadecenoic acid	*cis,cis*-,9, 12- Octadecadienoic acid (Δ-)9, 12-Octadecadienoic acid $\Delta^{9,12}$-Octadecadienoic acid ω-6,9-Octadecadienoic acid (*n*-6, *n*-9)-Octadecadienoic acid
Symbols	18:1 18:1;9 $C_{18:1}$ $\Delta^9C_{18:1}$ $\Delta^9 - 18:1$ $18:1^{\Delta 9}$ *ω-9-18:1 *18:1ω9 *9-18:1 *18:1(*n*-9)	18:2 18:2; 9, 12 $C_{18:2}$ $\Delta^{9,12}C_{18:2}$ Δ9, 12 – 18:2 $18:2^{\Delta 9,12}$ ω-6, 9-18:2 18:2ω6,9 6,9-18:2 18:2(*n*-6)

* Based on older terminology where the numbering is from the ω-carbon (the terminal methyl group) and *n* denotes the number of carbon atoms from the last double bond to the ω-carbon. More acceptable nomenclature designates the carboxyl carbon as carbon 1.

SIMPLE LIPIDS

Waxes

Waxes are esters of fatty acids with long chain primary alcohols. The fatty acid is usually straight chain which may be saturated or mono-unsaturated although occasionally branched chain or hydroxy acids are found. They are extremely non-polar compounds and are relatively inert chemically but they can be hydrolysed using a strong alkali, such as potassium hydroxide, a process called saponification.

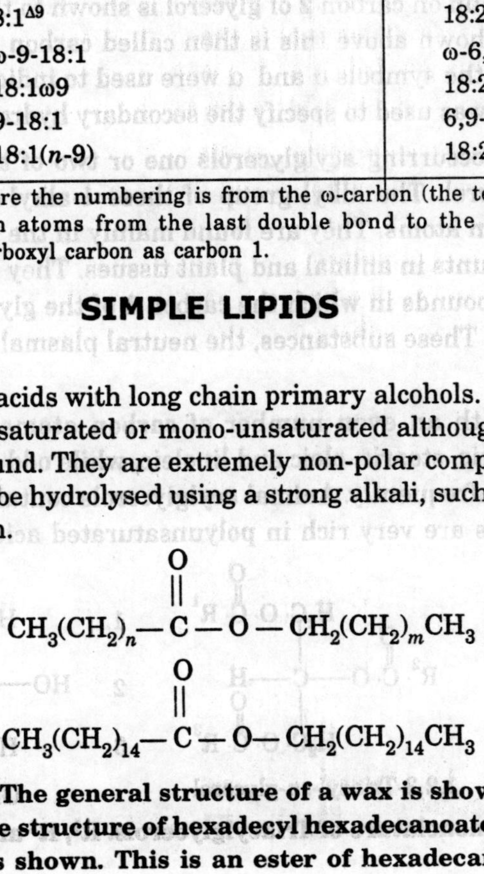

$$CH_3(CH_2)_n - \overset{\displaystyle O}{\overset{\|}{C}} - O - CH_2(CH_2)_mCH_3$$

$$CH_3(CH_2)_{14} - \overset{\displaystyle O}{\overset{\|}{C}} - O - CH_2(CH_2)_{14}CH_3$$

Fig. 2.2. Structure of waxes. The general structure of a wax is shown in which *n* and *m* are usually between 8 and 18. The structure of hexadecyl hexadecanoate (cetyl palmitate), a wax found in sperm whale oil, is shown. This is an ester of hexadecanoic (palmitic) acid and hexadecanol (cetyl alcohol).

Waxes are found in animal and insect secretions and the cell walls of some bacteria. The stored fat of marine animals has a high wax component which forms an energy reserve. Waxes are extracted and used commercially in the preparation of creams, cosmetics, polishes, lubricants and protective coatings for surfaces.

Acylglycerols

Acylglycerols are fatty acid esters of the trihydric alcohol, glycerol, and are often called glycerides. A prefix (mono-, di-, tri-) indicates the number of fatty acids esterified to glycerol which, in triacylglycerols, are nearly always different. Triacylglycerols are probably the most important single class of lipid and are major components of fats and oils of both animals and plants, whereas monoacylglycerols and diacylglycerols appear in only trace amounts in these tissues.

If positions 1 and 3 on the glycerol molecule contain different fatty acids a centre of symmetry is created and the triacylglycerol will exist in different enantiomorphic forms. Several systems of nomenclature have been used in attempts to describe the positioning of the fatty acids on the glycerol molecule but the stereospecific numbering system *(sn)* is now favoured. In this system, which gives an unambiguous description of glycerol derivatives, the secondary hydroxyl group on carbon 2 of glycerol is shown to the left in the Fischer projection and the carbon atom shown above this is then called carbon 1 and the one below, carbon 3. In older nomenclature the symbols α and α were used to indicate the position of the primary hydroxyl groups and β was used to specify the secondary hydroxyl group.

In some naturally occurring acylglycerols one or two of the fatty acids are esterified to an alkyl ether of glycerol. The alkyl group of these 1-alkyl-2,3-diacyl-*sn*-glycerols usually contains 16 or 18 carbon atoms. They are found mainly in the liver oil of marine animals and only in very small amounts in animal and plant tissues. They sometimes occur together with a related group of compounds in which the carbon 1 of the glycerol is linked by a vinyl ether bond to an alkyl group. These substances, the neutral plasmalogens, are l-alkenyl-2,3-diacyl-*sn*-glycerols.

The fatty acids with an even number of carbon atoms most frequently involved in acylglycerols are palmitic, stearic, oleic and linoleic, while odd-numbered and branched chain fatty acids only occur infrequently. Animal acylglycerols contain mainly saturated fatty acids although many fish oils are very rich in polyunsaturated acids. Plant acylglycerols contain

$$
\begin{array}{ccc}
 & & H_2C\cdot O\cdot \overset{O}{\overset{\|}{C}}\cdot R^1 \\
 & & | \\
R^2\cdot \overset{O}{\overset{\|}{C}}\cdot O & — & C — H \\
 & & | \\
 & & H_2C\cdot O\cdot \overset{O}{\overset{\|}{C}}\cdot R^3
\end{array}
\qquad
\begin{array}{lccc}
1. & & H_2COH & \\
 & & | & \\
2. & HO — & C & — H \\
 & & | & \\
3. & & H_2COH &
\end{array}
$$

1,2,3-Triacyl-*sn*-glycerol

Glycerol

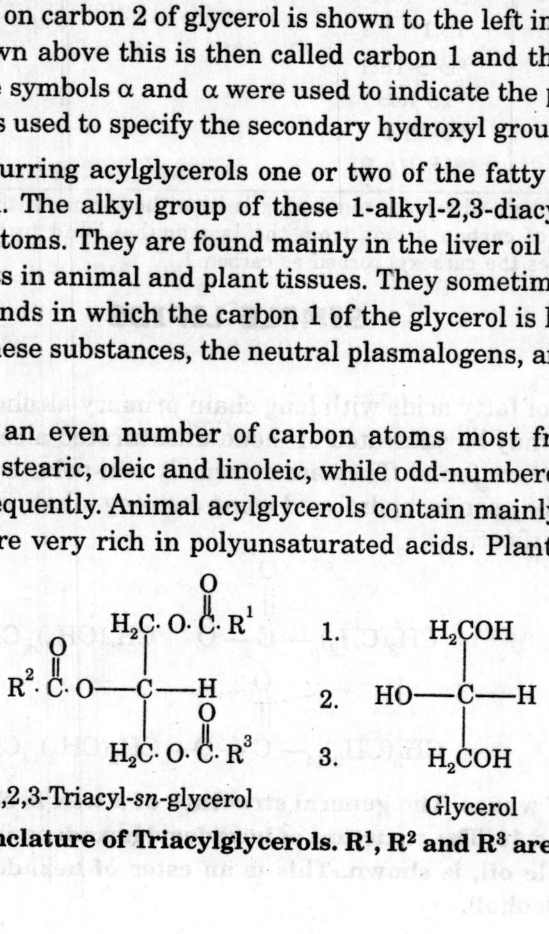

Fig. 2.3. *sn* Nomenclature of Triacylglycerols. R^1, R^2 and R^3 are acyl chains.

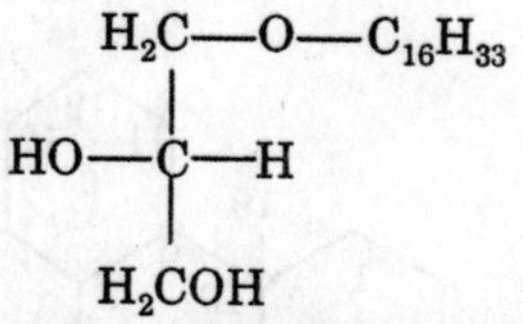

$$\begin{array}{l} H_2C{-}O{-}C_{16}H_{33} \\ HO{-}C{-}H \\ H_2COH \end{array}$$

1-Hexadecyl-*sn*-glycerol
(chimyl alcohol)

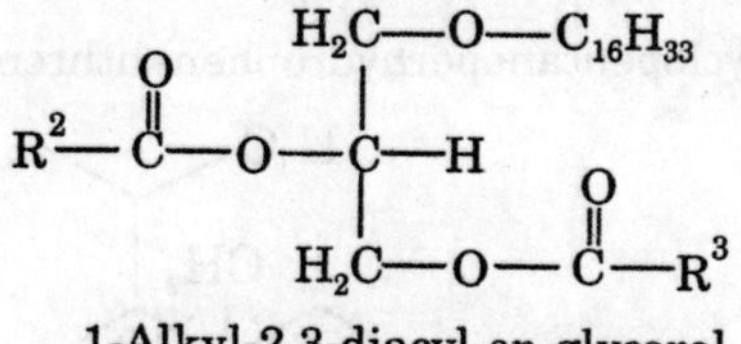

$$\begin{array}{l} \qquad\qquad\quad H_2C{-}O{-}C_{16}H_{33} \\ R^2{-}C(=O){-}O{-}C{-}H \\ \qquad\qquad\quad H_2C{-}O{-}C(=O){-}R^3 \end{array}$$

1-Alkyl-2,3-diacyl-*sn*-glycerol

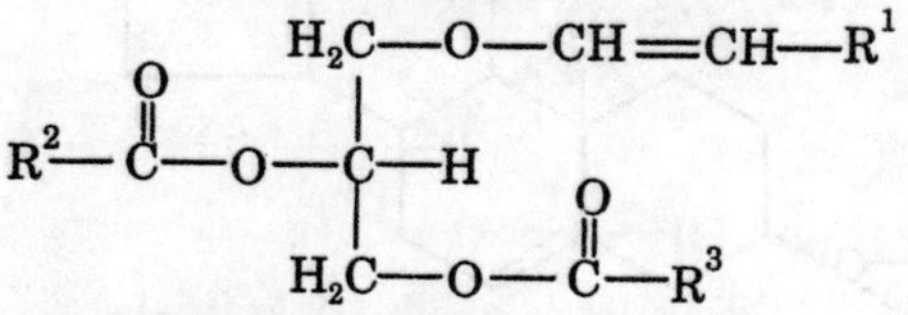

$$\begin{array}{l} \qquad\qquad\quad H_2C{-}O{-}CH{=}CH{-}R^1 \\ R^2{-}C(=O){-}O{-}C{-}H \\ \qquad\qquad\quad H_2C{-}O{-}C(=O){-}R^3 \end{array}$$

1-Alkyl-2,3-diacyl-*sn*-glycerol
(a neutral plasmalogen)

Fig. 2.4. Glyceryl ethers and their esters. R^1, R^2 and R^3 are acyl chains.

slightly increased proportions of unsaturated fatty acids, many of which show higher degrees of unsaturation. Milk fats and some seed and nut oils have a higher proportion of short chain fatty acids (C-4-C-10).

Sterol Esters

Sterols are compounds that have a steroid structure and contain a hydroxyl group, which is capable of forming an ester. They are widespread in nature and commonly exist as mixtures of the free sterol and esters with fatty acids. Cholesterol is by far the commonest sterol in animals, a high concentration being found in brain, nervous tissue and membranes. Plants contain the related steroids, β-sitosterol, ergosterol and stigmasterol but bacteria cannot synthesize sterols. Many biologically important substances have similar structures and the steroid sex hormones and adrenocortical hormones, bile acids and cardiac glycosides fall into this category. The different attachments to the steroid nucleus result in compounds with a wide range of biological actions.

The D vitamins are a group of compounds that are derived from a steroid nucleus. They are involved in calcium and phosphate metabolism.

The active vitamins are produced by converseion of provitamins by ultraviolet light. Ergosterol, a yeast sterol, is converted to its active form, ergocalciferol (vitamin D_2), and 7-dehydrocholesterol, which is found in many nature foods and is also synthesized in man, is converted to cholecalciferol (vitamin D_3). Fish liver oils are virtually the only source of vitamin D_3 in nature. The most active form of vitamin D_3 is 1, 25-dihydroxycholecalciferol and this is produced by the hydroxylation of cholecalciferol at position 25 in the liver and then at position 1 in the kidney.

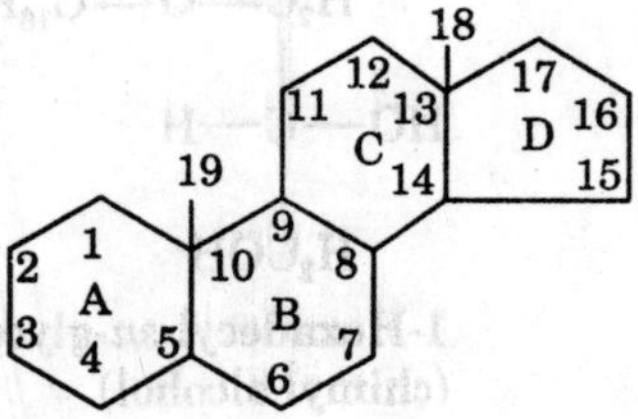

Cyclopentanoperhydrophenanthrene nucleus

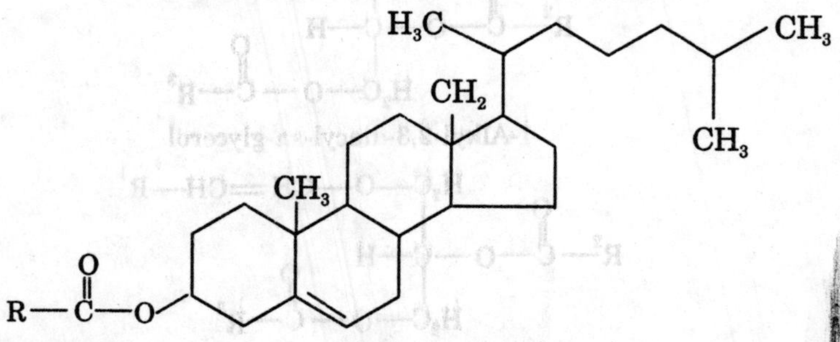

Cholesterol ester

Fig. 2.5. Cholesterol ester. Cholesterol is a derivative of the cyclopentanoperhydrophenanthrene condensed ring system and is therefore a member of the group of compounds called steroids. It often occurs esterified at position 3 with a long chain fatty acid, which may be saturated or unsaturated and commonly contains 16 or 18 carbon atoms. Steroids that form esters in this way are more correctly called sterols.

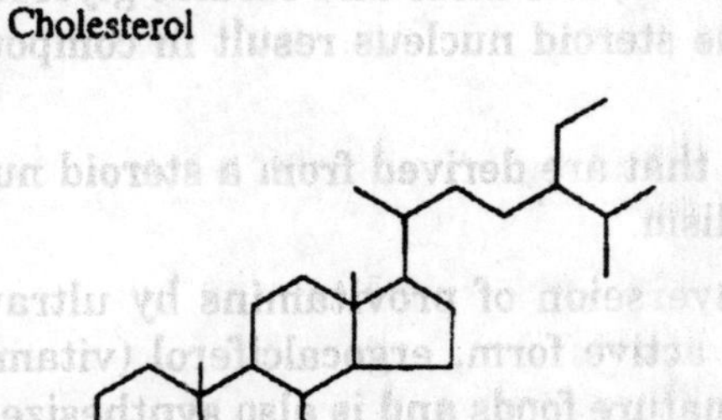

Cholesterol

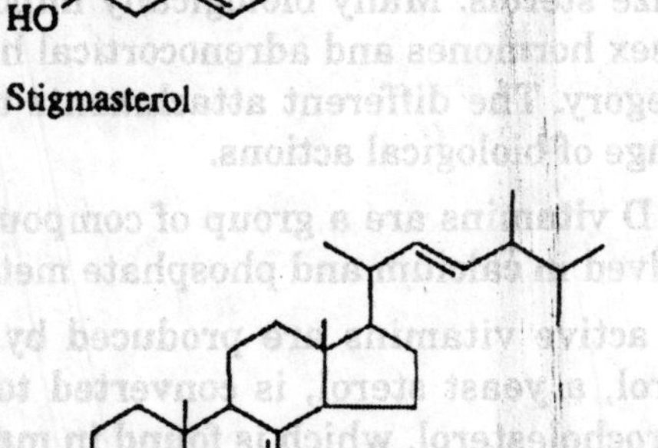

Stigmasterol

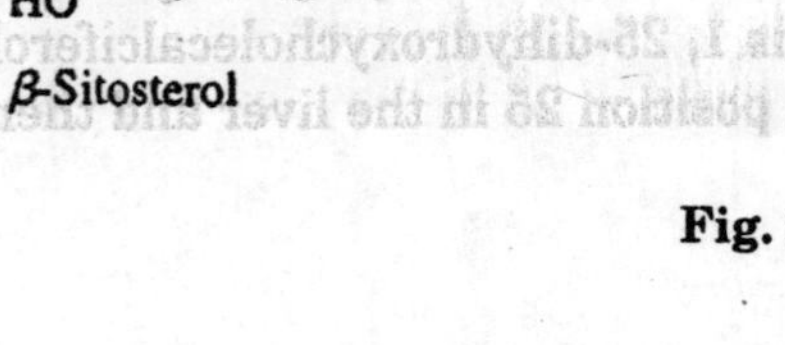

β-Sitosterol

Ergosterol

Fig. 2.6. Sterols.

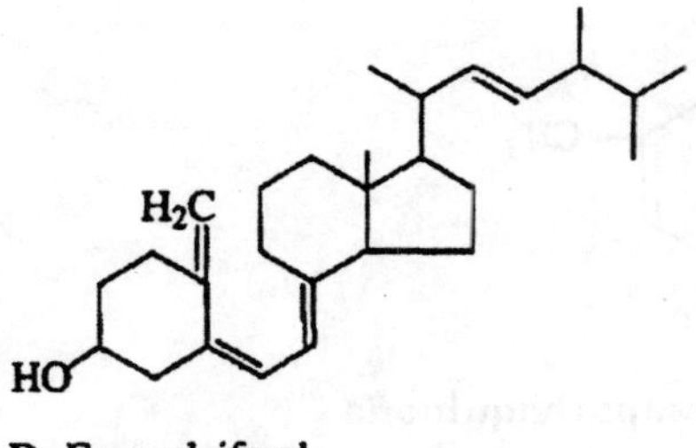

D$_2$ Ergocalciferol

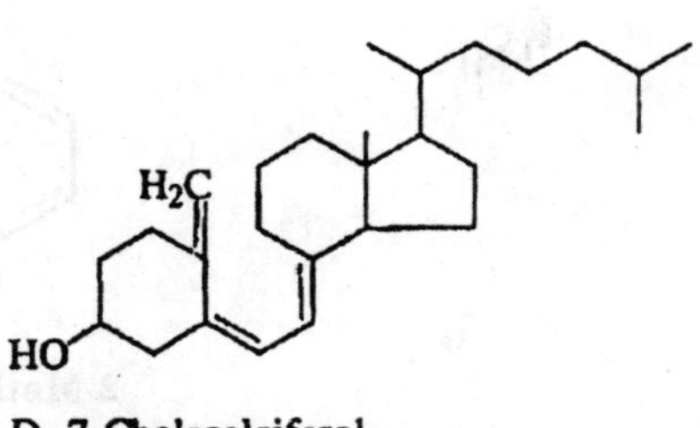

D$_3$ 7-Cholecalciferol

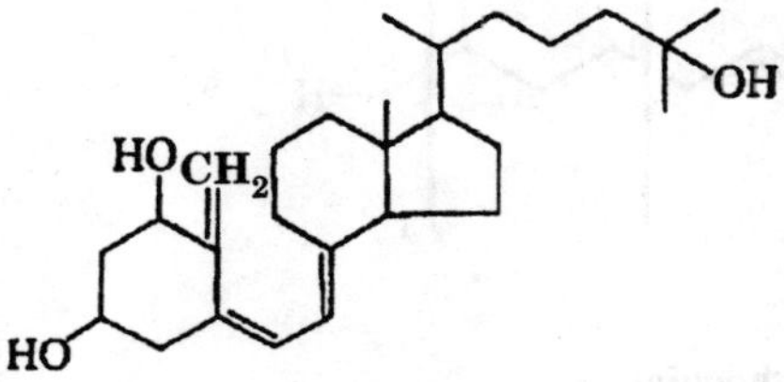

1,25-Dihydroxycholecalciferol

Fig. 2.7. Active forms of the D vitamins.

Vitamins A, E and K

The α-, β- and γ-carotenes, which are found in most plants, are vitamin A provitamins and are converted to vitamin A alcohol (all-*trans*-retinol), which is usually called vitamin A_1 by oxidative mid-point cleavage. Retinol and its fatty acid esters are the main forms in which vitamin A is stored in animals and humans, and its oxidation product, 11-*cis*-retinal (vitamin A_1 aldehyde), is required for the visual process.

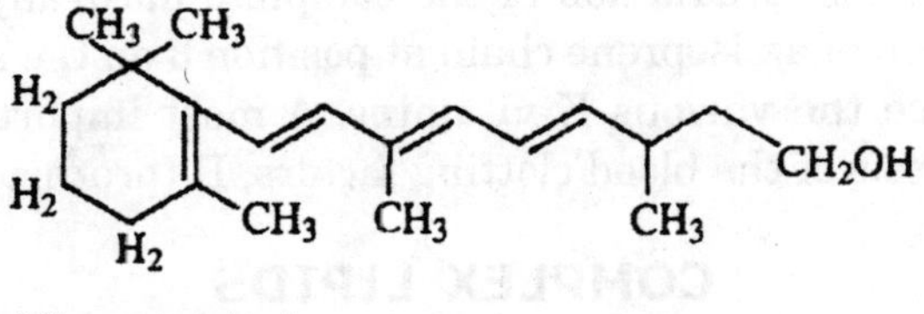

All-*trans*-retinol

Fig. 2.8. Vitamin A_1. All-*trans*-retinol (Vitamin A alcohol) is usually referred to as Vitamin A_1.

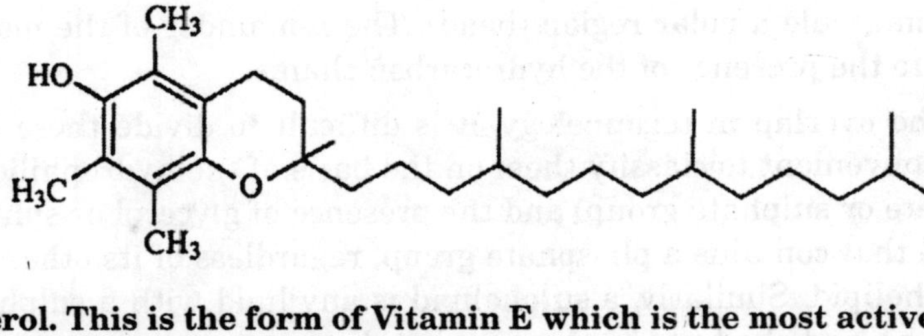

Fig. 2.9. α-Tocopherol. This is the form of Vitamin E which is the most active biologically, the β, γ and δ tocopherols being less potent.

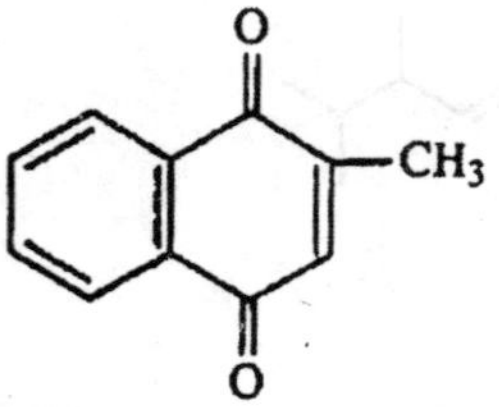

2-Methyl-1, 4-naphthoquinone

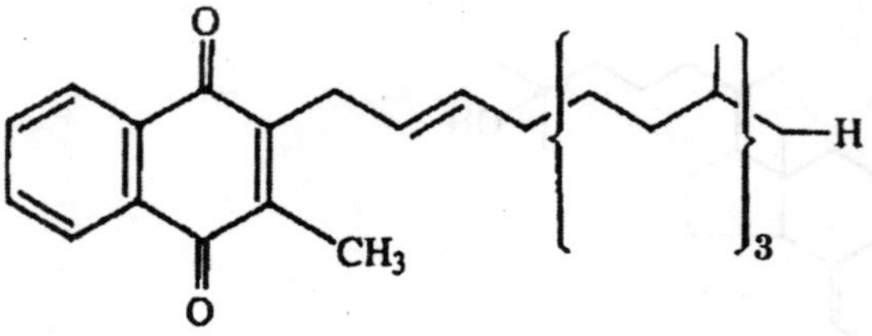

2-Methyl-3-phytyl-1,4-naphthoquinone

Fig. 2.10. Active forms of vitamin K. Although 2-methyl-1,4-naphthoquinone has vitamin K activity, the more important compounds have an isoprene chain at position 3.

Several different tocopherols are known to have vitamin E activity, but α-tocopherol, a trimethyltocol is the most biologically active. Other less potent forms are the β-, γ- and δ-tocopherols, which contain fewer methyl groups. They all have antioxidant properties and a deficiency results in a lack of protection of the unsaturated fatty acids in the membrane phospholipids against oxidation by molecular oxygen.

Compounds showing vitamin K activity are substituted naphthoquinones. The parent compound, 2-methyl-1,4-naphthoquinone, does show some biological activity as do other similar but synthetic compounds. The production of the complete naturally active forms is thought to depend upon the addition of an isoprene chain at position 3 on the aromatic ring. Differences in this side chain produce the various K vitamins. A most important physiological role of vitamin K is in the synthesis of the blood clotting factors, II (prothrombin), VII, IX and X.

COMPLEX LIPIDS

Complex lipids are considered here to be acyl esters of glycerol or sphingosine which also contain a hydrophilic group. They are sometimes called conjugated, compound or polar lipids. The latter name stems from the fact that they all contain at least one hydrophilic group, which gives the molecule a polar region (head). The remainder of the molecule, however, is non-polar owing to the presence of the hydrocarbon chains.

Because of the overlap in terminology, it is difficult to divide these lipids into distinct groups but it is convenient to classify them on the basis of the hydrophilic group (phosphate ester, carbohydrate or sulphate group) and the presence of glycerol or sphingosine. In such a system, any lipid that contains a phosphate group, regardless of its other polar components, is called a phospholipid. Similarly, a sulpholipid is any lipid with a sulphate group, while a glycolipid contains a carbohydrate but has no phosphate group.

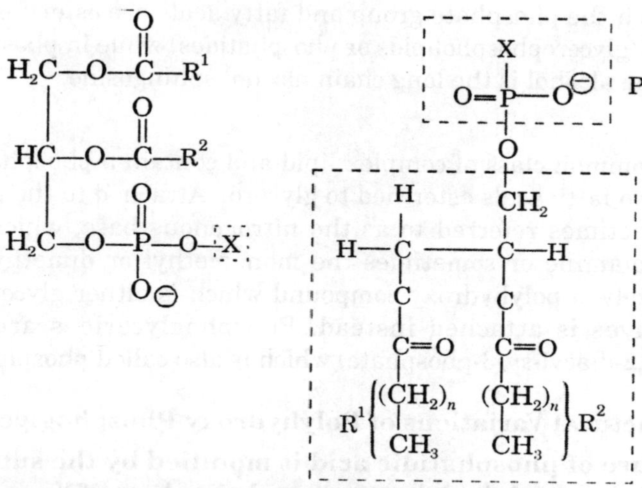

Polar head

Non polar tail

Fig. 2.11. Phosphoglyceride structure. The members of this group are derivatives of the parent compound, 1, 2-diacyl-in-glycerol-3-phosphate (phosphatidic acid) in which X is a hydrogen atom. This is replaced by either an amino alcohol or a polyhydroxy residue. In phosphoglycerides derived from animal tissues R^1 is usually a saturated acyl chain of between 16 and 20 carbon atoms and R^2 is usually unsaturated. Polyunsaturated acyl chains containing 16 or 18 carbon atoms predominate in leaf phosphoglycerides and those of bacterial origin are often more complex.

Table 2.4. Nitrogen-Containing Phosphoglycerides

Name	*Nitrogenous base**	*Major occurrence*	*Comments*
Phosphatidyl cholines	$-CH_2CH_2\overset{+}{N}(CH_3)_3$	Very abundant in animal tissues. Also found in plants and microorganisms. Important in lung alveoli	Commonly termed lecithins
Phosphatidyl ethanolamines	$-CH_2CH_2\overset{+}{N}H_3$	Large amounts in animals, plants and microorganisms	Old trivial name is cephalin. Methyl and dimethyl ethanolamine derivatives also occur
Phosphatidyl serines (or threonine)	$-CH_2CH(\overset{+}{N}H_3).COOH$	Widely distributed but in small amounts. Is a major component of brain red cell lipids	Usually isolated as salts with K^+, Ca^{2+}, Na^+ or Mg^{2+}

* Represented by X in Fig. 2.11.

Phospholipids

Phospholipids in which the phosphate group and fatty acids are esterified to glycerol are named phosphoglycerides (glycerophospholipids or phosphatides) while in phosphosphingolipids (sphingophospholipids) the alcohol is the long chain alcohol sphingosine.

Phosphoglycerides

These are the most common class of complex lipid and contain a phosphoric acid residue (phosphate group) and two fatty acids esterified to glycerol. Attached to the phosphate group is an amino alcohol, sometimes referred to as the nitrogenous base, which may be either serine, choline or ethanolamine or sometimes the monomethyl or dimethyl derivatives of ethanolamine. Alternatively, a polyhydroxy compound which is either glycerol, myo-inositol or one of their derivatives is attached instead. Phosphoglycerides are derivatives of glycerophosphoric acid (1,2-diacyl-*sn*-3-phosphate) which is also called phosphatidic acid.

Table 2.5. Structural Variations of Polyhydroxy Phosphoglycerides. The basic structure of phosphatidic acid is modified by the substitution of various polyhydric compounds in place of X.

Glycerol
Glycerol × 2
Glycerol-amino acid
Glycerol-3-phosphate
Myo-inositol
Myo-inositol-4-phosphate
Myo-inositol-4,5-diphosphate

The different phosphoglycerides are often named by placing the constituent attached to the phosphate group after 'phosphatidyl', *e.g.* phosphatidyl choline (3-*sn*-phosphatidylcholine or 1,2-diacyl-*sn*-glycero-3-phosphorylcholine). There are many phosphoglycerides because of the possible variation in the fatty acid chains, and when the full chemical structure is known, it should be used (*e.g.*, 1-palmitoyl-2-oleoyl-phosphatidylcholine). Nomenclature that entails the use of the DL system should be avoided.

The phosphoglycerides are good emulsifying agents because of the polar phosphate group and the presence of a group attached to it which may also be charged, making one end of the molecule highly hydrophilic while the rest of the molecule is composed of large hydrophobic groups. They are significant components of cell membranes and plasma lipoproteins and also have very important physiological roles including the prevention of the collapse of lung alveoli, which is made possible by the surfactant properties of mainly phosphatidyl choline.

Phospholipids are also found which have similar structures. These include lysophospholipids, which have only one of the two possible positions of glycerol esterified, almost invariably at carbon 1, and the plasmalogens, in which there is a long chain vinyl ether at carbon 1 instead of a fatty acid ester. These compounds also contain an amino alcohol, which may be either serine, ethanolamine or choline. Other rarer phospholipids are the monoacyl monoether, the diether and the phosphono forms.

Table 2.6. Variations in Phoshoglyceride Structure

Structural form	*Trival name*	*Structure*	*Common nature of X*
Monoacyl	Lysophospholipids	CH_2—O—CO—R^1 │ CH_2—OH │ O ‖ CH_2—O—P—O—X │ $O^{\ominus}$	Serine Ethanolamine Choline
Monoacyl, 1-alkenyl	Plasmalogens	CH_2—O—CH=CH—R^1 │ CH—O—CO—R^2 │ O ‖ CH_2—O—P—O—X │ $O^{\ominus}$	Serine Ethanolamine Choline
Monoacyl monoether	—	CH_2—O—R^1 │ CH—O—CO—R^2 │ O ‖ CH_2—O—P—O—X │ $O^{\ominus}$	Ethanolamine Serine
Diether	—	CH_2—O—R^1 │ CH—O—R^2 │ O ‖ CH_2—O—P—O—X │ $O^{\ominus}$	Glgcerol Glycerol-3-phosphate
Cabon-phosphorus bond	Phosphonolipids	CH_2—O—CO—R^1 │ CH—O—CO—R^2 │ O ‖ CH_2—O—P—CH_2—CH_2—$NH_3^{\oplus}$ │ $O^{\ominus}$	Ethanolamine

R^1 and R^2 represent saturated or unsaturated acyl chains.

Phosphosphingolipids

Sphingolipids contain the long chain alcohol sphingosine, or one of its derivatives. The sphingolipids that contain a phosphate group, *i.e.*, phosphosphingolipids, are normally referred to as sphingomyelins. They are esters of a ceramide and phosphoryl choline and named ceramide-1-phosphorylcholines. The ceramide is usually an amide of a fatty acid and the long chain amino alcohol, sphingosine (4-sphingenine), although derivatives of sphingosine, *e.g.*, dihydrosphingosine (sphinganine), are sometimes found. The fatty acid may be saturated or monoenoic and is often a hydroxy acid.

Sphingomyelins were first isolated from brain and nervous tissue, where they are abundant, but they also occur in many other animal tissues.

Glycolipids

Glycolipids may also be classified according to whether they contain glycerol or sphingosine. However, the latter group (glycosphingolipids) have varied and complex structures requiring subdivision into smaller groups.

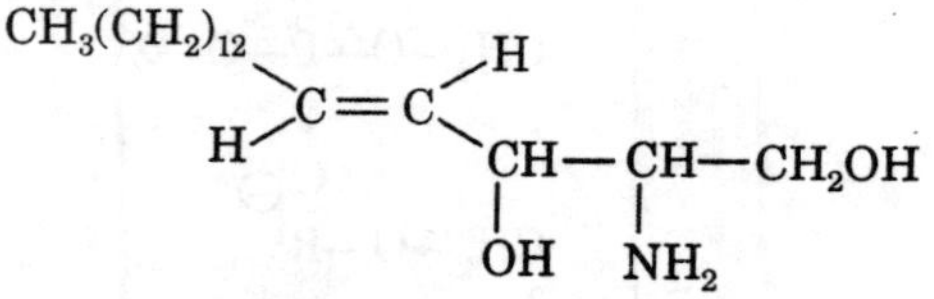

$CH_3(CH_2)_{12}$—CH=CH—CH(OH)—CH(NH_2)—CH_2OH

Sphingosine

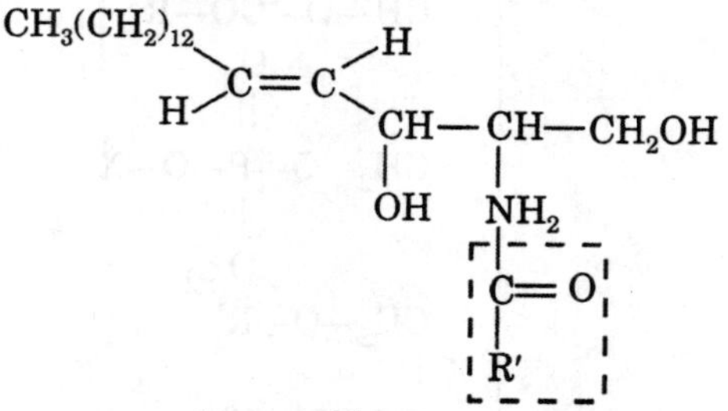

$CH_3(CH_2)_{12}$—CH=CH—CH(OH)—CH(NH—C(=O)—R′)—CH_2OH

A ceramide

$CH_3(CH_2)_{12}$—CH=CH—CH(OH)—CH(NH—C(=O)—R′)—CH_2O—P(=O)($O^{\ominus}$)—O_2 —CH_2—CH_2—$\overset{\oplus}{N}(CH_3)_3$

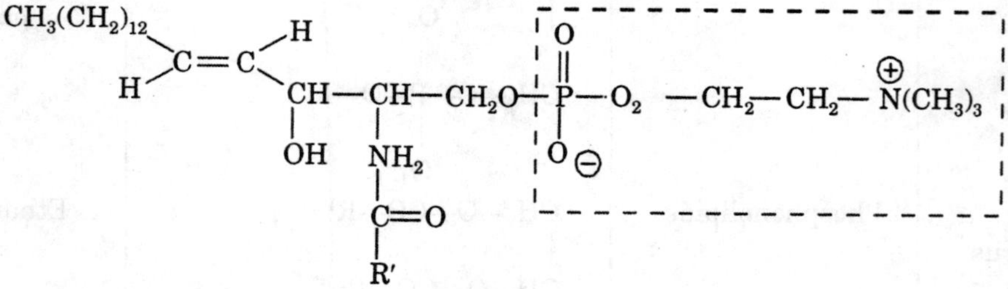

A sphingomyelin

Fig. 2.12. *Sphingomyelins.* Sphingomyelins are esters of a ceramide and phosphoryl choline. However, similar compounds are ceramide-1-phosphoryl ethanolamines and phosphono forms of sphingolipids. Ceramide (*N*-acyl-sphingosines) are amides of a long chain di- or trihydroxy base containing 12 to 22 carbon atoms, of which sphingosine (4-sphingenine) is the commonest, and a long chain fatty acid whose acyl chain is shown by R[1]. This may contain up to 26 carbon atoms.

Glycosyl Diacylglycerols

These are major components of plants and micro-organisms although small amounts can be detected in some animal tissues. They contain glycerol with two fatty acids esterified at positions 1 and 2, and one or more carbohydrate residues linked by a glycosidic bond to position 3. They are named 1, 2-di-acyl, 3-glycosyl-*sn*-glycerols. Those found in plants often contain a highly unsaturated fatty acid and one or two molecules of galactose. Bacterial glycosyl diacylglycerols often contain a branched chain fatty acid and up to four carbohydrates, which may include glucose, mannose, galactose, rhamnose and glucuronic acid.

Glycosphingolipids

Glycosphingolipids in which a ceramide unit (*N*-acyl sphingosine) is linked through position 1 on the long chain base to either glucose or galactose are called monoglycosyl ceramides (cerebrosides). Compounds containing more than one carbohydrate residue should be named appropriately as di-, tri- or tetra-oligoglycosylceramides and not called cerebrosides. Glycosyl

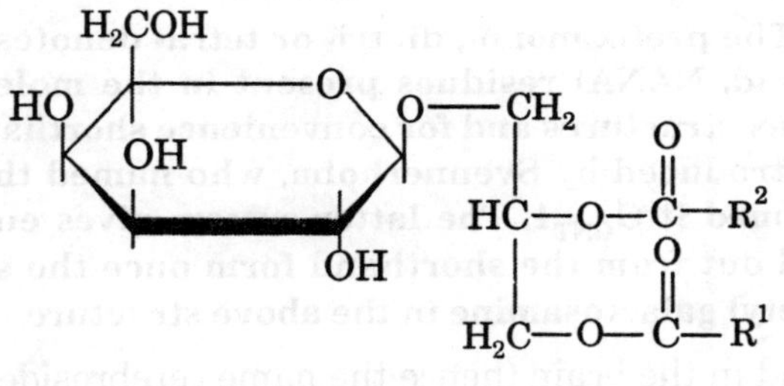

H_2COH, HO, O, OH, OH, O—CH_2, HC—O—C(=O)—R^2, H_2C—O—C(=O)—R^1

Fig. 2.13. Glycosyl diacylglycerol structure. The carbohydrate, which may be either a monosaccharide or a disaccharide residue, is attached by a glycosidic linkage to the sn-3 position of glycerol.

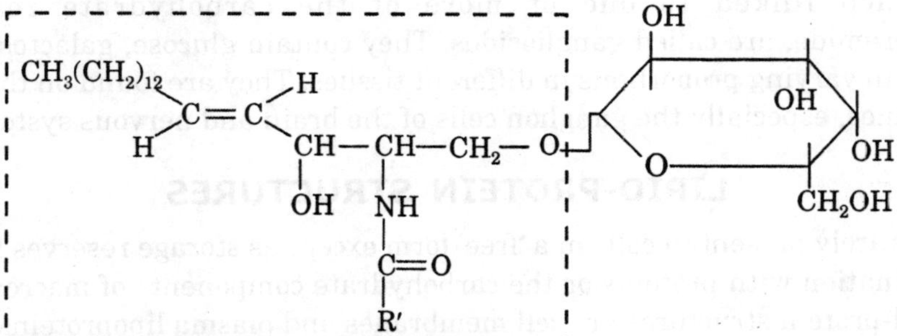

$CH_3(CH_2)_{12}$, H, C=C, H, CH—CH—CH_2—O, OH, NH, C=O, R′, OH, OH, OH, O, CH_2OH

Ceramide Carbohydrate

Fig. 2.14. Monoglycosyl ceramides (cerebrosides). R′ is an acyl chain. The carbohydrate residue of galactose is shown although this may be replaced by another hexose. Glycosyl ceramides containing between two and six carbohydrate residues are also found but only the mono derivatives are referred to as cerebrosides.

Monosialogaglioside (G_{M_1}) or ($G_{GNT}{}^1$)

$$\text{cer-glc}\left(4 \xleftarrow{\beta} 1\right) \text{gal}\left(4 \xleftarrow{\beta} 1\right) \text{galNAc}\left(3 \xleftarrow{\beta} 1\right)\text{gal}$$

$$\begin{pmatrix} 3 \\ \uparrow \\ 2 \end{pmatrix}$$

NANA

Disialoganglioside (G_{DIa}) or ($G_{GNT}{}^{2a}$)

$$\text{cer-glc}\left(4 \xleftarrow{\beta} 1\right) \text{gal}\left(4 \xleftarrow{\beta} 1\right) \text{galNAc}\left(3 \xleftarrow{\beta} 1\right)\text{gal}$$

$$\begin{pmatrix} 3 \\ \uparrow \\ 2 \end{pmatrix} \qquad \begin{pmatrix} 3 \\ \uparrow \\ 2 \end{pmatrix}$$

NANA NANA

Fig. 2.15. Gangliosides. The prefix mono-, di, tri- or tetra- denotes the number of sialic acid (*N*-acetyl neuraminic acid, NANA) residues present in the molecule. The many different gangliosides have complex structures and for convenience shorthand notations are used. The two commonest were introduced by Svennerholm, who named the 'parent' compound G_{M1}, and Wiegnandt, who named it $G_{GNT}1$. The latter system gives enough information for the structures to be worked out from the shorthand form once the symbols have been learnt. GalNAc represents *N*-acetyl galactosamine in the above structure.

ceramides were first found in the brain (hence the name cerebroside) and these mainly contain galactose, whereas ceramide glucosides are more common in other animal tissues and in plants. Cerebroside sulphates (sulphatides) are found mostly in the brain and in these a sulphate group is attached to position 3 of the galactose.

Glycosphingolipids which contain one or more *N*-acetylneuraminic acid (sialic acid) molecules, each linked to one or more of the carbohydrate residues of an oligoglycosylceramide, are called gangliosides. They contain glucose, galactose and *N*-acetyl galactosamine in varying proportions in different tissues. They are found on the outer surface of cell membranes, especially the ganglion cells of the brain and nervous system.

LIPID-PROTEIN STRUCTURES

Lipids are rarely present in cells in a 'free' form except as storage reserves but are usually found in combination with proteins or the carbohydrate components of macromolecules. The two major lipid-protein structures are cell membranes and plasma lipoproteins.

Cell Membranes

All biological membranes contain at least 25% lipid in association with protein in a complex, organized structure. Several different proteins may be present in a particular membrane and while some have structural roles others possess enzyrriic properties. The distribution of lipids throughout the membrane may be variable but generalizations can be made about the proportions and type of lipids present. The lipoprotein from myelin sheath of nervous tissue contains about 80% lipid, whereas the protein content is significantly higher in cells that are involved in active metabolism. The membranes of animal cells contain cholesterol, phosphoglycerides, sphingolipids and glycolipids. Bacterial cells do not contain cholesterol and neither animal nor bacterial cells contain sulpholipids, which are found in high concentrations in plant chloroplasts together with glycolipids. There is a predominance of phospholipids in animal cell membranes but the proportions of different lipids vary considerably from one membrane to another.

The general structure is that of a bilayer in which the hydrophobic sections of a variety of lipids are orientated towards the centre of the bilayer, with the hydrophilic areas towards the aqueous medium on the inside and outside of the cell. In the accepted fluid mosaic model, the lipids have lateral movement within the structure.

Plasma Lipoproteins

The lipoproteins that circulate in blood plasma are complexes of varying proportions of lipids and proteins. It is the presence of phospholipids and proteins that makes these aggregates water-soluble and it is in this form that otherwise insoluble lipids are carried in an aqueous environment between tissues. They are micellar and consist of hydrophobic centres of acylglycerols

Table 2.7. Plasma Lipoprotein Fractions

Fraction	*Precentage Lipid*	*Percentage composition by dry weigh*				*Transport role*
		Triacylglycerols	*Cholestrol and cholesterol esters*	*Phospholipids*	*Proteins*	
Chylomicron	90	80–95	2–6	3–9	~1	Transport of fat absorbed in the intestine to the liver and adipose tissers (exogenous triacylglycerol)
Very low density lipoprotein (VLDL)	90	40–80	10–40	15–20	5–10	Transport of fat synthesized in the liver (endogenous triacylglycerol)
Low density lipoprotein (LDL)	78	10	45–50	~20	20–25	Transport of cholesterol, mostly as the acytl ester, synthesized in the liver
High density lipoprotein (HDL)	50	1–5	~20	~30	45–55	Transport of cholesterol from peripheral tissues to the liver for catabolism

and cholesterol esters surrounded by hydrophilic coats of protein and phos-pholipids. The lipoproteins are usually grouped into classes based on their density. Each class includes lipoproteins whose physical and chemical characteristics and lipid composition are similar and whose densities, which are related to the lipid : protein ratio, fall within a specified range. The variations in density permit the different classes to be separated, if not completely, by ultracentrifugation. Alternative methods include precipitation by various salts and, more frequently, electrophoresis. The electrophoretic mobilities are comparable to some plasma proteins (α-, β- and pre-β-globulins) and this has been used for classification purposes. The investigation of patients with suspected lipoprotein abnormalities may include the analysis of the lipid content of a particular lipoprotein class, *e.g.*, HDL cholesterol levels.

Table 2.8. Plasma lipoproteins

Fraction	*Density range (gl^{-1})*	*Svedberg flotation rate (S_f)**	*Approximate diameter (nm)***
Chylomicrons	<0.95	>400	100—1000
VLDL (pre-β)	0.95—1.006	20—400	30—80
LDL (β)	1.006—1.063	0—20	15—25
HDL (α)	1.063—1.210	—	7.5—10.0
***Albumin:NEFA	>1.281	—	Smallest

* Svedberg flotation rate (analogous to a negative sedimentation coefficient) is the rate at which each lipoprotein floats up through a solution of sodium chloride of specific gravity 1.063.

** Diameters vary extensively in each group and if the extreme values of each group were given they would demonstrate the overlap between the groups.

***The albumin and non-esterified fatty acid complexes (99:1) constitute only about 5% of all plasma lipoprotein.

SAMPLE PREPARATION AND HANDLING

Quantitation of lipids may require an initial extraction step. This should neither degrade the lipids nor extract any non-lipid components, such as carbohydrates, amino acids, etc. Individual requirements will dictate how rigorous any extraction and purification procedure must be but several fundamental precautions must always be taken in order to minimize the possibility of errors.

Ideally samples from living organisms should be extracted without any delay to prevent autoxidative or enzymic deterioration of their lipid constituents. If this is not feasiable the sample should be frozen immediately and stored at –20 °C in a glass container under nitrogen. Often lipids will be extracted into an organic solvent and during this and subsequent steps in the analytical procedure minimal exposure of the lipids to air, light and heat is very important to prevent oxidation or destruction of the lipids.

The general rules that should therefore be observed include the use of a blanket of nitrogen whenever possible and evaporation of solvents at the lowest feasible temperatures, which must not exceed 50°C. The addition of an antioxidant such as butylated hydroxytoluene

(2,6-di-*t*-butyl-4-methylphenol) to the extraction solvents (0.1 gl^{-1}) might be necessary to prevent deterioration of unsaturated lipids but it is essential for storage of lipid extracts at about 0.1 % of the weight of lipid. Inactivation of lipolytic enzymes may usually be achieved by addition of an alcohol such as methanol or, in some cases, isopropanol. The latter is recommended for some more stable enzymes sometimes found in plant tissues. Alternatively the plant may be briefly immersed in boiling water.

Contamination is a major problem in lipid analysis and the use of plastic containers and stoppers, rubber bungs or tubing, and any grease on stop-cocks, etc., must be avoided. All solvents used should be of the purest grade and be peroxide-free and all glassware should be scrupulously clean.

Extraction Procedures

Lipids from liquid samples or cell suspensions may be extracted directly into organic solvents but solid samples will require prior treatment, such as homogenization or ultrasonication. The solvents are chosen with reference to the nature of the lipids present and to some extent, the type of sample, *e.g.* animal or vegetable tissue. Ethanol-diethyl ether (3:1) with subsequent extraction into a petroleum hydrocarbon is useful for non-polar lipids but variations of the Folch methanol-chloroform-water mixture are more widely used. The methanol disrupts the lipid-protein membrane complexes and inactivates the lipases, and the chloroform dissolves the lipids. A salt solution (KCl, NaCl or $CaCl_2$) is often added and the non-lipid substances are taken up into this aqueous phase by vigorous shaking. They can be removed in the upper aqueous layer following centrifugation to separate the phases. Repeated extraction and washing will improve the isolation of uncontaminated lipids in the lower chloroform layer, which can then be analysed by the method of choice.

QUANTITATIVE METHODS

Total Lipids

The total lipid content of a sample can be assessed gravimetrically if the lipid is first extracted and then completely dried until a constant weight is recorded. For accurate results, the whole procedure must be designed to minimize the loss of lipids at every stage. This includes using highly efficient extraction and purification techniques under appropriate analytical conditions.

Other methods, while being technically less demanding, also provide questionable results because they may only measure certain lipid components. Thus the pink colour produced by unsaturated lipids with a reagent containing sulphuric acid, phosphoric acid and vanillin may be used as a screening procedure to assess the total lipid content without preliminary extraction but the result will only be a reflection of the unsaturated lipid content of the sample. The sample is heated with concentrated sulphuric acid and after cooling a phospho-vanillin reagent is added, which contains 1.2 g vanillin in 1 litre of 80% phosphoric acid. The absorbance is read at 530 nm after 30 min and compared with known concentrations of lipid treated in the same way. Exact quantitation is not feasible, although a factor, derived with reference to the particular lipid used for preparation of the standard curve, may be useful in some circumstances.

Cholesterol

In many tissues cholesterol and other sterols exist as a mixture of the free alchohol and its long chain fatty acid ester (esterified at position 3 of the steroid nucleus). The determination of the cholesterol content of a sample may involve the measurement of either of these two fractions individually or the total cholesterol. It is possible to precipitate free cholesterol by adding an equal volume of digitonin ($1\ gl^{-1}$ in 95% ethanol), a naturally occurring glucoside, to form a complex that is insoluble in most solvents, including water.

The chemical methods for the quantitation of cholesterol measure total cholesterol, *i.e.*, free and esterified, and so a digitonin precipitate must be prepared if free cholesterol is to be measured. Enzymic methods do not measure the esters and a hydrolysis stage, either chemical or enzymic (using cholesterol ester hydrolase, is necessary for the measurement of total cholesterol.

Colorimetric Method

The colour reaction of cholesterol and cholesterol esters with acetic anhydride and concentrated sulphuric acid provides the basis of the method attributed to Liebermann and Burchard. This reaction in not entirely specific for cholesterol or its esters because other sterols will also react. In its original form the reagent consists of acetic anhydride, concentrated sulphuric acid and glacial acetic acid and the intensity of the green colour is affected by the proportions of the reagents and the amount of water present. It is possible to achieve an increase in sensitivity if the reagent contains ferric ions. Various modifications of reagent composition have been used and some methods are fluorimetric.

Enzymic Method

The enzymic methods show improved accuracy and reliability compared with the colorimetric procedures. The assay using cholesterol oxidase is analogous to the measurement of glucose using glucose oxidase but is not completely specific for cholesterol, and all sterols with a 3-β-hydroxyl group and a double bond in the 4-5 or 5-6 position will react. The method is based on the oxidation of cholesterol to cholest-4-ene-3-one with the simultaneous production of hydrogen peroxide. Total cholesterol can be determined if preliminary hydrolysis of the esters is undertaken. The concentration of the reaction products will be proportional to the amount of cholesterol originally present and either may be measured. Cholest-4-ene-3-one has an absorption maximum at 240 nm and its concentration can be calculated from the molar absorption coefficient, thus obviating the need for cholesterol standards. The alternative approach is to measure the hydrogen peroxide produced and although this may be achieved in a variety of ways, the usual procedure employs a peroxidase and a chromogenic oxygen acceptor:

$$\text{cholesterol} + O_2 \xrightarrow[\text{pH 7–8}]{\text{cholesterol oxidase}} \text{cholest-4-3-one} + H_2O_2$$

$$H_2O_2 + \text{4-aminoantipyrene} + \text{phenol} \xrightarrow{\text{peroxidase}} \text{quinoneimine} + H_2O$$

Acylglycerols

Mono-, di- and triacylglycerols may all be measured by determination of the amount of glycerol released by hydrolysis. The lipid is first extracted into chloroform-methanol (2:1) and saponification is performed under conditions that will not affect any phosphate ester bonds, otherwise glycerol originating from phosphoglycerides would also be measured. Heating at 70°C for 30 min with alcoholic potassium hydroxide (0.5 mol l^{-1}) has been shown to be satisfactory. However, the phospholipids may be removed prior to saponification either by extraction or by adsorption on activated silicic acid.

To compensate for any free glycerol present in the sample, the assay is performed both before and after the chosen hydrolysis step, the difference in results giving the glycerol derived from the acylglycerols.

Colorimetric Method

Glycerol can be oxidized to formaldehyde by periodic acid and the formaldehyde measured spectrophotometrically at 570 nm after reaction with chromotropic acid, or fluorimetrically after the addition of diacetylacetone and ammonia. The chromotropic acid reagent consists of 8-dihydroxynaphthalene-3,6-disulphonic acid dissolved in 50% sulphuric acid.

Enzymic Method

Enzymic assays have been developed for the measurement of glycerol and commercial kits are available, some of which include a microbial lipase to effect enzymic hydrolysis of acylglycerols. In a coupled assay method glycerol kinase is used to convert glycerol to glycerol-3-phosphate, which is coupled to the oxidation of NADH using pyruvate kinase and lactate dehydrogenase. The decrease in absorbance at 340 nm is related to the concentration of glycerol and therefore to the amount of acylglycerol originally present:

$$\text{glycerol} + \text{ATP} \xrightarrow{\text{glycerol kinase}} \text{glycerol-3-phosphate} + \text{ADP}$$

$$\text{ADP} + \text{phosphoenolypruvate} \xrightarrow{\text{pyruvate kinase}} \text{pyruvate} + \text{ATP}$$

$$\text{pyruvate} + \text{NADH} \xrightarrow{\text{lactate dehydrogenase}} \text{lactate} + \text{NAD}^+$$

An alternative method uses glycerol dehydrogenase to produce NADH. This is measured either by its absorbance at 340 nm or by using a diaphorase which catalyses the reduction of a dye, *p*-iodonitrotetrazolium violet (INT) to produce a coloured complex with an absorption maximum at 540 nm:

$$\text{glycerol} + \text{NAD}^+ \xrightarrow{\text{glycerol dehydrogenase}} \text{NADH} + \text{dihydroxyacetone}$$

$$\text{NADH} + \text{INT} \xrightarrow{\text{diaphorase}} \text{NAD}^+ + \text{INT (H)}$$

SEPARATION OF LIPID MIXTURES

Solvent Extraction

Lipids can be extracted from biological samples using a variety of organic solvents. A chloroform-methanol solvent is suitable for all lipids but it is possible to extract different classes of lipid selectively on the basis of their solubility in different organic solvents. This may be achieved by the addition of a solvent that will effect either the precipitation or the extraction of the lipids of interest. An example of the former is the precipitation of high concentrations of phospholipids with cold, dry acetone, and of the latter, the extraction of fatty acids into ether or heptane at an acid pH. However, like all solvent extraction procedures these are not entirely specific.

The partition of different lipids between two immiscible solvents (countercurrent distribution) is useful for crude fractionation of lipid classes with greatly differing polarities. Repeated extractions in a carefully chosen solvent pair increase the effectiveness of the separation but in practice mixtures of lipids are still found in each fraction. A petroleum ether-ethanol-water system can be used to remove polar contaminants (into the alcoholic phase) when interest lies in the subsequent analysis of neutral glycerides, which may be recovered from the ether phase. Carbon tetrachloride-methanol-water is useful in the analysis of plant lipids for separating non-polar carotenes and chlorophylls from more polar phospholipids and glycolipids. The lipid content of each phase can be easily monitored using thin-layer chromatography and the extraction process repeated until the desired separation is achieved.

Table 2.9. Lipid classes in order of increasing polarity

Least polar	
Saturated hydrocarbons	Neutral lipids
Unsaturated hydrocarbons	
Wax esters	
Steryl esters	
Long chain aldehydes	
Triacylglycerols	
Long chain alcohols	
Free fatty acids	
Quinone	
Sterols	
Diacylglycerols	
Monoacylglycerols	
Cerebrosides	
Glycosyl diacylglycerols	
Sulpholipids	
Acid glycerophosphatides	
Phosphatidyl ethanolamine	
Lecithin	
Sphingomyelin	
Lysolecithin	
Most polar	

Column Techniques

Adsorption chromatography on a column of silicic acid (silica gel) may be used to separate mixtures of lipids on a preparative scale. Alternative adsorbents are Florisil (magnesium silicate), acid-washed Florisil or alumina and these offer advantages in some cases, *e.g.*, Florisil is useful for the isolation of glycolipids. The extent of the separation generally achieved will depend on many factors including the complexity of the lipid mixture, the dimensions of the column, the type of adsorbent used and the choice of eluting solvents. Thin-layer techniques can be used to purify further the collected fractions as well as being useful to monitor the composition of the column effluent.

The lipids are bound to the fine particles of the adsorbent by polar, ionic and van der Waals forces, the most polar being held most tightly, and separation takes place according to the relative polarities of the individual lipids. It is usual to elute the lipids from the column with solvents of increasing polarity, when the least polar lipids will emerge in the early fractions. The solvents and the sequence in which they are used will be dictated by the components of the sample. Initial separation of a diverse mixture into the three broad groups of neutral lipids, glycolipids and phosphatides can be effected using a chloroform-acetone-methanol solvent sequence. The collected fractions may be then separated into their individual components by column or preparative thin-layer techniques.

Silicic acid columns can be used for the separation of mixtures of neutral lipids by elution with hexane containing increasing proportions of diethyl ether (0-100%). The lipids are eluted in the following order:

Hydrocarbons

Cholesterol esters

Triacylglycerols

Free fatty acids

Cholesterol

Diacylglycerols

Monoacylglycerols.

Samples that are rich in glycolipids and sulpholipids (*e.g.* plant extracts) can be separated by eluting the glycolipids with chloroform-acetone (1:1) followed by the sulpholipids using acetone. Fractionation of extracts containing high concentrations of several phospholipids (*e.g.*, animal tissues) can be achieved using chloroform to which increasing proportions of methanol have been added (95:5 increasing to 50:50) resulting in the following typical sequence:

Phosphatidic acid

Phosphatidyl ethanolamine

Phosphatidyl serine

Phosphatidyl choline

Phosphatidyl inositol

Sphingomyelin.

Ion-exchange cellulose chromatography with DEAE (diethylaminoethyl) or TEAE (triethylaminoethyl) is a useful technique for the separation of complex lipids on a large scale, the former having the widest range of application. It can sometimes be used after silicic acid chromatography of total lipids to resolve the lipids that are eluted together, *e.g.*, phosphatides and glycolipids or glycolipids and sulpholipids. Separation takes place primarily on the basis of ion-exchange of those lipids with ionic groups, although non-ionic polar groups, *e.g.*, hydroxyl, also exert an influence owing to adsorption by the cellulose. The DEAE cellulose is used in the acetate form, and charged, non-ionic lipids (*e.g.* phosphatidyl ethanolamine or choline, sphingomyelin, cerebrosides) are eluted with chloroform containing varying proportions of methanol. Weakly acidic lipids (*e.g.* phosphatidyl serine) can be eluted with glacial acetic acid and strongly acidic or highly polar lipids (*e.g.* phosphatidyl glycerol, phosphatidyl inositol, sulphatides and sulpholipids) with chloroform-methanol containing ammonium acetate and dilute ammonia.

Thin-layer Chromatography

Silica gel thin-layer plates may be used to separate lipids on either a preparative or an analytical scale. They are sometimes used to fractionate the lipids into classes prior to removal from the plate and further analysis by GLC. In this case the appropriate area of silica gel is scraped off and the lipid extracted into chloroform or diethyl ether containing 1-2% methanol for simple lipids or into chloroform-methanol-water (5:5:1) for polar lipids.

The polarity of the lipid determines the degree of adsorption to the silica, the most polar being most strongly held. The mobility of the lipids during chromatography will be affected by the polarity of the solvent and increasingly polar solvents will break the adsorptive bonds with greater efficiency, resulting in greater R_F values. When the sample contains compounds of widely varying polarities, a particular solvent may only be able to separate the lipids into general classes, *e.g.* fatty acids, acylglycerols, cholesterol, phospholipid. It may be possible to resolve the components of such groups by careful selection of solvents.

Double development in the same direction using the same or two different solvents sometimes aids the separation of non-polar lipids. However, two-dimensional techniques are often necessary to resolve mixtures of complex polar lipids (*e.g.* phospholipids, glycolipids, sphingolipids). This involves carrying out the chromatography in one solvent and after drying the plate and turning it through 90°, running it in another solvent.

Silica gel G contains a binder, calcium sulphate, to help it to adhere to the plate, but silica gel H does not and is preferable for some lipid separations, particularly polar mixtures. Some commercially prepared plates contain an alternative organic binder that does not interfere in the same way as calcium sulphate but can present some difficulty with location methods, particularly charring. The degree of hydration of the adsorbent and the particle size will effect the separation and because these cannot be guaranteed, authentic standards should always be run at the same time as the samples.

Silver nitrate may be incorporated in the adsorbent slurry (25 gl^{-1}) giving a final concentration of about 5% in the dry plate. The silver ions bind reversibly with the double bonds in the unsaturated compounds, resulting in selective retardation, and the lipids are separated according to the number and configuration (*cis* or *trans)* of their double bonds. This technique is extremely useful in fatty acids, mono-, di- and particularly triacylglycerol analyses when even positional isomers may be resolved. Borate ions may also be incorporated in the silica gel and these plates are used to separate compounds with adjacent free hydroxyl groups.

Non-polar lipids

Various combinations of hexane or light petroleum (40-60°C, bp) and diethyl ether, usually with a small amount of acetic acid (*e.g.,* 90:10 :1) or diisopropyl ether and acetic acid (98.5:1.5) are commonly used. The greater mobility is demonstrated by cholesterol esters followed by triacylglycerols, free fatty acids, cholesterol, diacylglycerols and monoacylglycerols, with complex polar lipids remaining unmoved. Double development in two solvents, *e.g.* diisopropyl ether-acetic acid (94:4) followed by hexane-diethyl ether-acetic acid (90:10:1) or diethyl ether-benzene-ethanol-glacial acetic acid (40:50:2:0.2) followed by diethyl ether-hexane (6:94) may improve the quality of separation.

Fatty Acids

Only a very small proportion of the fatty acids are present in the free, unesterified form and the vast majority are components of other lipids. Nevertheless it is important to be able to measure and identify the free fatty acids present in either form and for this they must be first extracted into an organic solvent and then usually converted to their methyl ester. The simplest method of methylation, which is applicable to both esterified and non-esterified fatty acids, is to heat the lipid sample for 2 h under a current of nitrogen at 80-90°C with 4% sulphuric acid in methanol. After cooling and the addition of water, the resulting methyl esters are extracted several times into hexane and the combined extracts are dried over sodium carbonate and anhydrous sodium sulphate. The solvent fraction is then reduced in volume by a stream of nitrogen.

Methylation of the fatty acids of acylglycerols may be achieved by dissolving the sample in dichloromethane and heating at 50°C with sodium methoxide in methanol in a stoppered vial for 15 min. This process effects both the release of the fatty acids from the acylglycerol and their methylation. It is known as transesterification. After acidifying with glacial acetic acid and diluting with water, the fatty acid methyl esters are extracted into diethyl ether. The extract is washed with sodium bicarbonate solution (20 gl^{-1}), dried and reduced in volume with nitrogen. Free fatty acids are not methylated by this procedure.

An alternative procedure involves the release of the fatty acids by alkaline hydrolysis (saponification) by refluxing the extracted sample with dilute alcoholic potassium hydroxide for 1 h. After cooling, adding water and acidifying, the fatty acids are extracted into diethyl ether. The methyl esters can then be prepared by treatment with diazomethane, which may also be used directly on free fatty acids. Saponification is less satisfactory, because it is a lengthy procedure and often results in the loss of lipid components.

Argentation thin-layer chromatography is an extemely useful procedure for the separation of methyl esters of fatty acids. Saturated fatty acids have the highest R_F values, which decrease with the increasing degree of unsaturation, and for a particular acid, the *trans* isomer usually travels ahead of its corresponding *cis* isomer. The solvents most commonly used contain hexane and diethyl ether (9 : 1) although a mixture of 4 : 6 is used to separate compounds with more than two double bonds. In order to separate positional isomers of the same acid, conditions must be carefully controlled and multiple development in toluene at low temperatures is often necessary.

Acylglycerols

A wide range of mono-, di- and triacylglycerols may be separated on silver nitrate silica plates using a chloroform-acetic acid solvent (99 : 0.5). For mixtures of monoacylglycerols and diacylglycerols, borate-impregnated plates offer better resolution and a chloroform-acetone solvent (96 : 4) can be used to separate monoacylglycerols from the diacylglycerols, 1,2 isomers from the 1,3 isomers and the 1 isomer from the 3 isomer in monoacylglycerols. Confusing in each compound.

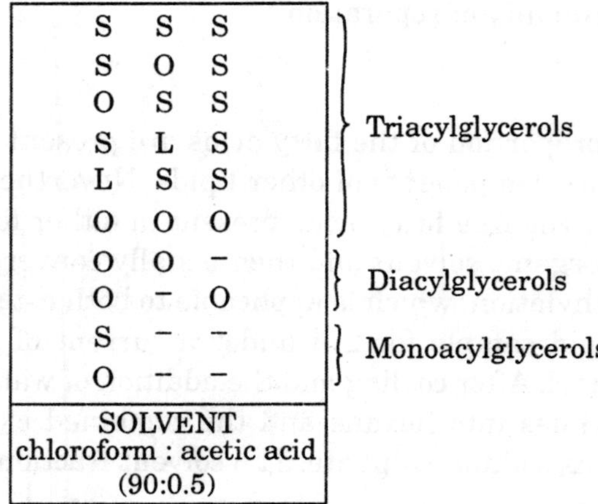

Fig. 2.16. Silver nitrate TLC of acylglycerols. S, O and L denote the fatty acids

S, stearyl, 18 :0

O, oleoyl, 18: $1^{\Delta 9}$

L, linoleyl, 18 : $2^{\Delta 9,12}$

Improved resolution and mobility of mixtures of monocylglycerols and diacylglycerols can be achieved using a choroform-acetone solvent and borate-impregnated plates.

results may be obtained owing to acyl migration occurring during the extraction even though care is taken to avoid the use of heat and alcoholic solvents. The production of the acetate derivatives prevents acyl migration and still permits separation in the usual way.

Triacylglycerols can be separated in a similar manner on silver nitrate-impregnated plates but different solvents may be required to resolve mixtures containing compounds with widely differing numbers of double bonds. Solvents such as hexane-diethyl ether (80 : 20) or chloroform-methanol (99:1) are applicable when up to four double bonds are present, whereas more polar solvents, *e.g.*, diethyl ether or chloroform-methanol (96 : 4) are required for higher degrees of unsaturation and even chloroform-methanol-water (65 : 25 : 4) for the highly unsaturated, *i.e.* 6 to 12 double bonds.

Complex lipids

It is difficult to separate the wide range of complex lipids in a single solvent system but the task is simplified if the sample has already been partially purified (*e.g.*, by a column technique) and only one class of lipid is present. Even so, it is often necessary to perform two-dimensional chromatography and silica gel without binder is often preferred.

Mixtures of phosphoglycerides can be separated using a chloroform-methanol-water mixture, the proportions of which may be varied to suit the sample constituents (*e.g.*, 65 : 25 : 4). Acetone is sometimes included in the solvent and silver nitrate-impregnated plates can be used. Acetic acid (1-4%) is also a useful additive to effect the separation of neutral phosphoglycerides

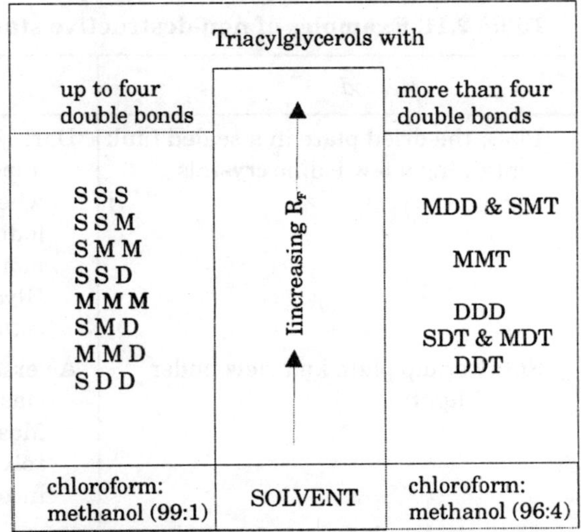

Fig. 2.17. Silver nitrate TLC of triacylglycerols.

S, saturated fatty acid

M, monoenoic fatty acid

D, dienoic fatty acid

T, trienoic fatty acid

Triacylglycerols with similar total numbers of double bonds can be separated using a chloroform-methanol solvent. The proportions of the solvent are adjusted to achieve resolution of compounds with either a total of up to four double bonds or more than four.

Table 2.10. Destructive staining procedures

Reagent	*Method*	*Comments*
30-50% sulphuric acid in water (v/v)	Spray the plate and heat at 180°C for 5 min on a not plate or 30 min in an oven	Detects all lipids by carbonization of organic material producing black charred areas. At lower temperatures (~80°C) sterols give a purple-red
Chromic acid $6gl^{-1}$ potassium dichromate in 55% sulphuric acid	Spray the plate and heat at 180 °C for 5 on a hot plate or 30 min in an oven	All lipids give black charred spots
50% orthophosphoric acid in water	After spraying, heat on a hot, plate at 300°C for 5 min	All lipids give black charred spots

from acidic phosphoglycerides (phosphatidyl serine and phosphatidyl inositol). Other modifications which may improve the quality of the separation include the addition of sodium carbonate or ammonium sulphate in the preparation of the silica gel or the preliminary removal of the simple lipids to the top of the plate by running it in acetone-hexane solvent (1 : 3). When the extract is rich in acidic phosphoglycerides it is preferable to use a solvent containing ammonia.

Table 2.11. Examples of non-destructive stains

Reagent	*Method*	*Comments*
Iodine vapour	Place the dried plate in a sealed tank containing a few iodine crystals	Dark yellow-brown spots appear within a few minutes where lipids have absorbed the iodine. Unsaturated lipids are more intensely stained. Glycolipids do not stain significantly
Rhodamine 6G (0.1 gl^{-1}in water)	Spray or dip plate and view under UV light	A versatile stain for all lipid classes. Pink-purple-red spots. Most useful with alkaline solvents. The dye may be incorporated in the plate
2',7'-dichloro-fluorescein (1-2 gl^{-1} in 95% ethanol)	Spray or dip plate and view under UV light	Does not detect phospholipids or glycolipids. Other lipids give yellow spots on purple background. Useful for acid solvents. Dye may be incorporated in the plate
Water	Spray the plate to saturate	Opaque spots appear on a translucent background. Not very sensitive and only applicable to preparative work

Two-dimensional techniques are usually employed if both phosphoglycerides and glycolipids are present, but it is possible to resolve members of both classes using a diisobutylketone-acetic acid-water mixture (40 : 25 : 5). A solvent composed of acetone, acetic acid and water (100 : 2:1) will separate the mono- and di-galactosyldiglycerides, which are particularly abundant in plant extracts, from phosphoglycerides, which remain at the origin.

Many solvent combinations have been described for two-dimensional separation and, in general, if an alkaline or neutral solvent is chosen for the first dimension then the second solvent should be acidic. Also it may be useful for one solvent to contain acetone, which will enhance the movement of glycolipids relative to the phosphoglycerides. Acetic acid should not be used in the first solvent, because it is difficult to remove completely and affects the quality of the separation in the second dimension. Difficulties are also encountered in the removal of butanol, which interferes with the charring process often used for the location of the spots.

Table 2.12. Specific locating reagents

Reagent	*Lipids detected*	*Comments*
Orcinol-sulphuric acid 2 g orcinol in a litre of 75% sulphuric acid	Glycolipids	Heat the plate at 120°C. Blue-violet spots on a white background Approximately 10 min
α-Naphthol 5 gl^{-1} α-Naphthol in methanol: water (1:1). Spray the plate, air dry and respray with 95% sulphuric acid	Glycolipids	Heat the plate at 120°C. Blue-purple spots. Other polar lipids give yellow spots and cholesterol gives a grey-red spot
Periodate-Schiff Spray with 10 gl^{-1} sodium periodate. Leave for 10 min for oxidation. Remove excess with sulphur dioxide. When plate is colourless, spray with 10 gl^{-1} *p*-rosaniline HCl in water previously decolorized by saturation with sulphur dioxide	Glycolipids Phosphatides with a vicinal hydroxyl group	Phosphatidyl glycerol gives a purple spot almost immediately and phosphatidyl inositol a yellow spot after about 40 min. Glycolipids give blue spots, the intensity increasing up to 24 hours.
Dragendorff stain A. Potassium iodide 400 gl^{-1} in water. B. Bismuth subnitrate 17 gl^{-1} in 20% acetic acid. Mix A and B and water (1:4:15) immediately before use	Choline containing	Orange-red spots within a few minutes with gentle warming
Ninhydrin 2 g l^{-1} in acetone	Containing free amino groups	Purple spots appear with phosphatidyl ethanolamine and phosphatidyl serine, for instance, after heating at 100°C
Molybdenum blue Sodium molybdate 100 gl^{-1} in 4 mol l^{-1} HCl. Hydrazine HCl 10 gl^{-1} in water. Mix 100 ml of each and heat in boiling water for 5 min. Cool and dilute to 1 litre	Containing phosphorus	Blue spots appear immediately
Resorcinol A. 20 g l^{-1} resorcinol in water B. Concentrated HCl C. 0.1 mol l^{-1} copper sulphate D. Water Mix A, B, C, D (10:80:0.5:10)	Gangliosides	Compounds containing sialic acid residues react to give blue-violet spots. Other glycolipids give yellow spots
Acidic ferric chloride reagent 0.5 g ferric chloride, 5 ml glacial acetic acid and 5 ml concentrated sulphuric acid in 1 litre of water	Cholesterol and its esters	Heat to 100°C. Cholesterol appears as a red-violet spot in a few minutes followed by the esters. Continued heating causes charring of all lipids

Detection and Quantitation

Lipids may be visualized on thin-layer plates using a general stain but this does not usually indicate the nature of the lipid present. Many of these are destructive and the lipid cannot be analysed further after elution from the plate, but other stains are non-destructive. There are specific reagents that will react with one of the chemical groups in the lipid *e.g.* ninhydrin for lipids containing a free amino group.

In some instances quantitation, or at least comparison of the proportions of the separated lipids, may be possible by scanning the plate directly by reflectance densitometry, *e.g.* after charring. Alternatively, the spot may be scraped off the plate and the colour or fluorescence of the eluted solution measured. The eluted compounds can be submitted to further analysis, by GLC for instance, to enable quantitation or identification. For reliable and meaningful results, known standards should always be run and analysed under identical conditions because of variable colour yields with different lipids. Even so, results obtained by different methods will often vary.

Gas-Liquid Chromatography

Gas-liquid chromatography is a very useful technique in lipid analysis, particularly for the separation of very similar compounds within classes. Because of the wide variations in structure and properties between classes it is not usually possible to resolve members of different classes on the same column. GLC is useful for both quantitative and qualitative analysis and also in the investigation of lipid structure.

However, full structural analysis of a lipid will often necessitate further analysis of the collected column effluent for a single GLC peak. Infrared and NMR spectroscopy and mass spectrometry are all useful techniques which will give information for identification purposes, including the position and configuration of any double bonds.

Acylglycerols

Acylglycerols may be separated as intact molecules by GLC after the preparation of acetate or trimethylsilyl (TMS) derivatives to reduce the polarity and prevent acyl migration. Acetylation can be carried out at room temperature by the addition of a solution of acetic anhydride in pyridine (5 : 1). After allowing the reaction to take place overnight, the excess reagent is removed with a stream of nitrogen and gentle warming. TMS derivatives may be prepared by treatment with excess N, O-bis(trimethylsilyl) acetamide (10% v/v in hexane) or with a mixture of hexamethyldisilazane and trimethylchlorosilane in hexane. Silylation is complete in a few minutes at room temperature.

TMS ether derivatives of monoacylglycerols are sufficiently volatile to be analysed on a polyester column (*e.g.*, EGA, PEGA), which permits separation on the basis of chain length and degree of saturation of the fatty acid component. Non-polar silicone stationary phase (*e.g.*, SE-30, OV-1), which are thermally stable up to 350°C are, however, necessary for diacylglycerols and triacylglycerols. The compounds are separated solely on the total number of fatty acid carbon atoms present, known as the carbon number, and those differing in their carbon number by two can be resolved. The temperature required for separation of diacylglycerols is approximately 270-310°C with temperature programming between 2 and 4°C per minute. A higher temperature may be required for triacylglycerols whereas monoacylglycerols, which may also be analysed on these columns, require much lower temperatures.

Qualitative analysis may involve the identification of an intact acylglycerol molecule or it may be a study of its composition and structure. GLC is particularly useful in the investigation of the fatty acids present in a triacylglycerol and their position in the molecule. Chemical hydrolysis can be used to release all the fatty acids whereas the enzyme pancreatic lipase will, in theory, only hydrolyse the fatty acids esterified at positions 1 and 3 leaving the 2-monoacylglycerol. In practice, hydrolysis at position 2 does eventually occur and the procedure must be carefully controlled or misleading results will be obtained. In a similar manner, specific phospholipases are also available which will cleave a particular bond in a phospholipid permitting an individual component to be analysed.

Wax Esters

Wax esters have similar relative molecular masses to diacylglycerols and are eluted under comparable gas chromatographic conditions. Alternatively, the alcohol and fatty acid moieties can be released by saponification and their methyl esters subjected to gas chromatography.

Cholesterol Esters

These can be analysed using the columns applicable to triacylglycerols. It may be possible with temperature programming to separate compounds that differ by only one carbon atom whereas isothermal analysis will normally resolve compounds with the same carbon number but which contain fatty acids with different degrees of unsaturation. These columns are also applicable to glycosyldiglyceride analysis.

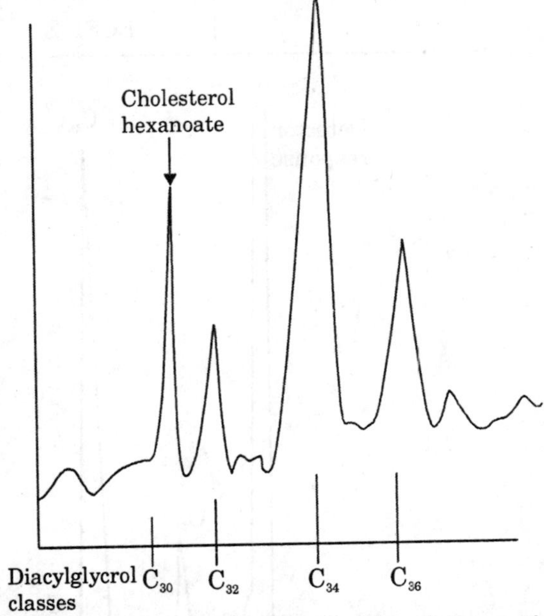

Fig. 2.18. GLC of diacylglycerols. Schematic representation of the separation of the trimethylsilyl derivatives of diacylglycerols on an SE-30 stationary phase with temperature programming. Cholesterol hexanoate is the internal standard.

Fatty Acids

It is usual to convert fatty acids to their methyl ester derivatives before separation by GLC, although it may be possible to analyse those with short chain lengths (two to eight carbon atoms) as the free fatty acids. Polar or non-polar stationary phases can be used and capillary (open-tubular) or SCOT columns will separate positional and geometric isomers. The *cis* isomers have shorter retention times than the corresponding *trans* isomers on a non-polar phase and visa versa on a polar phase.

The non-polar columns have either saturated paraffin hydrocarbon (Apiezon grease) or silicone greases as the liquid phase. Some suitable silicone greases include SE-30, OV-101, JXR 9 (all methyl silicones) and OV-1 (dimethyl silicone). Separation takes place largely on the basis of relative molecular mass (chain length) and these greases are useful for resolving mixtures of saturated fatty acids, usually isothermally. They can, however, differentiate between saturated and unsaturated components with the same chain length, unsaturated acids having shorter retention times than the corresponding saturated acids, as is also the case with branched chain structures compared with straight chains.

Table 2.13. Polar Stationary Phases for GLC of Fatty Acid (Methyl Esters)

Polarity	*Stationary phase*	
Highly polar	EGS	Ethyleneglycol succinate
	DECS	Diethyleneglycol succinate
	EGSS-X	Copolymer of EGS with methyl silicone
Medium polarity	EGA	Ethyleneglycol adipate
	BDS	Butanediol succinate
	EGSS-Y	Copolymer of EGS with a higher proportion of methyl silicone than EGSS-X
Low polarity	NPGS	Neopentylglycol succinate
•	EGSP-Z	Copolymer of EGS and phenyl silicone

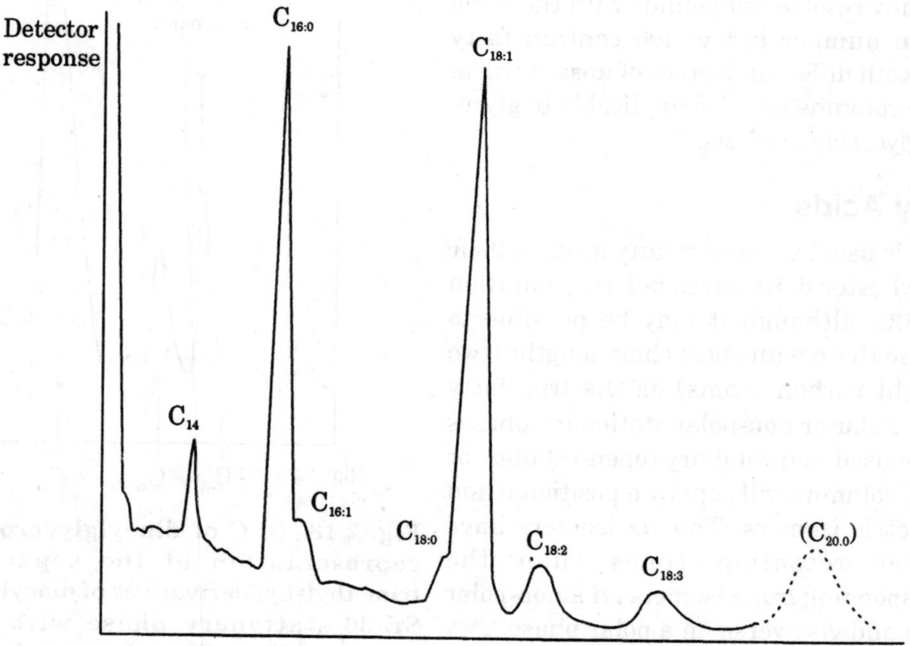

Fig. 2.19. GLC of fatty acid methyl esters. Schematic representation of isothermal separation on a PEGA stationary phase.

The liquid phases of polar columns are usually the heat-stable polymers of ethyleneglycol and the dibasic acids, succinic or adipic. Fatty acids are separated on the basis of both chain length and the degree of unsaturation and some columns are capable of resolving fatty acids with the same chain length but different numbers of double bonds (0-6). The saturated fatty acids show the shortest retention times followed by the monoenoic, dienoic, etc.

In some instances it may be possible to identify an unknown compound with some degree of certainty by showing that it has a retention time identical to that of a known fatty acid. The two substances must undergo co-chromatography on at least two columns which have stationary phases of different polarity, *e.g.*, EGSS-X and EGSS-Y. When this is not feasible, the chain length of the test substance may be determined using its relative retention time compared with the relative retention times of known fatty acids. The relative retention time of a substance is its own retention time compared with that of a chosen standard of known chain length, *e.g.* methyl palmitate (16:0).

$$\text{Relative retention time of X} = \frac{\text{Retention time of methyl ester X}}{\text{Retention time of methyl palmitate}}$$

To determine the chain length of saturated acids, the relative retention times of several fatty acids of known chain length are found. A plot of the logarithm of the relative retention time against chain length results in a straight line graph which can be used to find the chain length of the test substance.

The identification of unsaturated compounds is more involved but a useful approach is to determine the relative retention times of several different long chain fatty acids with known numbers of double bonds on both a polar and a non-polar column. A plot of the relative retention times of each reference compound on the polar phase against those demonstrated on the non-polar phase results in a series of parallel straight lines, each line corresponding to a homologous series of saturates, monoenes, dienes, trienes, etc. This graph can then be used to give information about the structure, chain length and number of double bonds present in the unknown acid.

Conversion of an unsaturated acid to the saturated compound permits the chain length to be more easily determined by GLC. This can be achieved by hydrogenation and if double bonds were originally present the resulting substance will alter its retention time to that of the saturated acid.

Unsaturated acids may be split chemically at their double bonds. Permanganate-periodate oxidation has been used to produce the corresponding carboxylic acids, while an alternative technique of ozonolysis results in the formation of aldehydes and aldehyde esters. All these reaction products may be identified by GLC and the information used to determine the position of the double bond in the original fatty acid.

CHAPTER 3

Oils and Fats

Phosphatides and mucilaginous substances are emulsifying agents and in excessive quantity may increase the loss of neutral oil when the oil is caustic refined. If the oil is going to be physical or steam refined, the content of phosphatides and metallic pro-oxidants must be lowered by a degumming process before being refined. Water degumming generally removes about 90 percent of the phospholipid material. There are, however, some non-hydratable phosphatides which do not react with water and are removed during caustic refining. Their presence tends to increase the neutral oil lost in the soapstock.

It is usually not economical to degum sunflower oil in a separate operation because the percentage of lecithin is so small. In addition to removing lecithin, degumming also removes some other foreign material. However, the quantity removed is usually so little that it can be easily removed during the alkali refining procedure. Additional equipment cost, labor expense and oil loss does not warrant degumming of sunflower oil if it is to be alkali refined.

NEUTRALIZATION OR REFINING

Neutralization removes nonglyceride impurities, which consist principally of free fatty acid (FFA), and significant quantities of mucilaginous substances, phosphatides, chlorophyll, and color bodies. The refining process removes the objectionable impurities in the oil with the least damage to either the glycerides, the tocopherols or other natural antioxidants.

Chemical (or caustic) refining converts the FFA to oil-insoluble soaps, removes all the substances, which become insoluble upon hydration, and removes some of the colour bodies present in the oil. Reactions are at a relatively low temperature, allowing for the refining equipment to be fabricated essentially from iron or mild steel.

The technology of caustic refining is concerned with: (*a*) the proper choice of alkali used, (*b*) the quantity of alkali used, and (*c*) the refining technique used to produce the desired purification (without excessive saponification of the neutral oil) and (*d*) the method used for efficient separation of the refined oil from the soapstock.

A disadvantage of the caustic refining method is the loss of a small portion of neutral oil, in addition to the FFA, as a result of saponification. This loss can be minimized, however, by utilizing good technology.

Sunflower oil can be alkali refined using either a batch method or a continuous method of refining. The batch method requires less capital investment, but; (*a*) the yield of refined oil is considerably less than via the continuous method, (*b*) the thruput rate is less and (*c*) there is a higher labor cost.

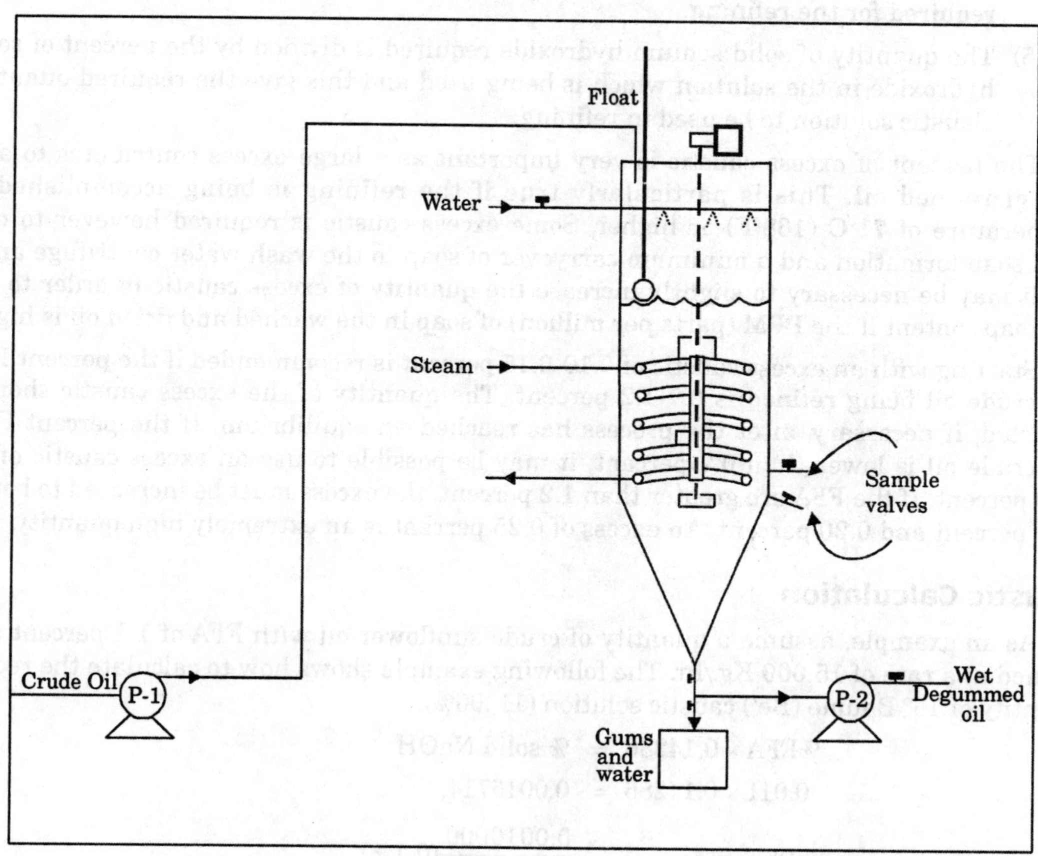

Fig. 3.1. Bath Degumming.

Calculation of Caustic

Caustic calculation is the method of determining the quantity and concentration of the caustic solution required for alkali refining as it is applicable to both the batch and continuous methods.

The quantity of caustic required for neutralization can be determined in the following manner:

(1) Multiply the percent FFA of the oil by 0.14286. This gives the percent of solid sodium hydroxide required for neutralization.

(2) To this percent, add a slight excess quantity, as the quantity calculated in Step 1 is the theoretical quantity. The excess usually ranges between 0.05 percent and 0.20 percent, depending upon the percent of FFA present in the oil and the quantity of non-oily constituents present in the crude oil.

(3) Add the percent calculated in (1) to the desired excess established in (2).

(4) The quantity of oil in pounds (or kilograms) is multiplied by the total percent arrived at in (3). This gives the number of pounds (or kilograms) of solid sodium hydroxide required for the refining.

(5) The quantity of solid sodium hydroxide required is divided by the percent of sodium hydroxide in the solution which is being used and this give the required quantity of Caustic solution to be used in refining.

The percent of excess caustic is very important as a large excess contributes to a high loss of refined oil. This is particularly true if the refining is being accomplished at a temperature of 71°C (168°F) or higher. Some excess caustic is required however to obtain good soap formation and a minimum carryover of soap to the wash water centrifuge and the oil. It may be necessary to slightly increase the quantity of excess caustic in order to lower the soap content if the PPM (parts per million) of soap in the washed and dried oil is high.

Starting with an excess caustic of 0.10-0.15 percent is recommended if the percent FFA of the crude oil being refined is 0.7-1.2 percent. The quantity of the excess caustic should be adjusted, if necessary, after the process has reached an equilibrium. If the percent FFA in the crude oil is lower than 0.7 percent, it may be possible to use an excess caustic of 0.05-0.10 percent. If the FFA are greater than 1.2 percent, the excess must be increased to between 0.15 percent and 0.20 percent. An excess of 0.25 percent is an extremely high quantity.

Caustic Calculation

As an example, assume a quantity of crude sunflower oil with FFA of 1.1 percent and is refined at a rate of 15,000 Kg./hr. The following example shows how to calculate the required quantity of 16° Baume (Be′) caustic solution (11 .06%):

$$\%\text{FFA} \times 0.14286 = \%\text{ solid NaOH}$$

$$0.011 \times 0.14286 = 0.0015714$$

$$\text{Plus excess of} = \frac{0.0010000}{0.0025714}(0.1\%)$$

$$15{,}000\ \text{Kgs/hr.} \times .0025714 = \text{solid NaOH}$$

$$\frac{38.571\ \text{Kgs. of solid NaOH}}{0.1106(16°\text{Be}')} = 384.74\ \text{Kgs. of } 16°\text{Be}'\text{ per hour}$$

It is suggested that a 16° Be′ caustic solution be used for refining sunflower oil. The concentration is not so high as to result in a significant amount of saponification of the neutral oil if the excess quantity was greater than the optimum. The Be′ of the caustic should be varied depending on various factors and conditions. If the percent FFA of the oil to be refined is greater than 1.5 and the crude oil is relatively free from non-oily constituents such as gums and lecithin, it would be advisable to use a caustic solution having a Be′ of 16.5-17.0. The reason for this is to avoid introducing an excessive amount of water which could inhibit the good separation.

If refining using the continuous method, adjust the quantity of caustic solution as the percent FFA changes or the quantity of non-oily constituents in the feed stock change. In a continuous refinery, a chart should be prepared which shows the quantity of caustic solution to be used as the percent FFA changes during the run.

If a phosphoric acid treat is used before caustic refining, additional caustic solution is required to neutralize the added phosphoric acid. The oil and acid should be mixed well, using good shearing agitation. The mixing should continue for a fairly long time (prior to the addition of the caustic solution) to prevent the formation of a sticky soap. Such soap is difficult to discharge and will foul the centrifuge discs.

Batch Refining

This method of refining is conducted in a mild steel, steep cone-bottom tank (about a 60° cone angle) equipped with an agitator (slow speed) having paddle-type blades which are pitched to lift the product. Also, it should have internal steam coils and a method of spraying the caustic solution and water on the surface of the oil. It is desirable to have a small draw-off valve for obtaining samples located about 24 cm (six inches) above the start of the straight side of the tank and another 24 cm. below the top of the cone bottom. A thermometer should be located about halfway up on the straight side of the tank.

Refining oil loss using the batch method is about six times the percentage of FFA in the crude oil.

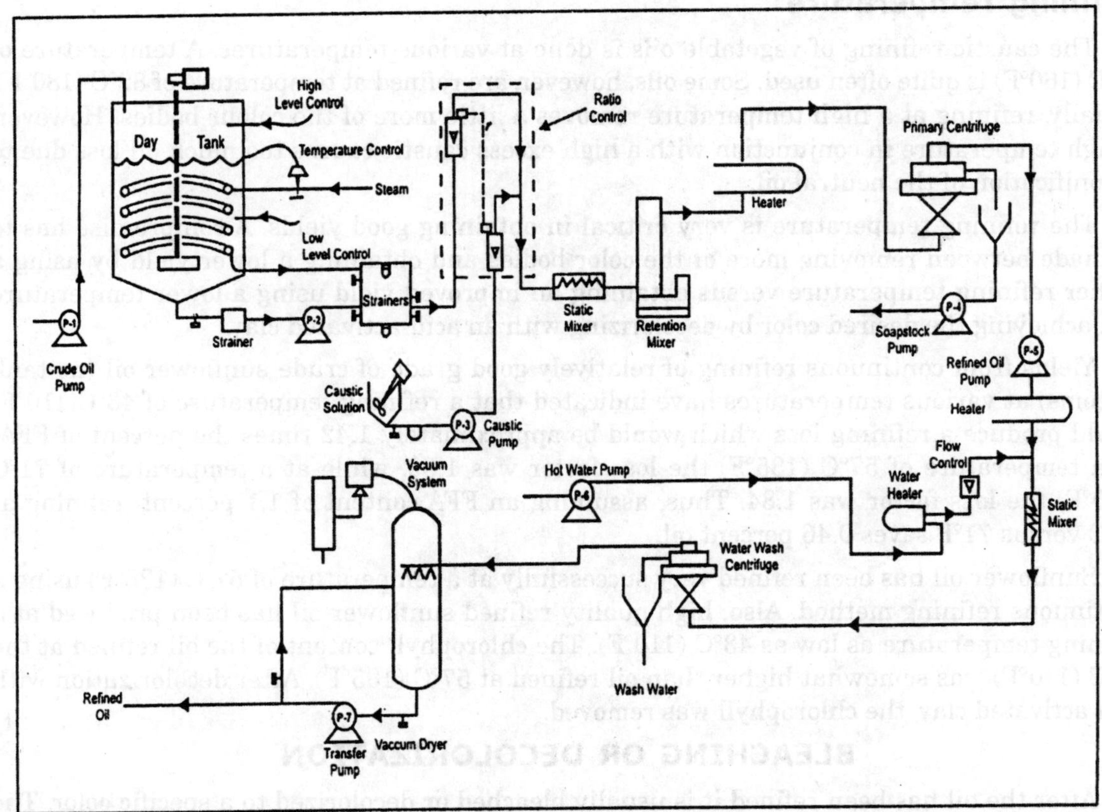

Fig. 3.2. Continuous Refining

Continuous Refining

A continuous method of refining requires a considerable investment in equipment. However, it is the most efficient and economical method.

The quantity of caustic soda solution to be used is calculated as previously shown for the batch method. The excess and °Be' for the continuous refinery is very similar to what is used in the batch refining. The excess alkali should be determined occasionally during the refining to ensure the process is under control. It is desirable to determine the soap content after the water washing and vacuum drying. This is so small changes can be made in the refining conditions to minimize the soap level before further processing the oil.

Addition of a small amount of weak acid solution (*e.g.* citric acid) prior to vacuum drying may be desirable if the oil is going to be deodorized or further processed without decolorization with an acid activated clay. However, if the oil is to be bleached with bleaching clay, the acidity in the clay usually neutralizes the slight alkalinity of the oil.

The continuous method of refining permits varying the refining temperature considerably to obtain a good separation of the heavy phase using the centrifuge. A temperature as low as 43°C (110°F) has been used to produce a good quality refined oil. A neutralization temperature of 57"C (135°F) gives both very good refined oil and good yield.

Refining Temperature

The caustic refining of vegetable oils is done at various temperatures. A temperature of 71°C (160°F) is quite often used. Some oils, however, are refined at temperature of 82°C (180°F). Usually, refining at a high temperature removes a little more of the colour bodies. However, a high temperature in conjunction with a high excess caustic causes too much oil loss due to saponification of the neutral oil.

The refining temperature is very critical in obtaining good yields. A compromise has to be made between removing more of the color bodies and obtaining a lower yield by using a higher refining temperature versus obtaining an improved yield using a lower temperature and achieving the desired color by decolorizing with an acid-activated clay.

Yields from continuous refining of relatively good grade of crude sunflower oil (no tank bottoms) at various temperatures have indicated that a refining temperature of 43°C(110°F) would produce a refining loss which would be approximately 1.42 times the percent of FFA. At a temperature of 57°C (135°F) the loss factor was 1.62, while at a temperature of 71°C (160°F) the loss factor was 1.84. Thus, assuming an FFA content of 1.1 percent, refining at 43°C versus 71°F saves 0.46 percent oil.

Sunflower oil has been refined very successfully at a temperature of 57°C (135°F) using a continuous refining method. Also, high quality refined sunflower oil has been produced at a refining temperature as low as 43°C (110°F). The chlorophyll content of the oil refined at the 43°C (110°F) was somewhat higher than oil refined at 57°C (135°F). After decolorization with acid activated clay, the chlorophyll was removed.

BLEACHING OR DECOLORIZATION

After the oil has been refined it is usually bleached or decolorized to a specific color. The bleaching process should be done under vacuum and at an approximate temperature of 77°C

(170°F). The use of acid activated clay is recommended as the residual acid in the clay neutralizes the traces of soap that may be present. The chlorophyll removal is an important step as otherwise a green tint may appear in the oil after deodorization.

The oil should be dry before the addition of the clay if an acid activated clay is used. Any moisture present in the oil causes the acid to leach out of the clay, which causes an increase in the FFA content of the refined oil. If the filters are well blown, the oil loss in the bleaching process is about equal to the quantity of bleaching clay used in the process.

DEWAXING

The dewaxing process removes the waxy-type of the non-oily constituents that are not removed either by the alkali refining process or the degumming process. There are various methods, which can be used for dewaxing the oils such as a batch method or various types of continuous methods.

Batch Dewaxing Method

The batch method is relatively simple. It consists of an well-insulated tank of a given capacity, equipped with some means of agitation. A relatively slow speed mechanical agitator is preferred, however, it is possible to use a pump to circulate the oil from the bottom to the top. Under no circumstance should air be used for agitation of oil. Air bubbles will form in the oil which (a) inhibit the heat transfer causing an extended processing time and (b) increase the peroxide value of the oil.

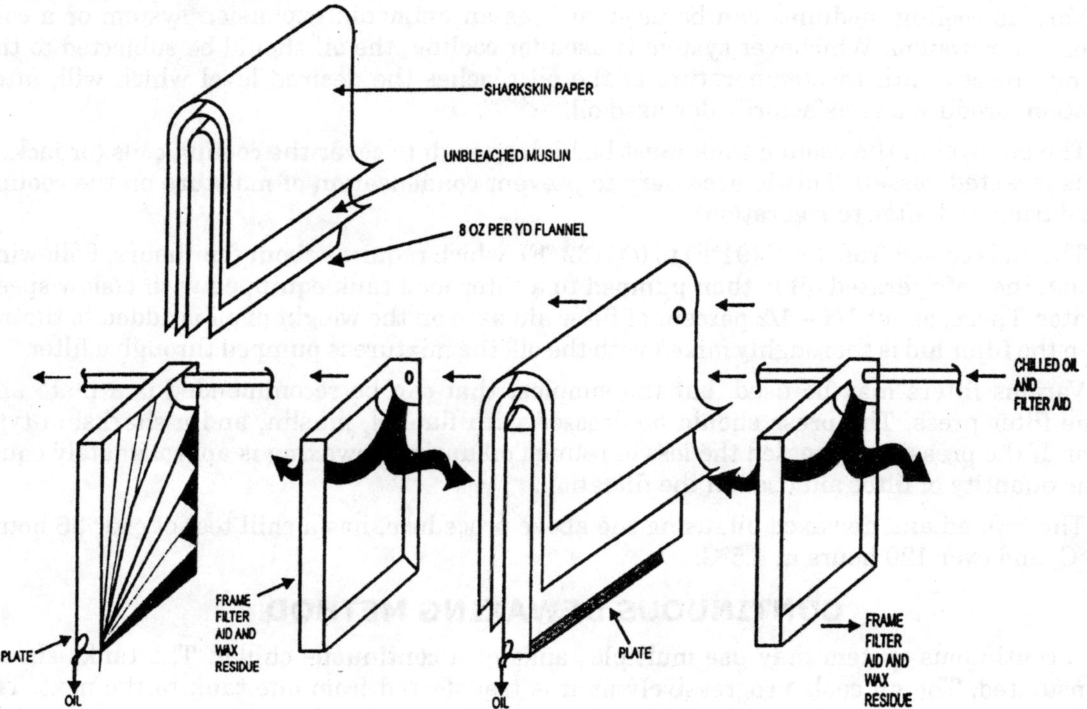

Fig. 3.3. Dressing the Filter Press-Dewaxing.

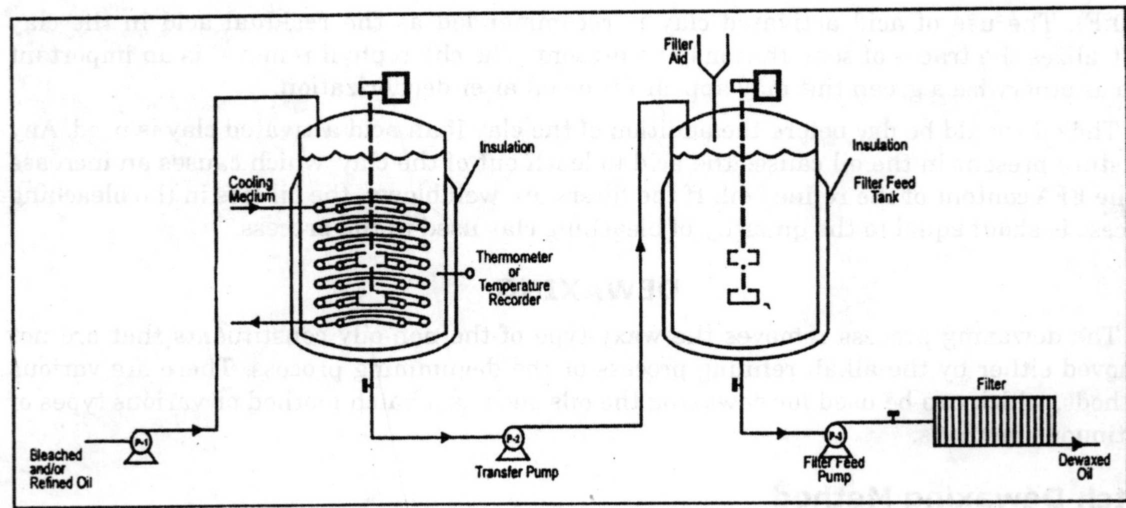

Fig. 3.4. Batch Dewaxing.

The tank used for cooling the oil should also be equipped with a large number of relatively small diameter coils in order to give a large cooling surface. It would be preferable to have a vertical-type tank rather than horizontal, though either is acceptable. However, the agitation should move the oil rather slowly across the cooling coils, yet it should be enough to keep the waxes from settling out of the oil.

Various cooling mediums can be used such as an ammonia expansion system or a cold glycol/water system. Whichever system is used for cooling, the oil should be subjected to the cooling process until the temperature of the oil reaches the desired level which will, after filtration, produce a satisfactorily dewaxed oil.

The oil level in the cooling tank must be high enough to cover the cooling coils (or jacket, if it is jacketed vessel). This is necessary to prevent condensation of moisture on the cooling coils during and after refrigeration.

The oil is cooled from 33°C (91°F) to 0°C (32 °F), which requires about four hours. Following cooling, the refrigerated oil is then pumped to a filter feed tank equipped with a slow speed agitator. There, about 1/4 – 1/2 percent of filter aid base on the weight of oil is added to the oil. When the filter aid is thoroughly mixed with the oil, the mixture is pumped through a filter.

Various filters may be used, but the simplest that can be recommended is a plate and frame filter press. The press should be dressed with flannel, muslin, and a sharkshin-type paper. If the press is so dressed the loss of refined oil during dewaxing is approximately equal to the quantity of filter aid used in the filtration.

The refined and dewaxed oil, using the above procedure, has a chill test of over 36 hours at 0°C, and over 120 hours at 4.5°C.

CONTINUOUS DEWAXING METHOD

A continuous system may use multiple tanks or a continuous chiller. The tanks should be insulated. The oil cools progressively as it is transferred from one tank to the next. The residence time in the system is long enough to permit reducing the temperature of the oil to the desired temperature before it is pumped from the final tank.

The chilled oil is then pumped to a filter feed tank where the filter aid is added and then pumped through a filter to remove the filter aid and waxes.

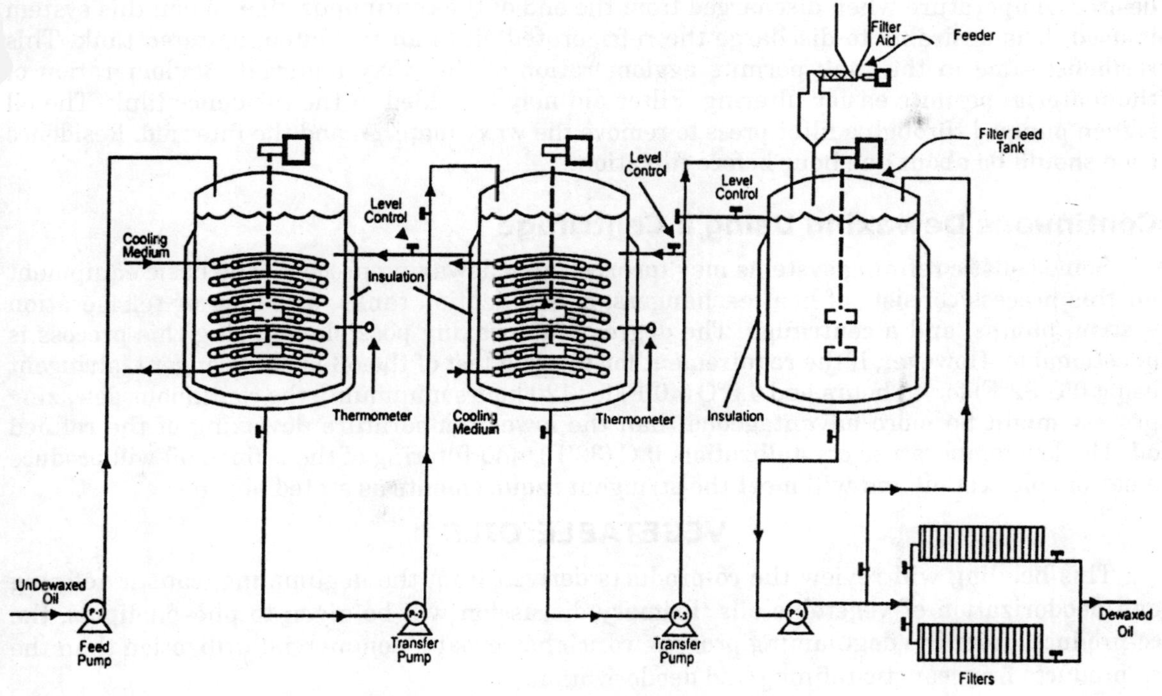

Fig. 3.5. Continuous Dewaxing.

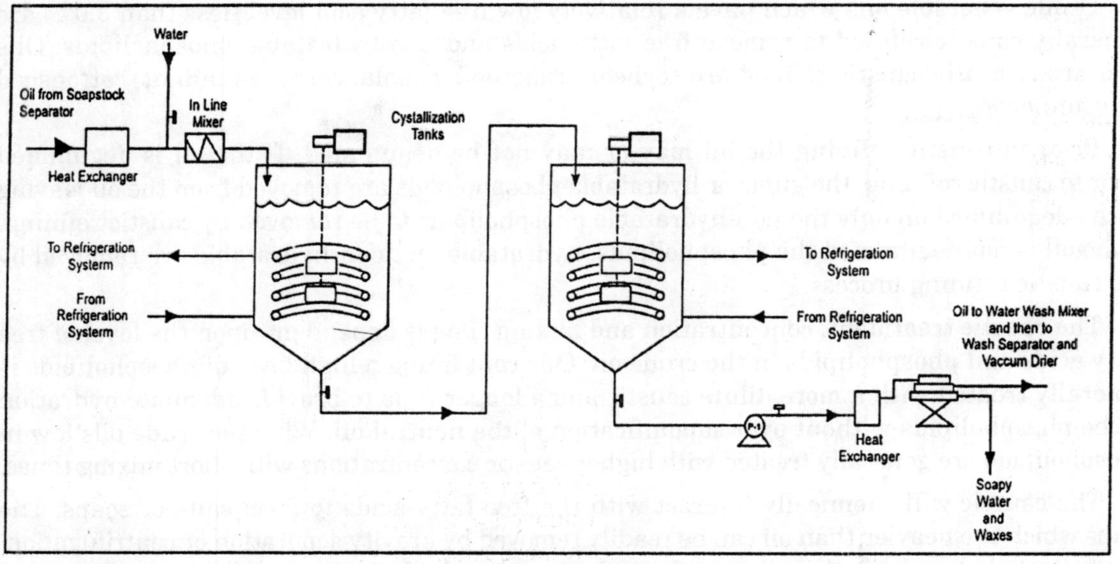

Fig. 3.6. Continuous Dewaxing (Using a centrifuge).

Alternatively, a continuous process may use a continuous chiller. With this process, the oil being chilled passes through a continuous pipe that has a cooling medium on the outside of a jacketed pipe. The flow rate is controlled so that the temperature of the oil is at the desired temperature when discharged from the end of the continuous pipe. When this system is used, it is desirable to discharge the refrigerated oil to an insulated, agitated tank. This residence time in the tank permits agglomeration of the waxy material. Agglomeration of the material permits easier filtering. Filter aid may be added in the residence tank. Th[illegible] is then pumped through a filter press to remove the waxy material and the filter aid. Residence time should be about one hour before filtration.

Continuous Dewaxing Using a Centrifuge

Some caustic refining systems may incorporate a dewaxing process. The basic equipment for this process consists of heat exchangers, crystallization tanks, a cooling or refrigeration system, pumps, and a centrifuge. The degree of de-waxing possible by using this precess is questionable. However, if the requirement for the cold test of the oil is not stringent [stringent being 0°C(32°F) for 48 hours and 4.5°C (40°F) for 120 hours minimum], the continuous dewaxing process might be more advantageous than the lower temperature dewaxing of the refined oil. The low temperature crystallization, 0°C (32°F), and filtering of the refined oil will produce a better cold test oil and will meet the stringent requirements as stated above.

VEGETABLE OILS

This heading will review the co-products derived from the degumming, caustic refining and deodorization of vegetable oils. Primary discussion will be given to phospholipids, the co-products from the degumming process, which has greater commercial utilization than the co-products from caustic refining and deodorization.

CO-PRODUCT FROM CAUSTIC REFINING-SOAPSTOCKS

Crude vegetable oils which have a relatively low free fatty acid level (less than 3.0%) are generally caustic refined to remove free fatty acids and non-hydratable phospholipids. Oils that are typically caustic refined are soybean, rapeseed, canola, corn, groundnut, cottonseed and sunflower.

Prior to caustic refining the oil may or may not be degummed. If the oil is degummed prior to caustic refining, the gums or hydratable phospholipids are removed from the oil leaving in the degummed oil only the non-hydratable phospholipids to be removed by caustic refining. If the oil is not degummed the phospholipids, hydratable and non-hydratable are removed by the caustic refining process.

The caustic treatment, concentration and mixing time is dependent upon the level of free fatty acids and phospholipids in the crude oil. Oils containing a high level of phospholipids is generally treated with a more dilute caustic and a longer time to provide adequate hydration of the phospholipids without over saponification of the neutral oil. Whereas crude oils low in phospholipids are generally treated with higher caustic concentrations with short mixing times.

The caustic will chemically interact with the free fatty acids to form salts or soaps. The soaps which are heavier than oil can be readily removed by gravity separation or centrifugation. The hydrated phospholipids and soaps being both more soluble in the water phase are removed collectively in water phase after refining and referred to as the soapstock.

The concentration of soaps (fatty acids) and phospholipids in the soapstock (generally total fatty acid content between 40 to 50%) is related to the strength of caustic (concentration of caustic in water) and additional water washing used for refining. The soapstock is a good source of non-edible fatty acids, which can be used to supplement animal feed or as a source for industrial fatty acids. Because of the high level of water and pH, soapstock can support microbial growth and will ferment.

The "wet" soapstock may be applied directly to animal feed, provided that the additional water from the soapstock does not raise the moisture level of the animal feed above the minimum level required to keep the animal feed stable.

To prevent microbial activity the soapstock can be dried or the pH lowered. Soapstock can be dried using a variety of methods. A common commercial practice of drying soapstock involves neutralizing the soapstock with sulphuric acid to a neutral or slightly acidic pH under mild heat, then removing the moisture via a thin film evaporator to less than 2%.

Soapstock is commonly acidulated with sulphuric acid to an acid pH whereupon the fatty acids are liberated from the metals to form free fatty acids. The acid salts and water are then separated from the free fatty acids via gravity separation or centrifugation to concentrate the total fatty acids to 85-95%.

Co-Products from Deodorization - Free Fatty Acids, Sterols, Tocopherols and Tocotrienols

The final processing step of either physical refining or caustic refining is a high temperature, high vacuum distillation, commonly referred to as deodorization. Of course in physical refining the primary objective of the distillation process is to remove a majority of the free fatty acids and flavor components, and in caustic refining the primary objective is to remove the flavor components and residual fatty acids.

In addition to the free fatty acids, sterols, tocopherols and tocotrienols are also removed by the distillation process along with pesticides, oxidized materials and truncated lipids (a result of oxidation or generated from previous processing steps). The materials along with the water used as sparging steam, are collected and commonly referred to a deodorizer distillate.

The components in deodorizer distillate having greater commercial value (sterols, tocopherols and tocotrienols) must be separated from the components of lessor commercial value by chromatography and additional distillation steps. While the isolated sterols, tocopherols and tocotrienols can command a high value in the market, the processing equipment required to isolate these materials is costly and is commercially practiced by a few oil refiners and oleo chemical manufacturers.

Co-Products from Degumming Phospholipids, Lecithin

Lecithin is a term that collectively refers to all the water-soluble materials or gums that are separated from a crude oil by the degumming process and are retained after the gums are dried.

The gums are removed from vegetable oils prior to caustic or physical refining because the gums will interfere with subsequent processing steps and degrade during the final distillation process imparting dark colors and off-flavors to the final processed oil.

Lecithin contains primarily the following components:

Phosphatidyl ethanol amine (PE), phosphatidyl choline (PC), phosphatidyl inositol (PI), phosphatidyl serine (PS), phosphatic acid (PA), glycolipids (GL) and free fatty acids (FFA), Proportions of these components will vary depending on the source, however the major constituents are generally PE, PC, PI and PA. All PL share a common chemistry. All contain two fatty acids attached at the 1 and 2 position on a glycerol molecule. A phosphate group is attached to the 3 position. If only a phosphate group is attached the molecule is PA. PE has an ethanolamine group attached to the PA; PC has choline attached to PA: PI has an inositol group attached to the PA, and PS a serine group attached to the PA. The functional properties of lecithin are due to the collective chemical properties of all the lecithin constituents.

Lecithins are surface active and have surface activity. The two fatty acid chains are very lipophilic and have a strong affinity to associate with oil. Whereas, the phosphatidyl groups are very hydrophobic, having a strong affinity to associate with water through hydrogen bonding. The surface activity of lecithin provides lecithin with very diverse functionality that is used in a variety of applications and will be discussed later in the paper.

Lecithins also contain small levels of carbohydrates and proteins. Generally the small levels do not affect the lecithin's functional properties. However, lecithin produced from poor quality beans or poor extraction practices will contain higher than normal carbohydrate levels that can alter functional properties.

Because the phospholipids (PL) are insoluble in acetone the purity of lecithin is measured on the basis of the percentage of acetone insolubles or % Al. Most commercial trading standards require at least 50% Al for a material to be traded as "lecithin." Edible fatty acids and vegetable oils are acceptable diluants to reduce the viscosity or commercial lecithins. Lecithins with greater than 65% Al are extremely viscous. Lecithin with 62% Al has a viscosity similar to honey or molasses.

THE PROCESS

Crude oil is mixed with water (de-mineralized or soft water) at generally an equivalent weight of water to PL's, at approximately 70°C. Because the phospholipids are amphoteric they have an affinity for water and will form gums under mild agitation that can be easily separated by centrifugation. If agitation is too severe the PL's will emulsify the water and oil making separation extremely difficult.

The wet gums are dried using thin film vacuum dryers. To reduce the color lecithin is reacted with hydrogen peroxide. Lecithins may have high peroxide values as consequence of hydrogen peroxide bleaching and is not an indication of excessive oxidation. The % of Al is adjusted using a combination of FFA and oil to obtain a desired viscosity and a lecithin that will not separate during storage. As a final step the lecithin is passed through a fine filter to provide a finished lecithin with less than 1% hexane insolubles and under 2% moisture.

Oil may be removed to make a solid or de-oiled lecithin with a minimum 97% Al by either acetone extraction or ultrafiltration. The de-oiled lecithin is ground or agglomerated to obtain a desired particle size distribution.

Damaged seed or poor crushing and extraction practices will affect the quality of lecithin. Excessive heat will develop non-bleachable colors, reduce oxidative stability and reduce functionality. PL from damaged seed can interact or complex with calcium or magnesium to form non-hydratable phospholipids NHPL will not be removed by the degumming process. Subsequently lecithin yields are reduced and additional effort is required from other oil processing steps (caustic refining or bleaching) to remove the NHPL from the oil.

MODIFICATION

To increase its hydrophobic properties lecithin is commercially modified by the following methods :

Hydroxylation—The fatty acid composition of lecithin is generally slightly more saturated than its source oil. Most commercial lecithins contain a high level of unsaturated fatty acids. The unsaturated double bonds are reacted with hydrogen peroxide using an organic acid catalyst to form hydroxyl groups at the double bond sites. The hydroxyl groups having a strong affinity for water increase the hydrophobic properties of the PL's providing lecithin with greater oil-in water emulsification properties. Approximately 25% of the available double bonds are hydroxylated to provide a functional effect.

Hydrolysis—A lipase enzyme is used to hydrolyze the middle fatty acid at the 2 position commonly referred to as lyso lecithin. Removal of the fatty acid decreases the level of fatty acids on the molecule, therefore reduce the relative ratio between the lipohilic (fatty acid) and hydrophilic (phosphate-based group) and increase the overall hydrophilic properties of the lecithin. A partial hydrolysis will not result in significant change in functionality. Nearly all the fatty acids from the 2 position must be hydrolyzed for lyso lecithin to have a meaningful change in functionality.

The heat stability of lecithin can be enhanced by acetylation of the ethanolamine group in PE. The ethanolamine group can interact with reducing sugars under mild heat to darken the lecithin. By treating the lecithin with acetic anhydride, the amine group is acetylated and will not interact with sugars, and increase the heat resistance of the lecithin. Acetylation also increases the lecithin's hydrophilic properties, increasing its oil-in-water emulsification properties.

Acetylation with hydroxylation produces a lecithin with greater heat resistance and additional hydrophilic properties.

APPLICATIONS OF LECITHIN

Chocolate—Lecithin is used in chocolate to reduce viscosity during tempering to improve the coating and filling properties of the chocolate. Because of lecithins effective lubricating properties the viscosity of chocolate can be significantly reduced with 0.5% lecithin (62% Al), that would otherwise require the addition of 3-5% cocoa butter to achieve an equivalent viscosity.

Margarines and spreads—Lecithin is used in margarines and spreads (margarine products with less than 80% fat-generally 20 to 52% fat) to improve the dispersion of water in fat. Because of lecithin's emulsification properties, lecithin can reduce the surface tension between the water phase and oil phase when they are added together during the production of margarines and spreads. In margarines 0.2 to 0.5% lecithin reduces the size of water droplets dispersed in fat prior to chilling. The reduced water droplets improve mouthfeel characteristics, improve flavor release and decrease microbial activity. In spreads 0.5 to 1.0% lecithin is generally used synergistically with 1 to 2.0% distilled monoglycerides to minimize water droplet size prior to chilling and stabilize the water-in-oil emulsions after chilling and through distribution.

Milk Replacers—The modified lecithins (hydroxylated and hydroxylated and acetylated) are effective oil-in-water emulsifiers to disperse and stabilize 10 to 30% oil-in-water emulsion used as milk replacers.

Baking Shortenings—One percent lecithin in baking shortenings can effectively enhance the emulsifying and aeration properties of other emulsifiers in the baking shortening (such as : distilled monoglycerides, propylene glycol monoesters).

Baking—Standard lecithins and deoiled lecithins are added to baking formulas to improve the lubricity properties of doughs and batters. Doughs with 0.5 to 1.0% lecithin are easier to mix, require less energy, reduce adhesion to equipment and have improved gas retention properties by improve the elasticity of the dough. 0.2 to 0.5% Lecithin in batters reduce batter viscosity improving the batters flow properties and consistent dispensing into baking pans.

Powdered Beverages—Lecithin can be sprayed on powdered beverages that are difficult to disperse in water or milk. Powders with a high fat content, such as a powdered cocoa or chocolate, will not readily disperse cold water or milk because of the high level of surface tension between the fat on the powder and water. Two percent lecithin (30 to 62% Al) sprayed onto the high fat powder will reduce the surface tension between the powder and water and allow easy dispersion of the powder in the water. The lipophilic fatty acids will adsorb on the fat surface, leaving the lecithin's hydrophilic surface to coat the powder and readily associate with the water allowing for easy dispersion.

In beverages that are very hydrophilic, such as, high protein powders, the lecithin is added to reduce the hydrophilic surface. Very hydrophilic powders have a tendency to absorb water very quickly forming capsules or "fish eyes" which are very difficult to disperse. Spraying two percent lecithin on the surface will reduce the hydrophilic surface of the protein powder, reducing water absorption and allowing for improve dispersion.

Anti-stick Sprays—Lecithin is commonly used in anti-stick sprays that are applied to baking pans, trays or belts to reduce the baked product from sticking to the metal. Generally a heat resistant, acetylated lecithin is used for these applications. The viscosity of the lecithin can be reduced for spraying by heating or adding additional vegetable oils.

Lecithins is also used as an anti-sticking agent between packaged cheese slices to prevent cheese slice from sticking to each other.

The anti-stick properties are also useful for mold release applications in the manufacture of ceramics, plastics and concrete.

Dry Mixes—Deoiled lecithins are used in dry mixes (sauces, batters, coatings) to improve dispersion of the powders in water, stabilize the dispersion of the solids in water and reduce sticking in batter dry mixes.

Paints—Lecithin is used to disperse pigments and reduce paint viscosity.

Drilling Muds—Lecithin is added as an effective "environmentally friendly" lubricant used in drilling for oil or water.

The unique chemical properties of lecithin provide lecithin with many multifunctional commercial applications.

BIOTECHNOLOGY APPROACHES TO MODIFICATIONS OF OILSEED AND VEGETABLE OILS

Increasing global human population and specific consumer demands have and will continue to have a direct impact on food supplies. As a major component of many diets definite concern needs to be directed toward oils and fats. This class of compounds not only serves as a major source of dietary energy but also has nutritional implications associated with specific lipids. Particular biological roles exist, for example, for essential fatty acids, sterols (especially cholesterol) and fat-soluble vitamins. Consequently it is imperative that oils and fats be available in both the desired quantity and quality.

There is an essentially universal reliance on plants as the major supplier of human food. In particular, certain herbaceous oilseed crops and cultivated trees are the primary source of edible oils and fats. Projections have been made, however, that greater supplies will be necessary to meet increasing future demands. In addition, health concerns are directing more immediate attention toward the actual composition of dietary oils and fats as related to nutritional requirements.

Several approaches may be adopted to meet the projected supply and demand for edible oils and fats. More efficient agronomic practices can be adopted to ensure greater crop productivity. Use of improved developed plant strains (crop cultivars) can be made to obtain increased oil yields. These improvements in plants can be sought not only from conventional breeding programmes, but also from the adaptation of many of the new techniques of biotechnology and molecular biology. Applications of cell culturing, genetic engineering and modification of enzyme systems have been made with varying degrees of success.

Table 3.1. The source and range of oil and protein yields in major crops (as percentage dry mass)

Crop	*Source*	*Oil*	*Protein*
Cottonseed	seed	18—20	19—24
Groundnut	seed	45—50	21—36
Linseed	seed	40—46	22—26
Rapeseed/canola	seed	40—45	25—29
Safflower	seed	30—35	14—17
Sesame	seed	50—55	18—22
Soybean	seed	19—23	36—44
Sunflower	seed	35—45	14—19
Coconut	endosperm	65—68	
Oilpalm	mesocarp	>60	
Olive	mesocarp	various	

DESIRED GOALS

Several major areas for possible crop improvements may be considered.

Plant

Application of better practices of agronomy should result in increased crop productivity. Cultivation of strains with optimal desired characteristics is, however, ultimately dependent upon the availability of seeds at a reasonable cost. In addition, a definite need exists for the development of further cultivars which can tolerate abnormal soil conditions of high acidity, alkalinity and salinity. Frost tolerance, drought resistance and other environmental conditions are also factors to be considered. Furthermore, scope exists for improvements in plant viability in such areas as germination, maturation and seed-pod shattering propensity.

High crop production is frequently governed by the availability of fertilizers that is accompanied by a significant cost component. A further cost can be incurred from the need to apply pesticides and herbicides which, in turn, can introduce problems of potential environmental concern.

Post-harvest practices can have a major impact on oilseed storage. Protection from conditions of extreme temperatures and moisture, fungal activity and rodent infestation is of prime importance and can be accomplished by procedures that do not require sophisticated approaches such as biotechnology.

Oil Yield

Most of the oilseed crops as presently grown appear to have reached a plateau in oil production. Crops such as rapeseed, sunflower, groundnut and linseed routinely yield seed with an oil content of approximately 40% with a general range of 38-48%. Soybean has a relatively low oil productivity of 19-23%. The very high oil output associated with the oil palm and coconut is associated with the mesocarp of the fruit of these trees.

Subject to a better fundamental understanding of crop physiology, plant genetics and the biochemistry involved with oil production, some definite improvement in the quantity of oil obtainable from oilseeds may be anticipated. Cell culturing techniques and various approaches of biotechnology can be fully expected to be applied in the relevant studies for the development of new crop strains.

Oil Quality

The fatty acid composition of a vegetable oil has a definite significance. Primary concern exists with nutritional requirements and their impact on human health. A second, but none the less important, consideration is with the keeping quality of the oil. Based on the general understanding of the involvement of saturated fatty acids in the aetilogy of cardiac vascular disease interests definitely exist in oils which are deficient or low in this class of fatty acids. Although olive oil and canola oils are characterized by very small contents of saturated fatty acids deliberate lowering of these acids in other commercial edible oils is a major topic of study.

Manipulation of the unsaturated fatty acid component of vegetable oils is variously sought to improve their nutritional value. Oils enriched in oleic acid and certain polyunsaturated fatty acids such as essential fatty acids as well as gamma linolenic acid are considered to be highly desirable.

The polyunsaturated fatty acid content of oils, on the other hand, due to their oxidative instability can present problems of storage and keeping qualities. As a consequence there is a demand by the food industry for the availability of oils with good keeping qualities obtainable from a designed reduction of the polyunsaturated fatty acid content of edible oils.

PLANT ENGINEERING

Trait Detection

Recognition of the occurrence within a plant of a desirable characteristic is essential if the intention is to incorporate that trait into some host crop. The older practice involving selective breeding is handicapped by an inefficient success rate and the length of time generally associated with this approach.

Plant stock evaluation, characterization of the diversity of germplasm and identification of quantitatively inherited traits can be made by one of several techniques of molecular biology. The practice of restriction fragment length polymorphisms (RFLPs) is being replaced by the faster procedures of randomly amplified polymorphic DNAs (RAPDs) or amplified fragment length polymorphisms (AFLPs). Application of expressed sequence tags (ESTs) generated by rapid-throughput automated cDNA sequencing has been made to facilitate the isolation of certain genes associated with the biosynthesis of plant oils.

The basis of these procedures is associated with electrophoretic separation of variously sized fragments of the plant DMA, obtained by the use of restriction endonucleases, and their visualisation by ethidium bromide. The procedures lend themselves to automation. Use of the polymerase chain reaction (PCR) is now being made routinely and at low cost to increase the number of copies of a specific DNA region under investigation. "Finger-printing" or probing of these segments is then made through hybridisation with DNA fragments recognized to be related to some desired trait.

Particular applications of RAPDs and ESTs are anticipated to be most useful for locating and manipulating particular genes as well as in the mapping of polygenic sources of variation such as plant characteristics and oil yields.

There has also been an increasing use of genomic and molecular techniques to assist in the rapid domestication of presently non-commercial plants such as coriander and cuphea.

Cell Culturing

Use of modern cell culturing techniques is continuing to have a major impact on the development of new plant strains. In particular, the time required for the introduction of new commercial cultivars can be shortened significantly from a minimum of approximately fifteen years to about five years. Some further delays, however, may be experienced in obtaining legislative approval for use as a food.

The ready totipotency of individual living cells when isolated in a suitable medium allows for cultured cells to be regenerated into whole plants possessing the desired genomic composition with the ultimate development of a new stable homozygous variety. The general practice is to culture explants (plant cells, tissue or organs) on a chemically defined medium under sterile conditions. Either microspore (immature pollen) or anther culture techniques are commonly used and are so effective that many plants can be produced eventually at a time. Manipulation of the genetics of these isolated haploid (single set of chromosomes) cells is relatively simple and can be attempted by one of several procedures. Diploidisation (creation of a double set of chromosomes) is assisted by colchicine addition or heat treatment and further cell fusion encouraged by electrical stimulation or by the presence of polyethylene glycol. Cell growth and division is then regulated to produce individual cells and callus (clumps of cells) showing little differentiation. Subsequent organised growth is accompanied by embryogenesis and organogenesis of shoots and roots resulting in the ultimate development of plantlets. Following a hardening period the plantlets may be planted in the open field.

The ease of regeneration of plants from cell cultures varies, but a particularly useful system has been established for rapeseed; the system for soybean is more labour-intensive while that for sunflower continues to be developed.

Genetic Modification

Use of haploid cell systems allows for simpler manipulation of the single set of chromosomes than would be the case with diploid cells. As a general practice protoplasts are created by the enzymatic removal of the cell wall to facilitate the introduction of the desired foreign DNA.

Several approaches have been undertaken to induce changes in the plant genome and the incorporation of novel traits. Commercially valuable successes have been limited and varied.

Mutagenesis

An earlier practice with some noted successes in mutation breeding involves the use of a variety of mutagenic agents. The process is associated with point mutations and the attempt to unmask traits normally hidden in the native strain. Compared to targeted procedures, the process trends to have a low success rate due to its random approach.

Exposure of explants, seeds or whole plants can be made to ionizing irradiation or chemical mutagens such as ethyl methanesulphonate (EMS) and sodium azide. The practice is heavily dependent upon the availability of good screening methods. For example, when modification of the fatty acid composition of an oilseed is the aim, good use can be made of gas liquid chromatography to analyse a portion of a developing cotyledon and the remainder can be allowed to develop into a mature plant if found to possess the desired characteristic.

Gene Transfer Technology

Several systems are available for the introduction of selected foreign DNA into plant cells or protoplasts.

1. *Transfection :* Certain bacteria and viruses can be used as vectors in which a desired DNA fragment has been incorporated and then used to infect host, cells leading to alteration of the host genome. Frequent application has been made with the Ti plasmid of *Agrobacterium tumefaciens* as vector to transfect dicots - rapeseed, soybean, sunflower, flax and cotton; this system is ineffective with monocot plants such as corn and oil palm.
2. *Electroporation :* The desired foreign DNA can be introduced into host cells where a transient increase in membrane permeability due to a brief voltage pulse allows for its uptake from the medium.

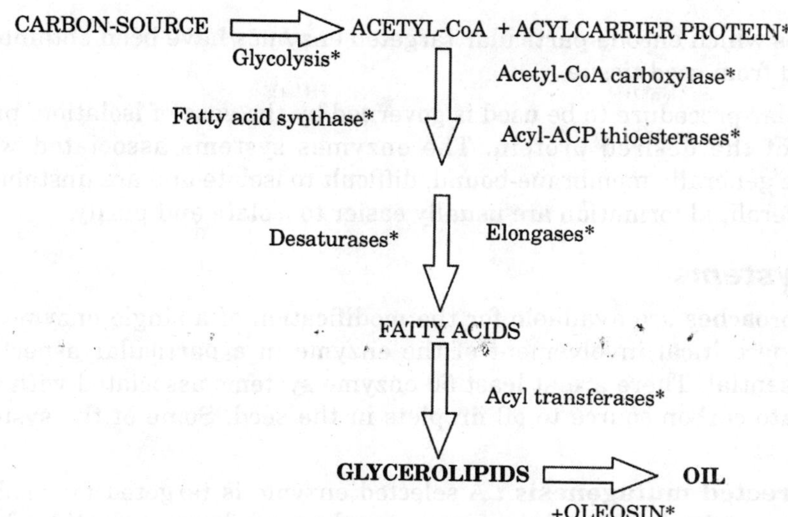

Fig. 3.7. Major target systems* for fatty acid, glycerolipid and oil formation in oilseeds.

3. *Microprojection :* Very small particles of tungsten or gold coated with the desired DNA can be shot into host cells at very high speed using the 'gene gun.'

These procedures are generally limited in their efficiency. Selection of protoplasts and cells where successful uptake of the foreign DNA has occurred frequently involves a known characteristic of the vector system where *e.g.*, the gene for resistance to some antibiotic such as ampicillin or tetracycline resides. Only the transformed systems will survive and grow in medium treated with the appropriate antibiotic.

Successful uptake of the desired genetic material resulting in the creation of a new allele does not guarantee that it will be either expressed or inherited.

cDNA

Specific isolation and .purification of a particular gene is seldom undertaken, although the successful cloning of many of the targeted enzymes of fatty acid biosynthesis and glycerolipid formation has been obtained. Two general practices can be used to prepare a fragment of DNA that contains the desired genetic information. Recombinant DNA libraries representing all of the mRNAs in a given cell can be prepared. Following the isolation and purification of the desired individual single-stranded mRNA double-stranded cDNA can be produced by the action of reverse transcriptase.

Alternatively the desired cDNA may be prepared from synthesis of the mRNA based on the knowledge of the full or partial amino acid sequence of the target protein. A frequently used approach involves the purification of the target protein (enzyme) and raising antibodies against them. Expression gene libraries are then screened, preferably with probes of plant origin, and the desired DNA located. Increased quantities of the desired cDNA can be obtained by cloning in some established system such as *Escherichia coli* or, more effectively, through application of the polymerase chain reaction (PCR). Genetic materials including partial and complete cDNAs which encode particular targeted enzymes have been obtained from several systems derived from seed tissue.

The particular procedure to be used is governed by the ease of isolation, purification and solubilisation of the desired protein. The enzymes systems associated with fatty acid biosynthesis are generally membrane-bound, difficult to isolate and are unstable; the enzymes involved in glycerolipid formation are usually easier to isolate and purify.

Targeted Systems

Several approaches are available for the modification of a single enzyme system. Exact knowledge of the critical involvement of the enzyme in a particular aspect of plant lipid formation is essential. There are at least 50 enzyme systems associated with the conversion of a carbohydrate carbon-source to oil droplets in the seed. Some of the systems that have been targeted.

1. **Site-directed mutagenesis :** A selected enzyme is targeted for modification of its activity by introduction of chemically synthesised deoxynucleotides having discrete changes in sequence from that normally present in the genome. Expression of this altered gene in the host plant is then expected to yield the required protein generally modified at its active site.
2. **Anti-sense RNA technology :** Gene expression for a particular system is effectively reduced by providing a chemically synthesised RNA fragment which is complementary to the mRNA for the targeted enzyme. Subsequent hybridisation of the sense and anti-sense RNA molecules is expected to reduce the quantity of sense mRNA available for translation.

   Application of this technology is particularly useful in obtaining mutants for specific genes when they are members of multigene families *e.g.*, the fatty acid synthase complex and hence more difficult to engineer. Problems, however, may arise when the desired reduction of targeted traits is accompanied with umpredicted accentuation of others. None the less is should be possible, at least in theory, to "fine time" the overall fatty acid composition of plane coils.
3. **Sense gene/co-suppression technology :** Introduction of a hormologous gene into the host plant system can result in suppression of subsequent genetic expression rather than an expected increase in the quantity of the targeted enzyme. Further application of this procedure may prove to be useful ior the modification of the unsaturated fatty acid composition of oils with improved nutritional characteristics.

All of these approaches to modification of the plant genome are heavinly dependent upon accurate knowledge of plant lipid biochemistry, genetics and factors influencing genetic expression. Introduction of foreign DNA into a host plant with the creation of a new allele does not guarantee its expression or its inheritance. Particular requirement exists for the availability of a strong piromoter to initiate the process of transcription.

TRANSGENIC OIL SEED CROPS

Successful modifications of oilseed crops have been made in two general areas and have allowed the introduction of cultivars with desired characteristics.

Table 3.2. Some Available Transgenic Crops and Their Trait Modification

Crop	*Altered Trait*
Canola *(Brassica napus)*	glufosinate ammonium tolerance* glyphosate tolerance ** high-lauric high-stearate/low-polyunsaturate high-oleic high-medium chain
Soybean *(Glycine max)*	glyphosate tolerance high-oleic/low-saturate
Corn *(Zeamays)*	Bt-resistant *** glufosinate ammonium tolerance glyphosate tolerance
Cotton *(Gossypium hirsutum)*	Bt-resistant glyphosate tolerance
Flax (Linum usitatissimum	sulphonylurea tolerance
*Liberty Link **Roundup Ready	***Bt = Bacillus thuringiensis

Plant Modifications

Plant protection has been achieved through genetic engineering to provide both tolerance to selected herbicides and resistance to insect infestation.

Strains of corn (maize), flax, rapeseed and soybean are commercially available which are herbicide-tolerant. The presence of weeds can have a major impact on crop yield where growth inhibition of *e.g.*, untreated canola can be as much as 35%. Theoretically plant growth of the crop can be expected to be favoured when weeds sprayed with the herbicide die off. The anticipation is that there will be a lower need to apply the herbicide with an accompanying reduction of costs and potential environmental problems. Preliminary results suggest that herbicide-tolerant cultivars show slightly enhanced crop yield with a lower need for spraying.

Similar increases in crop yields have been obtained with both corn and cotton which have been engineered to contain the gene from Bacillus thuringiensis responsible for the production of a toxin fatal to certain insects.

Selective breeding of cultivars and the application of good agronomic practices have resulted in an apparently optimal range of total oil content for the major oilseed crops.

At present there does not appear to have been any success in developing transgenic crops with enhanced oil productivity. Based on the fact that certain (non-domesticated) plants are capable of accumulating, up to 70% (w/w) oil in the seed an understanding of the mechanism(s) involved might conceivably allow development of higher oil-yielding crops.

Oil Modifications

Some very notable successes have been obtained in the development of transgenic plants in which the seed oil has been modified to yield increased and/or decreased contents of particular fatty acids. Certain of these cultivars are available for commercial use, subject to legislative approval, while others are still in field trials. The impact, if any, of modified fatty acid composition of seed oil on the viability of the plant is not generally known. An engineered high stearic acid content in Brassica rapa seeds resulted in low oil content and low germination.

Further progress in the tailoring of seed oils can be fully anticipated.

COMMENTARY

Successful use of genetic engineering has been made with the incorporation of some desirable traits into commercial crops, particularly canola. The techniques and protocols are available, but more definite applications are heavily dependent upon the availability of more advanced knowledge of plant lipid metabolism-total fatty acid synthesis, unsaturated fatty acid synthesis and glycerolipid (triacylglycerol formation)-gene expression and regulation, and all factors that impact on plant viability.

The first generation of transgenic crops has been primarily associated with input traits associated with protection-herbicide-tolerance and insect-resistance-and with certain modifications of the fatty acid composition of the seed oil. Rapeseed (canola) has proved itself to be particularly amenable to manipulation and further advantages lie in the fact that it is an established crop.

The second generation it is believed will focus on output traits associated with the use of crops modified to produce a wide variety of consumer-desirable products such as nutraceuticals, pharmaceuticals and renewable feedstocks for the oleochemical industry.

Great care will need to be exercised with the use of crop engineering to avoid any possibly negative impacts on the environment and human health. A Luddite philosophy exists in certain segments of society that will have to be overcome by good solid reasoning if further successes with the technology is to be achieved.

Claims that transgenic plants will go a long way to solving the global lack of available food have yet to be substantiated. Emerging nations which generally lack adequate supplies of food may look toward biotechnology and genetic engineering as a solution, but the cost may be prohibitive. It may well be prudent for them to give greater attention to other approaches such as better agronomic practices and post-harvest technologies.

SEPARATION OF OILS AND FATS

Centrifuges are machines from mechanical process engineering for the separation of solids from liquids, the separation of two immiscible liquids or two such liquids and a solid component. The pre-condition is that the media to be separated from each other have a different specific gravity.

Centrifuges are subdivided into different types depending on their application. Disk-type centrifuges are almost exclusively used for oil and fat refining which are in turn classified

use of hydrogel silica, such as Trisyl, can reduce soap, trace metals, phospholipids and free radicals that are present in the oil and improved its stability.

The final step in oil processing is deodorization. In this process the bleached oil is steam distilled under vacuum .The object is to remove the volatile impurities from the oil. The red color of the oil is reduced due to heat bleaching of the carotenes

In the deodorization process it is necessary that the absolute pressure in the deodorizer system is maintained as specified by the manufacturer. The author has experienced in the United States and around the world that the production personnel as well as the management in an oil processing operation tend to operate the deodorizers as described below:

1. A throughput far beyond the designed limit.
2. Maintain an absolute pressure in the deodorizer much higher than the designed operating limit.
3. Deodorize oils at a temperature that is too high in order to compensate for poor vacuum in the deodorizer and/or higher deodorizer throughput.

All of the above practices produce either incompletely deodorized oil or damage the oil in deodorization process.

A higher absolute pressure in the system, generally, indicates presence of air. At a deodorizer temperature of 242° to 260°C, the oil can undergo very rapid oxidation and polymerization if the system is not maintained at the designed operating pressure. Researchers have shown that as dimer content in the deodorized oil goes up, flavor acceptability in fresh deodorized oil goes down. At the same time, the oil containing higher levels of dimers, demonstrates poor flavor stability. Poor vacuum in the deodorizer can produce high amounts of oxidation polymers in the oil. Some of these polymers contain as much as nine times the amount of oxygen present in a triglyceride molecule. These compounds break down in the absence of light, oxygen and even at very low temperatures, releasing more free radicals that cause flavor degradation in the oil in storage or in the food processed with this oil. This is known as hidden oxidation in the oil. Standard analytical tools that are normally used to measure oil oxidation, such as peroxide value, cannot explain this phenomenon.

The oil in the deodorizer must be cooled down to 150°C or less, before it is taken out of the deodorizer for external cooling. Addition of citric acid is critical to chelate the metals that can cause autoxidation of the oil in storage. Citric acid must be added at a temperature less than 145°C to prevent its decomposition.

The oil, leaving the deodorizer must be cooled down to a temperature not more than 5°C above its melt point before it is stored. The liquid oils must be cooled down to less than 40°C. It is highly recommended that all deodorized oils must be saturated with nitrogen as they leave the deodorizer. The oils must be stored under nitrogen atmosphere with a maiximum oxygen content of 0.5 per cent in the head-space.

Oils stored under these conditions have shown no rise in the peroxide value for 6-8 weeks, without using antioxidants.

Photooxidation

It has been mentioned earlier that there are two types of photooxidation reactions:

1. Photochemical reaction

2. Photosensitized reaction

Photochemical Reaction

This reaction is initiated by the ultraviolet (UV) light. Unsaturated fatty acids form free radicals in presence of the UV light. The rest of the reaction proceeds in the same manner as the autoxidation process. Photochemical reaction proceeds at the same rate as autoxidation.

Photosensitized Reaction

Photosensitized reaction occurs when the oil is exposed to the visible light in presence of a photosensitizer. This reaction proceeds 1500 times faster than autoxidation or photochemical reaction.

Chlorophylls, naturally present in soybean and canola oils, are photosensitizers. This is why excess chlorophylls, present in these oils must be removed in order to reduce photosensitivity of these oils in following applications:

1. When the oils are packaged and stored in clear bottles.
2. Processed foods are prepared with these oils and packaged in clear bags or containers.

Chlorophylls (A and B) are reduced in the bleaching process. Acid activated clay is used to remove chlorophylls from vegetable oils. In the United States and Canada, the chlorophyll content is maintained at less than 30 parts per billion in both soybean and canola oils.

It has been found that chlorophylls undergo decomposition under acidic pH in presence of oxygen. The breakdown products of chlorophylls are known as pheophytins, pheophorbides and pyropheophorbides. These components are ten times stronger photosensitizers than their parent compounds. Therefore, measurement of only chlorophylls in the bleached oil may not provide total information regarding its photosensitivity. It has also been demonstrated that chlorophyll B and their decomposition products are stronger photosensitizers compared to chlorophyll A. Therefore it is important to measure the breakdown products of chlorophylls in the bleached oil to establish its photosensitivity.

Atmospheric bleaching process provides an ideal opportunity for the chlorophylls to break down and make the oil more prone to photooxidation.

Silica treatment of oil in both conventional refining process and modified chemical refining process (28) has shown improved oil stability. This is because silica is a very good adsorbent for soap, trace impurities and including free radicals. This can improve the autoxidative stability of the oil. In addition, use of silica reduced the dosage of bleaching clay required to achieve the desired green color in soybean and canola oils. This may provide the opportunity to reduce chlorophyll breakdown during bleaching and improve photo-oxidative stability of the oil.

1. Atmospheric bleaching process is damaging to the oil. Vacuum bleaching process is a preferred option for oil bleaching.
2. Silica pre-treatment, prior to the acid activated clay treatment of the oil is recommended for bleaching all types of vegetable oils.
3. In addition to the chlorophyll measurement, it is prudent to check for the chlorophyll break down compounds in soybean and canola oils. This can provide a better idea about photo-oxidative stability of the oil.

HYDROLYSIS

When a molecule of water reacts with a molecule of triglyceride, the resultant products are diglyceride and fatty acid (commonly termed as free fatty acid). As more water molecules react with the oil, more free fatty acids are formed together with diglycerides and monoglycerides. Eventually the reaction products are glycerol and fatty acids. This reaction can occur only when the oil and water are in perfect solution. Oil and water do not form a solution under normal processing conditions, because both temperature and pressure are insufficient to create a solution. However, the presence of a small amount of surfactant in the oil increases inter-solubility between the two liquids and promotes hydrolysis of the oil during processing. This occurs because the surfactant reduces the interfacial tension between oil and water.

In commercial fat-splitting reactors, a small amount of caustic is used to produce soap through the reaction between the oil and caustic. The soap thus produced act as surfactants for the reaction. The crude oil contains natural surfactants, such as phospholipids. The other source of surfactants are:

1. Soap left in the oil from the caustic refining process or from neutralized (unbleached) oil.
2. Some products of oil oxidation have surface activity that can promote hydrolysis in the oil.
3. Mono and diglycerides produced in the oil due to over refining of the crude, or excess caustic treat needed for refining poor quality crude oil.

In chemical refining process of seed oils, the final phosphorus content in the oil is recommended to be less than 1 part per million, it has been experienced by the author that when the phosphorus content in the vegetable oil is 3 parts per million or higher, the free fatty acid in the finished oil begins to rise while the oil is in storage. This is commonly experienced in physically refined palm oil or palmolein, where due to the distance between the source and the end-users, the oil has to be purchased in large quantities and stored for months at a time. In many instances, the oil needs to be re-deodorized to reduce the increased free fatty acids produced during storage. Physically refined palm oil or palmolein contains 3 to 5 parts per million of phosphorus in the finished oil. This is believed to be one of the major drivers for the rise of free fatty acid in these oils during storage.

High phosphorus, left in the oil, can cause some hidden losses in oil yield. High phosphorus in the oil poisons the nickel catalyst in hydrogenation. This increases the consumption of catalyst. Additional amount of diatomaceous earth is needed to remove the extra nickel catalyst

Hydrogenated oil, produced with poisoned catalyst tends to contain high level of colloidal nickel. This is removed from the oil in the post-bleaching process where the hydrogenated oil is treated with acid activated bleaching clay together with citric or phosphoric acid under vacuum, post-bleaching process always increases the free fatty acid content in hydrogenated oil, which increases oil loss in the deodorization. Process, A higher free fatty acid in deodorizer feed reduces the rate of deodorization. This means reduced productivity due to slower deodorizer throughput. Although the excess free fatty acids are removed in the deodorizer, there is always some additional loss of neutral oil in the distillate when the deodorizer feed contains high concentration of free fatty acids. If the phosphorus content is high in the deodorizer feed, the oil can undergo hydrolysis during deodorization. This can be verified by measuring the diglyceride content of the feed to the deodorizer and the deodorized oil. The additional diglycerides formed in the deodorized oil do not get distilled and can be analyzed.

There is a tendency to deodorize physically refined palmolein at a temperature of 260°C to 265°C. This is generally done to reduce the deodorization time and increase deodorizer throughput. However, at this higher temperature the oil can undergo hydrolysis, creating a condition where reduction of the free fatty acid to less than 0.05 per cent becomes difficult. One must also recognize that surfactants, such as phospholipids reduce interfacial tension between the oil and air. Therefore, under poor vacuum and with high level of phosphorus, there is an opportunity for the oxygen to react with the oil in the deodorizer and form free radicals that lead to the formation of oxidative dimers and polmers.

Soap can be left in the refined oil due to inadequate bleaching. This can produce similar reaction as phospholipids in the oil . Some cottonseed oil processors believe that prime Bleach summer yellow (PBSY) cottonseed oil does not require bleaching because of the low color of the oil. This is not correct because PBSY cottonseed oil contains some residual soap that must be removed from the oil to prevent hydrolysis and oxidation of the oil in subsequent processing and handling. Neutrallized oils also suffer form the same risk as the PBSY cottonseed oil since it contains soap.

Crude palm oil may contain soap. This normally comes from improperly cleaned shipping vessels. Even a small amount of soap content must be eliminated from the oil before it is deodorized. Some unscrupulous palm oil processors have also delivered crude palm oil that has been partially neutralized in order to meet the 5 per cent limit on the free fatty acid, set forth by PORAM. This practice causes long-term instability in the oil.

Table 3.11. Recommended Oil Quality Parameters for Seed Oils

Analysis	*Crude Oil*	*Refined Oil*	*Water Washed Oil*	*Bleached Oil*
% FFA	—	0.01—0.03	0.02—0.05	0.04—0.1
Soap, ppm	—	< 500 Not > 1000	< 50 Not > 100	0
PV, maximum	4*	4	4	0
PAV	4	4	4	6
Phosphorus, ppm	—	3	3	< 1
Calcium, pm	—	—	—	< 0.2
Magnesium, ppm	—	—	—	<0.2
Iron, ppm	—	—	—	< 0.2

*At PV higher than 8, seed oil may produce unacceptable flavor stability.

It is recommended that the soap content in all palm oil and all types of partially neutralized oil receipts is measured and the oil is bleached to remove the soap before the oil is further processed. The oil can be treated either with silica or bleaching clay to accomplish this. The recommended oil analytical parameters for crude, refined, water washed and bleached oils for best performance.

THERMAL DECOMPOSITION

Vegetable oils undergo decomposition when they are subjected to high heat. When the oil is heated to 150°C in air, even the saturated fatty acids break down to produce compounds, such as homologous series of carboxylic acid, 2-alkanones, 2-alkanals, lactones, *n*-alkanes and

n-alkenes. When triglycerides containing even numbered fatty acids, ranging from 6 to 16 carbon atoms are heated in air to 180°C to 250°C for 1 hour, they all produce homologous hydrocarbons, aldehydes, keton lactones.

Thermolytic non-oxidative products from saturated fatty acids can be detected even when the oil is heated to 180°C under vacuum. Volatile compounds detected under this condition include normal alkanes, fatty acid, oxypropyl ester, propene, propanediol esterdiglycerol and acrolein.

In case of unsaturated fatty acids, formations of dimeric and cyclic compunds appear to be the predominant thermolytic breakdown product. Degrees of complexity of these compounds increase with the degree of unsaturation of the fatty acids present. Heat treatment also leads to the formation of several aromatic compounds. Thermal dimers and cyclic monomers have been isolated from several types of heated vegetable oils.

It can be seen from the above dicussion that exposure of oil to high heat can lead to both oxidative and thermolytic degradation of the oil.

RECOMMENDATIONS

The recommended analytical parameters for deodorized seed oil is listed. The general information provided in this table is applicable to all types of seed oils and palm oil (palmolein).

Table 3.12. Recommended Anlysis for Various Vegetable Oils.

Analysis	*Target*	*Maximum / minimum*	*Remarkes*
%FFA	< 0.03	0.05, max.	Higher FFA may indicate lower self-life stability
PV	0	0.5 max... 1.0 max...	As loaded in trucks As received by Customer
AV	2	4	Never to exceed 6
%Dimers	< 0.2	0.4 max	Never to exceed 0.5%
%Polymers	< 1	2 max.	Higher polymers may impart bitter flavor in prepared foods
Chloropyhll	—	30 max.	Increases photosensitivity in the oil
Phosphorus, ppm	< 0.5	1.0 max.	Baking shortening and margaring can tolerate higher phosphorus level
Calcium, ppm	—	0.2 max.	Higher than 0.5 ppm implies poor quality oil
Iron, ppm	—	0.2 max.	Higher than 0.5 ppm implies poor quality oil
Magnesium, ppm	—	0.2 max.	Higher than 0.5 ppm implies Poor Quality Oil
Lovibond Red color	< 0.1, for Soybean Oil, Canola Oil, Sunflower Oil	1.5 max. for Soybean OIl, Canola Oil, Sunflower Oil	> 1.5 Lovibod Red color Implies Poor Quality Crude or Poor Oil Processing
% Triglycerides	98 minimum	-	A Lower Value Implies Poor Quality Crude In Case of Seed Oils
% Monoglyceride*	—	0.3	A Lower Value Implies Poor Quality Crude or Over Refining In Case of Seed Oils
% Diglyceride**	—	0.6	A Lower Value Implies Poor Quality Crude or Over Refining In Case of Seed Oils

*Typical values in commerical palm and palmolein : 0.5 – 1% monoglycerides

**Typical values in commerical palm oil and palmolein : 3 – 11% diglyceride

Lovibond Red color is different for different oils. Mono and diglyceride contents in palm oil and palmolein are always higher than in the seed oils.

PALM OIL INDUSTRY

Composition

Palm oil is obtained from the flesh surrounding the seed, simply by cgoking, meshing and pressing. In this process the seeds are separated and after cracking and removing the shell, the kernel can be processed to yield palm kernel oil. Palm kernel oil forms about 10% of the total oil yield and has different composition and uses from palm oil.

As regards to fatty acid composition, palm oil occupies an intermediate position among natural fats. Its saturated fatty acid content is 50% and the rest is monounsaturated and polyunsaturated. By virtue of the fatty acid composition, palm oil is semi solid containing mainly palmitic, oleic and linoleic acids, which are the most common components of all food fats. Palm kernel oil, on the other hand is rich in lauric acid and is interchangeable with coconut oil in most of its applications but is more flexible because it is more unsaturated compared to coconut oil.

Crude palm oil, besides its fatty acid composition also contains 1% minor components. A diagrammatic presentation of crude palm oil minor components.

CAROTENOIDS AND TOCOLS

Carotenoids form a class of C40 polyunsaturated hydrocarbons (carotenes) and their oxygenated derivatives (xanthophylls). These compounds impart a rich orange-red colour to the oil.

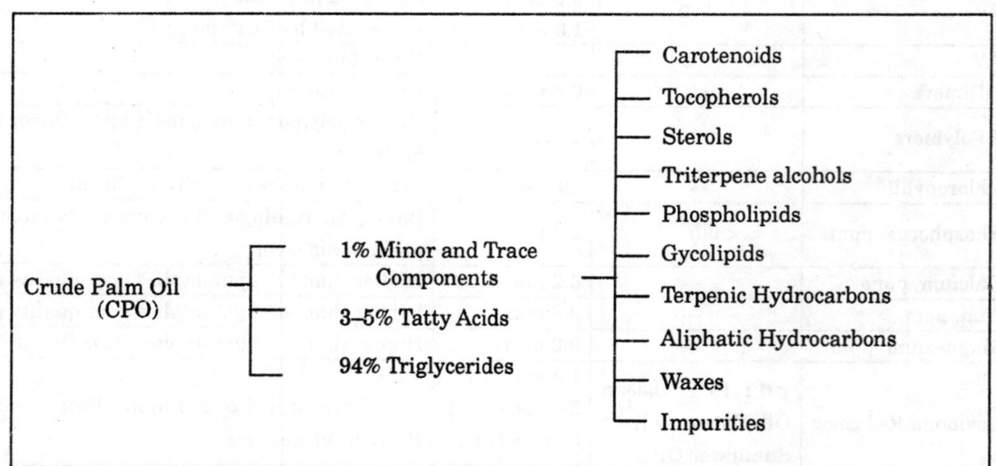

Fig. 3.18. Constituents of Crude Palm Oil

Crude palm oil is one of the richest natural plant sources of carotenes with the concentration of 500-700 ppm. It has 15 times more retinol equivalents than carrots and 300 times more than tomatoes. Analysis shows that alpha and beta carotenes constitute approximately 90% of the total carotenoid content and the rest are delta carotenes, gamma carotene, phytofluene, phytoene, zea-carotene, lycopene, neurosporene, alpha and beta zeacarotenes.

The production of palm oil from Malaysia is projected, to reach ever 8 million tonnes by the year 2000. Hence, the potential availability of carotenes in the year 2000 is about 4800 tonnes.

The tocopherols and tocotrienols are found in cude palm oil at 600-1000 ppm concentrations. The main constituents are 44% gamma tocotrienol, 22% alpha tocophercel, 12% delta tocotrienol, the rest being alpha and beta tocotrienols, gamma and delta tocopherols.

NUTRITIONAL BENEFITS OF CAROTENOIDS AND TOCOLS IN PALM OIL

In human nutrition, carotenoids play important role as a source of provitamin A. Beta-carotene, especially, is the most nutritionally active carotene as provitamin A. It is converted in the gastrointestinal tract into vitamin A. Vitamin A of course, plays an important role in regulating vision growth and reproduction. It is essential for the normal cellular differentiation of most epitilia, including those of the skin, bronchi and trachea, stomach, intestine, uterus, kidney and other organs.

The FAO/WHO recommendations on fats and oils in human nutrition stated that "in countries where vitamin A deficiency is a health problem, the use of red palm oil, wherever readily available should be encouraged."

Human studies have also indicated strongly that carotenoids possess antioxidant properties and may protect against certain types of cancer. It was postulated that beta carotene has significant protection against cancer. This has sparked much interest and numerous reports have now appeared to shed new light on this question. An inverse association between serum beta carotene and the risk of squamous cell carcinoma of the lung has been found. It was also reported that beta carotene inhibited oral tumors when applied topically or given as a dietary supplement.

The antitumor effects of palm carotenes had been studied on the developments of chemically-induced skin cancer and viral-induced lymph cancer in HRS/J female hairless mice. Palm carotene was found to possess antitumorigenic effects and the inhibition of subcutaneous lymphoma was concentration dependent: 0:3% palm carotenes (PC) > 0.1% PC > 0.05% PC. All the palm carotenes group inhibited epithelial pappilloma formation as well.

Vitamin E functions primarily as an antioxidant. especially to prevent oxidation and peroxidation of polyunsaturated fatty acid units of membrance phospholipid (within and on the plasma membrane of cells). This works against injury to cell membranes as in red blood cell fragility in man. It is also implicated in preventing photooxidative damage to the lens and thus in preventing cataract formation. The antioxidant properties of vitamin E have been implicated to prevent/delay the onset of various degenerative diseases such as aging, cancer, arthritis, damage to cells caused by air pollution and also in preventing coronary heart diseases.

The nutritional benefits of palm vitamin E (palm vitee) has been studied on human as well as animal models. The hypocholesterolemic effect of palmvitee has been studied by various researchers. It was found in one study that the concentrations of serum total cholesterol (–15%), LDL cholesterol (–8%), apoprotein B(–10%) and thromboxane (–25%) decreased significantly in 15 hypercholesterolemic subjects fed 200 mg palmvitee capsules/day . Another study on 9 healthy volunteers taking one capsule of palmvitee perday also showed a reduction of serum cholesterol (reduction rage –5% to –35.9%) and LDL cholesterol (reduction range 0.9% to 37%).

The effect of palm oil vitamin E on isolated Langendorff perfused rat hearts has also been studied and results indicated that palmvitee containing a mixtue of tocotrienols and tocopherols may be more efficient than alpha tocopherol alone in the protection of the heart against oxidative stress induced by ischemia-reperfusion.

Large population studies have analysed the relationship between antioxidant nutrients such as carotenoids and vitamin E and degenerative diseases. A continuing study among 22,000 male physicians in the US yielded an interim finding that those taking a supplement of beta carotene reduced the occurrence of major cardiovascular events by almost 50%. It was reported that an inverse correlation between plasma vitamin E levels and mortality from ischemic heart disease in a cross cultural epidemiology.

CAROTENOIDS

Sources of Carotenoids

Various analytical methods have been used for the determination of the carotenoid profile of crude palm oil. Column chromatography has been used in earlier studies using different adsorbents and reverse phase HPLC has been shown to have several advantages for the separation of carotenoids in the oil.

Table 3.13. Carotenoid contents of Palm Oil Extracts

Palm Oil Extract	*Carotenoid Content (ppm)*
Crude palm oil	630—700
Crude palm olein	680—760
Crude palm stearin	380—540
Second pressed oil	1800—2400
Residual oil from fiber	4000—6000
Total carotenoids estimated at 446 nm	

The carotenoid contents of various palm oil extracts. A carotenoid rich oil can be obtained from palm pressed fiber, a palm oil by product which is presently burnt as a fuel in palm oil mills. The pressed fiber was found to contain about 5-6% of residual oil with a carotenoid concentration of 4000-6000 ppm, and the carotenoid content of the residual oil in the pressed fiber of the hybrid oil palm fruit is even higher, at 6000-7000 ppm.

Carotenoids from the commercial crude palm oil are concentrated during the extraction and fractionation process. A system of palm oil extraction based on a double pressing technique has been implemented by several mills in Malaysia. The expected advantages of double pressing over the conventional single-stage pressing are lower oil loss in fiber, higher kernel extraction rate, less wear on the screw worm and cages, and reduction of contamination of the kernel oil with crude palm oil. More interestingly, the oil produced from the second pressing has a higher concentration of carotenoids. This could be due to the fact that the first pressing in a double pressing process is carried out at a lower pressure (to avoid cracking of the nut) and oil relatively higher in carotenoids from the second pressing is extracted. After removal of the nuts the fiber is then subjected to a higher pressure and more carotenoids are extracted out of the mesocarp, together with the residual oil from the first pressing.

Carotenoids are also being concentrated in the industrial process called fractionation. Palm oil is a semi-solid fat at ordinary room temperature. Fractionation is carried out to extend the uses of palm oil and the products obtained are the liquid oil (olein, 70-80%) and the solid fat (stearin, 20-30%). The carotenoid content in the palm olein (lower melting) fraction is enriched by 10-20%.

Recovery and Extraction of Carotenoids

The carotenes on the market at present are chemically synthetic beta carotene, carotene extracted from carrots and carotene extracted from algae Dunaliella. The carotenoid composition of the various carotenes. The demand for natural carotenes is such that customers are willing to pay a premium for these, which they value for their perceived safety. The major commercial uses of beta carotene in the global market in which they are widely used as a food colorant.

Table 3.14. Carotenoid Compositions of Various Carotenes.

	alpha-C	*beta-C*	*Others*
Carrot carotene	30—50%	40—60%	10—15%
Synthetic beta-carotene		100%	
Dunaliella cartonene	6—10%	84—90%	
Palm oil carotene	30—35%	60—65%	5—10%

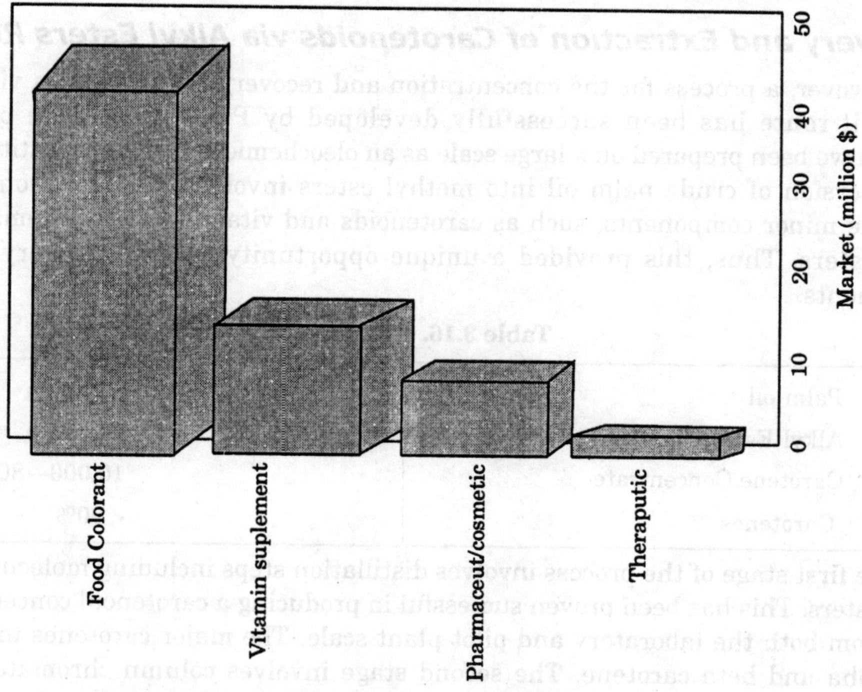

Fig. 3.19. Major commercial uses of beta carotene.

Because carotenoids are likely to grow in importance and value, the recovery of carotenoids from palm oil and its by-products is important. In addition, some studies are also being carried out to recover and concentrate palm-based carotenoids for pharmaceutical purposes.

Numerous extraction methods have been developed to recover the carotenoids from crude palm oil. Most of the reported methods are difficult, costly or inefficient.

Table 3.15. Carotenoid Concentrates from Various Methods.

Method	*Carotenoid content (ppm)*	*Recovery*
Vacuum distillation*	>20,784	<46
Molecular distillation*	>80,000	>80
Adsorption*		
C18	8000—9000	>90
Carbon	5000—7000	<50
Molecular distillation of crude palm oil	1290—1990	
Adsorption from crude palm oil (activated carbon)	3700—5600	<80

* through methyl ester route
Total carotenoids estimated at 440 nm

Recovery and Extraction of Carotenoids via Alkyl Esters Route

However, a process for the concentration and recovery of carotenoids via alkyl esters of palm oil route has been successfully developed by PORIM. Volatile palm oil methyl esters have been prepared on a large scale as an oleochemical or diesel substitute. This process of conversion of crude palm oil into methyl esters involves mild reaction conditions, the valuable minor components, such as carotenoids and vitamin E would remain intact in the alkyl esters. Thus, this provided a unique opportunity for the recovery of these minor components.

Table 3.16. Carotene Content

Palm oil	600 ppm
Alkyl Esters	600 ppm
Carotene Concentrate	10,000—80,000 ppm
Carotenes	< 90%

The first stage of the process involves distillation steps including molecular distillation of alkyl esters. This has been proven successful in producing a carotenoid concentrate of >80,000 ppm from both the laboratory and pilot plant scale. The major carotenes in the concentrate are alpha and beta carotene. The second stage involves column chromatography with an absorbent to give a carotene fraction. The concentration of this carotene fraction ranged from

90-100%. The carotene fraction can then be diluted to produce the desired commercial carotene concentrate in various concentrations (1-30%) for various applications.

Table 3.17. Carotene composition (%)

Carotene	*Concentrate*	*Crude Palm Oil*
Phytoene	1.5	1.3
Phytofluene	0.3	0.1
Cis-β-Carotene	0.9	0.7
β-Carotene	49.9	56.0
α-Carotene	33.3	35.1
Cis-α-Carotene	5.5	2.5
γ-Carotene	1.7	0.7
ζ-Carotene	1.3	0.3
δ-Carotene	0.6	0.8
Neurosporene	0.1	0.3
β-Zeacarotene	1.3	0.7
α-Zeacarotene	0.4	0.2
Lycopene	3.4	1.3
Total (ppm)	80560	673

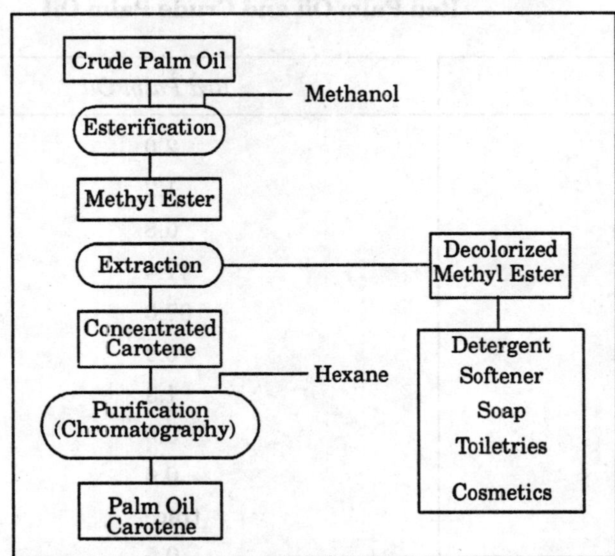

Fig. 3.20. Lion Procedure for Producing Palm Oil Carotene

Lion Corporation of Japan has made efforts to recover carotene from palm oil since the 1970s. They have developed a process which is also via the methyl ester route. Crude palm oil is converted into methyl esters by transesterification with methanol at low temperature which results in little loss of carotenes, if any. In the second step methyl esters are mixed in an organic solvent, and then the mixture is separated into a carotene rich layer and a decolourizsed methyl ester layer. The methyl ester is used for oleochemical purposes while the carotene may be further concentrated from the carotene rich layer by ordinary means. Finally, the carotene is purified by liquid chromatography with hexane. The palm oil carotens contains more than 95% of carotenoids (60-70% of beta carotenes, 30-35% alpha carotenes and small amounts of gamma and delta carotenes).

Carotech has also developed an integrated propess for recovery of carotenoids and tocotrienols from crude palm oil, again via the alkyl esters route. Fatty acids in a vegetable oil are subjected to alcoholic esterification to form an ester-rich layer including fatty acid alkyl esters, carotenoids and tocotrienols. The ester rich layer is exposed to solvolytic micellization to form a carotenoid rich layer. The ester rich layer is separated from the carotenoid-rich layer. The carotenoids in the carotenoid-rich layer are concentrated and can be adsorptively separated from the carotenoid-rich layer. Fatty acid alkyl esters are separated from the ester-rich layer to form a tocotrienol-rich layer.

Red Palm Oil

A process has also been developed by PORIM to produce deacidified and deodorized red palm oil from the degummed palm oil with more than 80% of the original carotenes still intact as well as the vitamin E. This red palm oil is of similar excellent quality to the refined, bleached and deodorized (RBD) palm oil that is normally traded but contains no carotenes.

Table 3.18. Carotene Composition (%) of Deacidified and Deodorized Red Palm Oil and Crude Palm Oil

Carotene	*Red Palm Oil*	*Crude Palm Oil*
Phytoene	2.0	1.3
Phytofluene	1.2	0.1
Cis-β-Carotene	0.8	0.7
β-Carotene	47.4	56.0
α-Carotene	37.0	35.1
Cis-α-Carotene	6.9	2.5
Carotene	1.3	0.7
ζ-Carotene	0.5	0.3
γ-Carotene	0.6	0.8
Neurosporene	trace	0.3
β-Zeacarotene	0.5	0.7
α-Zeacarotene	0.3	0.2
Lycopene	1.5	1.3
Total (ppm)	545	673

The process involves two stages *i.e.*, pretreatment of the crude palm oil followed by deodorization and deacidification by molecular distillation. The pretreatment is carried out in a conventional manner using phosphoric acid followed by bleaching earth. This allows impurities and oxidative products in the crude oil to be removed. The deodorization and deacidification is carried out using molecular distillation unit at temperatures <165°C and a pressure of 20-35 X 10-3 Torr.

Analysis by HPLC shows that the profile of carotenes are not destroyed during the process and the quality of red palm oil has also been found to be good. Sensory evaluation carried out on red palm oil showed that it is of very good quality and comparable to RBD palm oil.

Table 3.19. Quality Parameters of Red Palm Oil

Carotenes	>80% intact
Tocopherols and Tocotrienols	>80% intact
Free fatty acids	<0.1%
Peroxide value	<0.2
Phosphorus content	<2 ppm
Moisture and impurities	<0.1%

The commercial production of red palm oil involves the capital investment in a molecular distillation unit, which includes a vacuum system. The other costs involved are the power supply, water supply for cooling and storage tanks. The cost of production, based on a 5-year investment plan is given below.

Plant capacity (T/yr)	10,000	30,000
Pretreatment (RM/T)	17.00	17.00
Molecular distillation (RM/T)	99.00	96.30
Total cost of Red Palm Oil (RM/kg) (includes cost of crude palm olein)	1.2529	1.2374

Unitata has also developed a process to produce carotene enriched red palm oil. This is a chemical process which involves chemical neutralisation of the free fatty acids, followed by pretreatment, then deodorization using conventional method to remove the odour. The quality of the product obtained from this process is also similar to the RBD palm oil, containing not less than 250 ppm of carotenes and 800 ppm of vitamin E. The oil produced has been marketed which bears the tradename of NUTROLEIN.

TOCOLS

Source of Tocols

Table 8 shows the various sources of tocopherols from palm oil and its products. Tocopherols accumulated in the palm fatty acid distillate (PFAD) after physical refining can be as high as

0.8% or greater so that the PFAD has been looked upon as a good source of natural tocopherols and tocotrienols. In fractionation or fractional crystallizations tocopherols tend to remain in the liquid or olein fractions thus rendering the solid fractions deficient in antioxidants.

Table 3.20. Tocopherols in Palm Oil (ppm)

Crude Palm Oil	600_1000
Crude Palm Olein	800—1000
Crude Palm Stearin	250—350
PFAD	150—8500
RBD Palm Oil	356—630
RBD Palm Olein	468—673
Palm Kernel Oil	(80—100)*

The vitamin E content of palm oil and other edible oils. Palm oil has an abundance of tocotrienols as compared to other oils.

Palm Vitamin E Isolation

The palm fatty acid distillate (PFAD) contains 0.4%-0.8% tocopherols and tocotrienols with the tocotrienols present as a major component. The bulk of the PFAD is free fatty acids.

The concept of isolating the tocopherols and tocotrienols (palm vitamin E) is based on the conversion of free fatty acids, partial glycerides and triglycerides into methyl esters after removal of which by distillation a concentrate of palmvitee is obtained. Purification of the concentrate will yield tocopherols and tocotrienols.

The purification of the crude mixture is effected by crystallization during which sterols are separated, followed by ion exchange chromatography, in which the tocopherols and tocotrienols are separated from squalene. Palmvitee is then washed dried and purified by molecular distillation. Finally it is deodorized before being dissolved in super olein prior to encapsulation.

The process flow of isolation of palmvitee. The concentration of vitamin E at different stages of the process.

This pilot plant is capable of producing vitamin E with the following results:

purity	95—99%
overall yield	65%
production	5—6 kg/batch

Table 3.21. Total Tocopherol and Tocotrienol of Oils and Fats

Oil and fats	*Tocopherol*					*Tocotrienol*					
	α − T (ppm)	*β − T (ppm)*	*γ − T (ppm)*	*δ − T (ppm)*	*%T (ppm)*	*α − T3 (ppm)*	*β − T3 (ppm)*	*γ − T3 (ppm)*	*δ − T3*	*%T3 T + T3*	*Total (ppm)*
Palm Oil	256	—	316	70	55%	143	32	286	69	45%	1172
Palm Oil	279	—	61	—	31%	274	_	38	69	69%	1081
Palm Oil	152	—	—	—	17%	205	—	439	94	83%	890
Soyabean Oil	101	—	593	264	100%	—	—	—	—	0%	958
Safflower Oil	387	—	174	240	100%	—	—	—	—	0%	801
Corn Oil	112	50	602	18	100%	—	—	—	—	0%	782
Cottonseed Oil	389	—	387		100%	—	—	—	—	0%	776
Sunflower Oil	487		51	8	100%	—	—	—	—	0%	546
Peanut Oil	130	—	216	21	100%	—	—	—	—	0%	367
Cocoa Butter	11	—	170	17	99%	2	—	—	—	1%	200
Olive Oil	51	—	—	—	100%	—	—	—	—	0%	51
Coconut Oil	5	—	—	6	31%	5	1	19	—	69%	36
Palm Kernel Oil	13	—	—	—	38%	21	—	—		62%	34
Lard	12	—	7	—	73%	7	—	—	—	27%	26

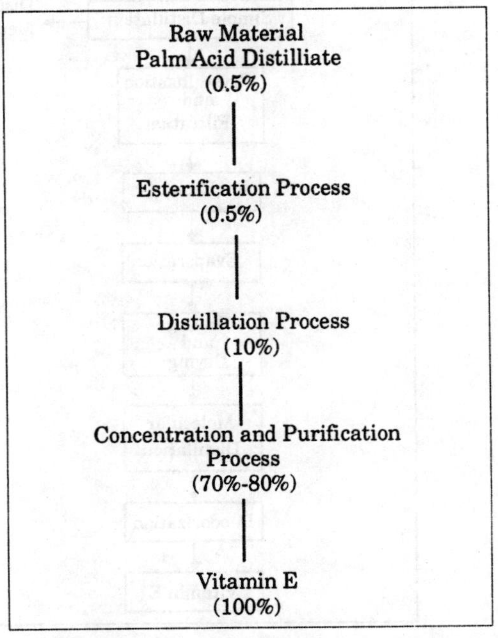

Fig. 3.21. Concentration of Vitamin E at Different Stages.

ECONOMIC VIABILITY EVALUATION

It has been shown that it is technically feasible to produce tocopherols and tocotrienols from PFAD and preliminary economic assessment on the process has been attempted.

Vitamin E extraction is a concentration process from the 0.5%. Vitamin E content of the feed material, PFAD to 99% purity. There are 8 processing steps, and based on two capacities of 25000 tonnes and 50000 tonnes PFAD per year, the plant investment amounts to RM 49.86 and RM 75.50 million respectively (1989 estimates).

The material, utility and labour costs are estimated on the pilot plant experience. The effect of variation of selling prices on project feasibility is analysed.

The economic analysis indicated that for 50000 tonnes of plant capacity for PFAD or 162500 kg of vitamin E, the project is feasible when the selling price is maintained at about RM200 per kilogram for an investment of 75.5 million ringgit. For 25000 tonnes of plant capacity for PFAD or 81250 kg of vitamin E the project is marginally feasible if selling price of vitamin E can be maintained at RM200 and more, some of the costs may be further reduced such as depreciation of plant and material (PFAD).

Nevertheless a detailed analysis, including market study is required before these processes could be fully commercialized.

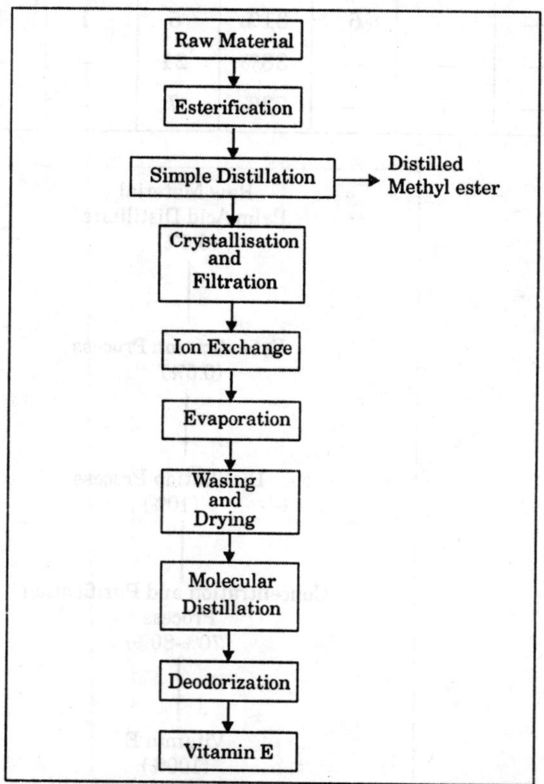

Fig. 3.22. Vitamin E pilot plant production process.

CHAPTER 4

Cuticular Lipids

The cuticular lipids of insects consist of aliphatic material which is present on the outer layer of the integument. These waxes serve the critical function of restricting water loss to prevent desiccation in most terrestrial insects, and also serve as sex attractants and aphrodisiacs, as species and caste recognition cues, and as kairomones in certain insect groups. The cuticular lipids affect the absorption of insecticides and other chemicals from the environment and have been suggested to serve as a barrier to the penetration of micro-organisms. Some in-sects, such as the scale insects and honeybees, produce much larger amounts of waxy material for protection and for raising brood and storing honey.

All insects which have been carefully studied possess hydrocarbons in their cuticular lipids, and they are the most abundant class of compounds in many species. Other types of waxes commonly found in cuticular extracts include wax esters, ketones, epoxides, secondary alcohols, primary alcohols, free fatty acids, sterols, sterol esters and triacylglycerols. In this context, wax is used in its broad sense to include a variety of lipids rather than in the strict chemical definition.

Cuticular lipids are synthesized by cells which are located near the integument or in the peripheral fat body, probably oenocytes. The biosynthetic pathways for the major cuticular components, including hydrocarbons, wax esters and secondary alcohols, have been established. Prior to deposition, the lipids must pass through the epidermal cells and then through the cuticle to reach the surface of the insect. The pore canals have been implicated in the process by which cuticular lipids cross the cuticle (Locke, M., 1965; Hepburn, H., this volume). The thickness of the cuticular lipid layer appears to vary considerably among different species, and values from a few molecules thick to over half a micrometer have been reported.

Various aspects of insect cuticular lipids have recently been reviewed. Discussions focusing on aspects of the chemistry and biochemistry of cuticular lipids have been presented by Blomquist, G. and Jackson, L. (1979); Nelson, D. (1978); Lockey, K. (1980*a*); Jackson, L. and Blomquist, G. (1976*a*); Blomquist, G. and de Renobales, M. (1982) and Jackson, L. and Baker, G. (1970). Analytical techniques used to characterize insect waxes have been reviewed by Jackson, L. and Armold, M. (1977) and Gilby, A. (1980*a*). The role of cuticular hydrocarbons in chemical communication has been dealt with by Howard, R. and Blomquist, G. (1982), and discussions of the role of cuticular lipids in limiting water loss have recently been presented by Hadley, N. (1980*a*, 1981); Gilby, A. (1980*b*); Machin, J. (1980); Edney, E. (1977) and Ebeling, W. (1974). This chapter summarizes the functions attributed to insect cuticular lipids, outlines analytical procedures used in characterizing wax components, reviews the types of chemicals commonly found in cuticular lipids, discusses the biosynthesis of cuticular lipids and reviews the physiological and endocrine regulation of cuticular lipid formation.

FUNCTIONS OF CUTICULAR LIPIDS

Prevent Desiccation

Perhaps the most critical of the functions attributed to insect cuticular lipids is the formation of a waterproof layer which prevents a lethal rate of desiccation. Many investigators have measured the transpiration of water from insects, and these studies have been reviewed periodically. The evidence is in agreement that the cuticle is permeable to water vapors, and that the cuticular lipids play a major role in reducing transpiration.

Major lines of evidence which support this view are:

(1) Cuticles from which the surface lipids are removed with organic solvents are relatively permeable to water. Vegetable oils, lecithin, and a series of wetting agents and detergents affect the permeability of the cuticle to water.

(2) The waterproofing observed with intact insects is closely duplicated when extracted cuticular lipids are deposited on collodium membranes or intact wing membranes.

(3) Rapid desiccation occurs as a result of scratching the outer surface of the cuticle with abrasive dusts, and this has been used in insect control. Adsorption of the lipids onto the dust may also play a role in increasing transpiration.

(4) If the abraded insect is kept in a moist atmosphere to prevent desiccation, the lipid layer is restored, and along with it, the resistance to desiccation returns.

(5) Positive correlations have been shown between the quantity of extracted cuticular wax and the measured or potential transpiration. In addition, a greater amount of cuticular lipid is often present when there is a greater need for water conservation.

(6) Insects which inhabit desiccating environments have a surface-wax composition that provides maximum impermeability to water vapor.

(7) The transpiration rate from an insect is found to increase rather abruptly at a temperature that corresponds to the transition point or change of phase point of the lipids on the cuticle of the particular species.

This latter point has become quite controversial. The discovery that the permeability of insect cuticle underwent abrupt increases at specific temperatures prompted Beament, J. (1945) and Wigglesworth, V. (1945) to propose that cuticular permeability is controlled by a monolayer of polar lipids oriented at an angle of 65° to the cuticle surface. It was proposed that at the transition temperature (Tc), sufficient thermal energy is available to disrupt this ordered state, leading to increased transpiration. Another model proposed to account for the permeability changes at Tc includes M. Locke's (1965) suggestion that the molecules in the wax canals undergo a phase change at Tc.

Machin, J. (1980) and Hadley, N. (1981) have summarized the recent evidence which argues against the oriented monolayer model proposed by Wigglesworth and Beament. The major lines of evidence against the model are:

(1) The cuticular lipids of most insects studied to date are not composed primarily of aliphatic alcohols and acids, but rather contain large amounts of hydrocarbon;

(2) Forced area curve data for cockroach cuticular lipid indicated that a monolayer was absent and that the molecules were only weakly attracted to the water surface;

(3) Toolson, E. *et al.* (1979), using spin-labeled molecules, showed that there was not a preferred orientation on the cuticle, and furthermore showed that the phase transitions take place with lipids associated with the surface rather than the pore canals;

(4) Toolson, E. (1978), using mathematical arguments, has shown that the abrupt permeability increases which occur at Tc when water loss is plotted against temperature are in reality artifacts created by incorrectly dividing the observed rate of water loss by the saturation deficit to correct for thermally induced changes in the diffusion gradient.

Using Toolson's method, Machin, J. (1980) calculated data for a number of insects, and suggests that the presence or absence of temperature transitions depends upon whether the insect has a "greasy" or "waxy" cuticle. A generally accepted Model for the waterproofing observed by cuticular lipids awaits development.

Effect on Entry of Insecticides and Micro-Organisms

The cuticular lipids play a role in controlling the entry of insecticides and other chemicals from the environment. Olson, W. (1970) suggests that the cuticular lipid layer constitutes a significant barrier to the penetration of lipoidal solutes through the integument. He concludes that partitioning of lipid-soluble insecticides such as DDT from the epicuticular lipid into lower phases is the rate-limiting step in penetration. Disruption of the cuticular lipids increases the rate of solute penetration into the cockroach. Other workers consider the underlying proteinacious part of the cuticle as the main barrier to insecticide penetration, and con-sider the surface lipid as facilitating the lateral movement of topically applied insecticides. Gerolt, P. (1969) has obtained evidence contradicting the widely held view that contact insecticides reach the site of action by penetrating through the integument. He suggests that topically applied dieldrin accumulates in the cuticular lipids, spreads laterally, and then reaches the site of action via the tracheal system.

The protection from micro-organisms provided by the cuticular lipids may be due to the non-wetting nature they provide the insect's surface, making it difficult for the organisms to establish themselves. It is also possible that the cuticular lipids provide chemical protection by being toxic to micro-organisms. Koidsumi, K. (1957) reported that the cuticular lipids of the larvae of *Bombyx mori L.* and *Chilo simplex* Butler were capable of restricting the growth of pathogenic fungi *in vitro,* and indicated thai the antifungal compounds were short-chain saturated fatty acids. However, no definitive structural identification of these compounds was made.

CHEMICAL COMMUNICATION

In addition to their critical role in preventing desiccation, insect cuticular lipids also serve in Chemical Communication. Major roles assigned to cuticular hydrocarbons include serving as sex pheromones, as species and caste recognition cues in several social insects, as thermoregulatory pheromones and as kairomonal cues for parasites. Table 1 lists the species reported to use cuticular components in chemical communication and lists the types of components involved.

Sex Phermones

Cuticular hydrocarbons have been widely implicated as dipteran sex pheromones. Rogoff, W. *et al.* (1964) reported that benzene extracts of female *Musca domestica* elicited both attraction and excitation mating behavior patterns in males. Carlson, D. *et al.* (1971) reported that (Z)-9-tricosene was a "sex attractant" of *Musca domestica* and that it could be extracted from the bodies of females or from their fecal deposits.

A low level of response to (Z)-9-tricosene, both in olfactometer and pseudofly bio assays, prompted several workers to re-examine the female housefly's surface lipid chemistry, and led to the identification of over 100 isomeric methylalkanes, (Z)-9,10-epoxytricosane, (Z)-14-tricosene-10-one, and a C_{36} wax ester. The methylalkanes are terminally and internally branched monomethylalkanes and 3,X-, 4,X- and internally branched dimethylalkanes. Uebel, E. *et al.* (1976) showed that addition of branched alkanes of 28-30 carbons to a pseudofly treated with (Z)-9-tricosene enhanced the mating strike activity of the males, although the alkanes by themselves had no activity. This stimulatory effect of branched alkanes was confirmed by Rogoff, W. *et al.* (1980). In addition, Uebel, E. *et al.* (1978*a*) showed that the epoxide and ketone both released strike activity by the males in the absence of (Z)-9-tricosene. The ester showed no biological activity. Combinations of all active and synergistic components still failed, however, to match a live virgin female as a target for a mating strike. Richter, I. *et al.* (1976) proposed that (Z)-9-tricosene may serve as a releaser mechanism for responsiveness to optical cues, leading to both mating strikes and aggregation.

Considerable effort has also been made at elucidating the importance of cuticular lipids as pheromones for the face-fly *Musca autumnalis.* Chaudhury, M. *et.al.* (1972, 1973, 1974) suggested the presence of a sex pheromone in this species from olfactometer bioassay data. Uebel, E. *et al.* (1975*c*) and Sonnet, P. *et al.* (1975) isolated and characterized several cuticular hydrocarbons which elicited strike behavior by males in a modified pseudofly bioassay. Active components included (Z)-14-nonacosene, (Z)-13-nonacosene and (Z)-13-heptacosene. These olefinic components were found in approximately equal quantities on both males and females. Males, however, contain greater quan-tities of heptacosane and nonacosane than do females, and the authors suggest that these alkanes attenuate the effects of the olefins. The overall hydrocarbon differences are however slight and a closer examination of this species likely to be fruitful.

An even more complex situation exists regarding the sex pheromone complex of the stable-fly, *Stomoxys calcitrans* (L). Muhammed, S. *et al.* (1975) reported that the female produced both "sex attractant" and "mating stimulation" pheromones. Uebel, E. *et al.* (1975*b*) showed that (Z)-9-hentriacontene, (Z)-9-tritriacontene and methyl branched hentria- and tritriacontenes incited males to attempt copulation, as did a series of monomethyl-and dimethylalkanes. Later work by Son-net, P. *et al.* (1977) on the saturated components showed that 15-methyl- and 15,19-dimethyltritria-contanes were most active. Harris, R. *et al.* (1976) showed that increasing amounts of these hydrocar-bons were produced from emergence until the female was 3 days old. Meola, R. *et al.* (1977) reported that this species required a blood-meal for sex pheromone production and successful sexual behavior. Sonnet, P. *et al.* (1979) further characterized the female pheromone complex which induces male copulatory behavior. The major methyl branched olefins were identified as 13-methyl-l-hentriacontene and 13-methyl-l-tritriacontene. The most active preparation was a combination of the normal alkenes, methylalkenes, and mono- and dimethylalkanes.

Three species of *Fannia* have been examined for sex pheromones using the pseudofly bioassay. (Z)-9-Pentacosene constitutes the major cuticular hydrocarbon component of female *Fannia canicularis,* and was reported to stimulate the male to initiate copulatory attempts. The mating stimulants from female *Fannia pusio* are normal mono-olefins of 31 and 33 carbons. Bioassay data indicate that (Z)-l 1-hentriacontenc is the most active component. Essentially the same mixture of alkenes is found on females of *Fannia femoralis.* For both species alkanes are also present, and Uebel, E. *et al.* (1978*a*) showed that addition of the alkanes of *F. femoralis* to its C_{31}, olefins increased mating strike frequency.

Mackley, J. *et al.* (1981) and Bolton, H. *et al.* (1980) investigated the possibility that the horn-fly *(Haematobia irritans)* courtship behavior involved chemical components. Alkenes, including (Z)-9-tricosene, (Z)-S-triccsene, (Z)-9-pentacosene and (Z)-9-heptacosene were identified from both sexes. Males had greater quantities of the (Z)-9-tricosene than did females, but the females contained greater quantities of all other olefins. Each alkane was individually active in releasing strike behavior, but only mixtures elicited the remaining steps of the courtship.

The fruit-fly *Drosophila melanogaster* also utilizes cuticular hydrocarbons as part of its sex pheromone. Newly emerged males and females have an almost identical hydrocarbon profile. A sexual dimorphism in long-chain hydrocarbons occurs starting on the second day after eclosion, and males lose sexual attractiveness. An aphrodisiac pheromone begins accumulating on females 1 day after eclosion. This hydrocarbon pheromone gives a dose-dependent wing vibration response in males.

Initial attraction of male tsetse flies *(Glossina* spp.) to females is known to involve visual cues (Swynnerton, C., 1936) with no evidence presented for olfactory sex attractants. Good evidence does exist, however, for the presence of a contact sex pheromone in *Glossina morsitans* which elicits copulatory behavior by the males. Similar evidence exists for *Glossina palpalis palpalis* and *Glossina pallidipes* (Carlson, personal communication). Carlson, D. *el al.* (1978) found that male behavior associated with copulation was elicited only by three high-molecular weight hydrocarbons. The most active hydrocarbon was identified as 15,19,23 trimethyl-heptatriacontane, with 15,19-dimethylheptatriacontane and 17, 21-dimethylheptatriacontane being less active. Female *G. morsitans* was shown to contain about 4.2/μg of the trimethylalkane

Table 4.1. Cuticular Lipids in Chemical Communication

Process / species	*Components involved*
Sex pheromone	
Musca domestica	(Z)-9-ticosene,(Z)-9,10-epoxytricosane (Z)-14-ticosene,(Z)-10-one, Methylalkanes
Musca aulumnalis	(Z)-9-nonacosene,(Z)-13 nonacosene, (Z)-13-Heptacosene
Stomoxys calcitrans	(Z)-9-hentriacontene,(Z)-9,-tritriacontene, methylakanes 15-methyl and 15, 19-dimethy-tritriacontanes 13-methyl 1-hentriacontene,13-methyl-1 tritricontene
Fannia canicularis	(Z)-9-pentacosene + alkanes
Fannia pusio	(Z)-11-Hentriacontene + alkanes
Fannia femoralis	(Z)-11-hentriacontene + alkanes
Haematobia irritans	(Z)-5-tricosene(Z)-9 pentacosene,(Z)-9-heptacosene
Glossina morsitans	15, 19, 23-timethyheptariacontane, 15, 19-and 17, 21-diamethyheptriacontane
Glossina sp. (seven other species)	methylakanes
Lucilia cuprina	cuticular lipids (unidentified)
Pikonema alaskensis	9, 19-alkadienes of C_{29}, C_{31}, C_{33}, C_{35}, and C_{37},
Lycoriella mali	*n*-heptadecane
Colias eurytheme	13-methylheptacosane and esters
Copulation release pheromone	
Callosobruchus chinensis	mono-and dimethylakanes of C_{26}-C_{33} and (E)-3. 7-dimethyl-2octene-1,8-dioic acid
Species and caste recognition	
Zooptermopsis angusticollis	hydrocarbon composition
Rcticulitermes flavipes	hydrocarbon composition
Reticulitermes virginicus	hydrocarbon composition
Trichopsenius frosti	hydrocarbon composition
Trichopsenius depressus	hydrocarbon composition
Philotermes howardi	hydrocarbon composition
Xenitusa hexagonalis	hydrocarbon composition
Kairomones	
Heliothis zea (attract parasites)	13-methylhentriacontane
Heliothis virescens (attract parasites)	11-methylhentriacontane, 16-methyl dotriacontane, 13-methyltroiacontane
Thermoregulatory	
Vespa crabro	(Z)-9-pentacosene
Oviposition markers	
Cardiochiles nigriceps	nC_{23}-nC_{33}

and 1.2/µg each of the dimethylalkanes. With all three compounds, male copulatory behavior was released only on physical contact with a surrogate female containing the chemicals (Carlson, D. *et al.*, 1978). Huyton, P. *et al.* (1980) showed that solvent-extracted dead males baited with the trimethylalkanes gave an ED_{95} of 2µg. No synergistic effects were found with the dimethylalkanes, and Huyton, P. *et al.* (1980) concluded that the trimethyl compound is the sex pheromone of this species. The natural pheromone appears on the pharate adult female about 2 days before emergence from the puparium and is present throughout her life. Male responsiveness to pheromone baited decoys increases to a maximum 3-4 days after emergence. These authors suggest that another possible function of this pheromone may be to ensure that the union of male and female is maintained throughout the lengthy copulation period which is required for successful ejaculation and spermatophore transfer (Pollock, J., 1970).

Huyton, P. *et al.* (1980) conducted interspecific mating tests with seven species of *Glossina* to evaluate the specificity of the contact pheromone isolated from *G. morsitans*. Their tests with either live or dead females as targets indicated that the responses of the males are largely the consequence of differences in composition of the female surface cuticular hydrocarbons and not of female behavior.

In addition to Diptera, other insects use long-chain hydrocarbons as sex pheromones. Several sibling species in the genus *Holomelina* (tiger moths) use 2-methylheptadecane, as do females of *Pyrrharctia isabella*. Conner, W. *et al.* (1980) noted that the female arc-tiid moth *Utetheisa ornatrix* emits (Z,Z,Z)-3,6,9-heneicosatriene plus small quantities of an unidentified C_{21} tetraene as her sex pheromone. These hydrocarbons appear gland-specific and are probably not cuticular components.

Two sulfur butterflies, *Colias eurytheme* and *Colias philodice* (Pieridae) coexist over much of North America and originally were thought to interbreed. Reproductive isolation is based upon female response to species-specific visual and chemical signals. Males of *C. philodice* contain on their wings three major esters (*n*-hexylmyristate, *n*-hexylpalmitate, and *n*-hexylstearate), a series of *n*-alkanes with odd carbon numbers (C_{23} to C_{29}), and small quantities of 13-methylheptacosane. Males of *C. eurytheme* contain on their wings the same *n*-alkanes, no esters, and a large amount of 13-methylheptacosane. The esters and 13-methylheptacosane all have significant electrophysiological activity. Preliminary behavioral data suggest that the esters are the important species recognition cues and that the branched hydrocarbon functions as an aphrodisiac. Grula, J. and Taylor, O. (1979) studied the inheritance of male pheromone production by gas chromatographic analysis of wing extracts of various genotypes derived from forced interspecific crosses. The X-chromosome carried most of the gene(s) controlling production of 13-methylheptacosane. Expression of this compound in hybrids displayed a codominant pattern, and was not influenced by diet.

Cuticular hydrocarbon components are part of the sex pheromone of the yellowheaded spruce saw-fly, *Pikonema alaskensis* Rohwer. A series of (Z,Z)-9,19-alkadienes of 29, 31, 33, 35 and 37 carbons are present on the surface of the female and possess pheromone activity.

Long-chain methyl branched alkanes are components of the copulation release pheromone, erectin, of the azuki bean weevil, *Callosobruchus chinen-sis*L. 3-Methyl-, internally branched mono-methyl- and dimethylalkanes of C_{26}-C_{35} along with a dicarboxylic acid comprise the copulation release pheromone, which is distinct from the sex attractant.

Species and Caste Recognition Cues

Social insects are well known for their ability to recognize conspecifics (Wilson, E., 1971), and to distinguish the caste and sex of the individuals they interact with. Because many social insects live either wholly in the soil or distributed between soil and wood (especially subterranean termites and many ants), much of their sensory perception must be either olfactory or tactile. The nests and galleries of these insects are often tightly closed, low-volume systems. Such systems place a premium on semiochemicals which are complex enough to have a high information content, but which are of low enough volatility to minimize sensory habituation. Howard, R. *et al.* (1978) and Blomquist, G. *et al.* (1979*a*) suggested that the cuticular hydrocarbons of termites might serve as semiochemical cues for caste and species recognition. They found that *Reticulitermes flavipes* (Rhinotermitidae) and *Zoopiermopsis angusticollis* (Hodotermitidae) possess drastically different hydrocarbon profiles, and both of these differ markedly from *Nasuititermes exitiosus* (Nasuititerrnitidae). In addition, R. Howard and co-workers completely characterized the cuticular hydrocarbons of *R. virginicus* and partially characterized by GLC the hydrocarbons of four other species of *Reticulitermes,* as well as those of *Coptotermes formosanus, Marginitermes hubbardi, Incistitermes minor, Pterotermes occidentis,* and *Tenuirostitermes tenuirostris* (unpublished data). In every case, the hydrocarbon composition is unique. Furthermore, examination of the hydrocarbon compositions by caste indicates caste-specific ratios of hydrocarbon components. Preliminary bioassay data have confirmed the role of cuticular hydrocarbons as species recognition cues for *Reticulitermes virginicus* and *Reticulitermes flavipes*.

Additional evidence that cuticular hydrocarbons serve as species recognition cues comes from the finding that the highly integrated, host-specific termitophilous beetle *Trichopsenius frosti* has an identical cuticular hydrocarbon profile as its host termite *R. flavipes*. In addition, the cuticular hydrocarbons of three species of termitophilous beetles (representing two subfamilies of Staphylinidae) associated with *R. virginicus* are identical to those of that termite (Howard, unpublished data). Biosynthetic studies showed that *T. frosti* produces its own hydrocarbons, rather than procuring them in some manner from its termite host.

The scarab beetle *Myrmecaphodius excavaticollis* associated with *Solenopsis* spp. ("fire ants") has a cuticular hydrocarbon composition that loosely mimics that of its current ant host. The mechanism by which the beetles achieve this is unknown. The cuticular hydrocarbons of two *Solenopsis* species *(Solenopsis invicta* and *Solenopsis richteri)* have been thoroughly characterized and those of two other species *(Solenopsis geminata* and *Solenopsis xyloni)* characterized by GLC (Vander Meer, personal communication). All four species have unique hydrocarbon profiles, suggesting that, like termites, ants may use these chemicals as species recognition cues.

Thermoregulatory Pheromones

Wasps and hornets are known to actively heat their brood combs, especially those cells containing pupae. They do this by positioning their respiratory spiracles over the silken dome of the pupal cell and alternately retracting and extending their abdominal segments up to 180 times per minute. Ishay showed by preliminary bioassays that it was the pupae themselves that triggered this brooding behavior, and that the stimulus involved was that of a "volatile" chemical which could be extracted with alcohol from the pupae. Veith, H. and Koeniger, N. (1978) identified the pheromone as (Z)-9-pentacosene.

Kairomonal Cues for Parasites

Parasites are such ubiquitous components of practically every ecosytem that it is not surprising that some of them have managed to evolve to utilize cuticular hydrocarbon cues produced by their hosts as primary cues in locating their hosts. Hymenop-teran parasites of *Heliothis* spp. exhibit intensified searching behavior for their hosts when exposed to either frass or scales of the ovipositing female moth. Jones, R. *et al.* (1971) showed that 13-methylhentriacontane was the major component of the hemolymph, cuticle, and frass of larvae *Heliothis zea* which was responsible for intense searching behavior by the larval parasite *Microplitis croceipes*. Vinson, S. *et al.* (1975) similarly showed that 11-methylhentriacontane, 16-methyl-dotriacontane and 13-methyltritriacontane, produced in the mandibular glands of *Heliothis virescens* and deposited on the leaf surface by feeding larvae, stimulate *Cardiochiles nigriceps* to search the immediate vicinity more efficiently. Jones, R. *et al.* (1973) reported that an egg parasite of *H. zea (Trichogramma evanescens)* was stimulated by extracts of scales left by the ovipositing *H. zea* female, and that the most active factor in these extracts was *n*-tricosane. Lewis and co-workers have since conducted extensive laboratory and field investigations on the role and practical utilization of *n*-tricosane in manipulating population levels of *H. zea*. Much of their findings regarding the behavioral sequence of host selection by parasites, the chemistry of the kairomones involved, their role in the host finding process, and the basis for their potential employment in pest management programs has recently been reviewed.

WAX OF HONEYBEES, SCALE INSECTS

In addition to a thin layer of cuticular lipid, some insects secrete large amounts of wax. These include the honeybees, which use the wax as structural material to build the honeycomb, and the scale insects, which use the wax for protection. The scale insects are so called because in many species the female is protected by a scale or shield consisting of a mixture of wax and cast skins. The amount of lipid covering the scale insects is much larger than that usually seen on the epicuticle of most insects; however, its chief function is still most likely to protect the insects against desiccation. The adult females of scale insects are degenerate and are attached or fastened to the host plant by the mouthparts. It is probably because they are immobile that the scale insects form their characteristic protective coat, which also protects them from insect predators and the weather.

ANALYTICAL PROCEDURES

Extraction and Separation

The analytical techniques used in insect wax chemistry have been reviewed comprehensively by Jackson, L. and Armold, M. (1977) and by Gilby, A. (1980*a*). The cuticular lipids from whole insects or cast exuviae can be conveniently extracted by immersion in organic solvents such as hexane or chloroform for short periods of time. Usually a few seconds to 15min is recommended for hexane, and shorter extraction times for chloroform. Chloroform is more penetrating and the chances of extracting internal lipids increase as the extraction time increases. Chloroform-methanol mixtures and diethylether are not commonly used except for very short extraction periods due to their interaction with water and the increased possibility of extracting internal lipids. Most often one immersion does not extract all of the cuticular lipid; two or more short immersions are usually better than one long immersion.

After extraction, the major lipid classes can usually be separated by column chromatography and/or thin-layer chromatography (TLC). The use of silver nitrate-impregnated TLC plates for unsaturated components and molecular sieve for branched components are often very helpful. Gas-liquid chromatography (GLC) is routinely used to separate and quantify the components of each class of lipid. Infrared spectrophotometry, nuclear magnetic resonance spectrometry, mass spectrometry (MS) and combined GLC-MS, including chemical ionization (CI)-MS techniques have proven very useful in structure determination. GLC-MS is the principal tool used in obtaining structure assignments for hydrocarbons.

Structure Determination of Hydrocarbons

n-Alkanes can be identified by their retention times on GLC and inclusion in 5 A molecular sieve. Structure assignments can be confirmed by GLC-MS.

The methyl branched alkanes have retention times on GLC which are somewhat less than an *n*-alkane of the same total carbons. Monomethylal-kanes with methyl branches on carbons 2, 3 or 4 elute about 0.3 or 0.4 carbon units in front of *n*-alkanes of the same total carbon number on packed columns. Internally branched monomethylalkanes elute about 0.6 to 0.7 carbon units in front of corre-sponding *n*-alkanes. Internally branched dimethylalkanes elute about 1.3 to 1.4 carbon units before corresponding *n*-alkanes, and dimethlyal-kanes with one methyl branch near the end and one in the middle (3,X-, 4,X-) elute about 1 carbon unit in front of corresponding *n*-alkanes.

When subjected to mass spectrometry each type of hydrocarbon yields a diagnostic spectrum. Discussions on the interpretation of spectra from insect hydrocarbons can be found in papers by Jackson, L. and Blomquist, G. (1976*a*), Nelson, D. and Sukkestad, D. (1970), Nelson, D. *et al.* (1972), Nelson, D. (1978), and Pomonis, J. *et al.* (1978). Recently, the application of chemical ionization-mass spectrometry to insect hydrocarbons has proven useful.

Most of the alkenes characterized from insects to date are *cis* / *(Z)*. This assignment is usually based on the presence of a peak at 730 cm^{-1} and the absence of a peak at 970 cm^{-1} (indicative of *trans* or E double bonds) in the infrared spectrum. Z and E isomers can be separated by silver nitrate-impregnated TLC. Oxidative cleavage followed by GLC or oxidation of the alkene to a diol with osmium tetroxide followed by silylation and mass spectrometry have been successfully used to determine the position of double bonds. More recently, methoxymercuration-demercuration followed by mass spectral analysis has been successfully applied to alkenes (Blomquist, G. *et al*, 1980*c*). This technique is especially advantageous with mixtures of components, as it eliminates the need for isolating each component by preparative GLC.

Characterization of Wax Esters, Fatty Alcohols and Sterols

Wax esters are usually characterized by saponification and characterization of the component fatty acids and fatty alcohols. However, some information can be obtained about the total number of carbon atoms in the wax ester chains by GLC on a high-temperature silicone column. Since GLC only gives peaks according to total number of carbons in the wax ester there may be a mixture of various alcohols esterified to a mixture of fatty acids in any one peak. In order to isolate the alcohols and fatty acids, the wax esters are usually either saponified or transesterified. Saponification has the advantage that the alcohols can be extracted with an organic solvent from the alkaline Saponification mixture and then extraction of the acidified solution will yield the fatty acids. Transesterification provides the methyl esters of the fatty acids which are easily separated from the alcohols by TLC or column chromatography and then the fatty acid methyl esters and fatty alcohols can be readily charac-terized by GLC. Free primary alcohols, secondary alcohols and sterols can usually be separated from the other lipid components by TLC or column chromatography. The purified fractions of the various alcohols are then submitted to GLC and GLC-MS for further characterization. Secondary alcohol wax esters require some special techniques since they do not saponify or transesterify well under normal conditions. Secondary alcohol wax esters have been characterized by GLC of the total fraction to determine the range of chain lengths. This step is followed by $LiAlH_4$ reduction of the ester linkage to yield secondary alcohols and primary alcohols. TLC separation of primary and secondary alcohols followed by GLC or GLC-MS can be used to characterize the components of the secondary alcohol wax esters.

Characterization of Free Fatty acids and Acylglycerols

The free fatty acid fraction is generally characterized by methylation followed by GLC on a polyester column which separates the methylesters according to chain length and number of double bonds. The triacylglycerols, diacylglycerols and monoacylglycerols can be separated by TLC or column chromatography and then each fraction can be analyzed by GLC on high-temperature silicone columns, or they can be simply saponified and methylated or transisterified and the resulting methylesters can be subjected to GLC analysis.

STRUCTURE AND COMPOSITION OF CUTICULAR LIPIDS

The cuticular lipids of many insect species consist of complex mixtures of various types of hydrocarbons and other components. The structures of the major types of insect cuticular lipids are presented.

Hydrocarbons

Hydrocarbons have been the most extensively studied cuticular components and frequently are the only class characterized. They are common components of surface extracts and are present on all insects which have been carefully examined. The insect surface lipids of some species are > 90% hydrocarbon whereas hydrocarbons account for only 0.5% of the surface lipid of the tobacco budworm pupae. Insect surface lipid hydrocarbons are usually a mixture of components which includes *n*-alkanes, *n*-alkenes, terminally branched monomethylalkanes, internally branched monomethylalkanes, dimethylalkanes, trimethylalkanes and others. The types of hydrocarbons present in over 80 species are presented. The hydrocarbons of some insects have a relatively simple composition, such as is present in *Periplaneta americana*. Other insects have exceedingly complex mixtures, such as the female housefly which has 134 methyl branched components.

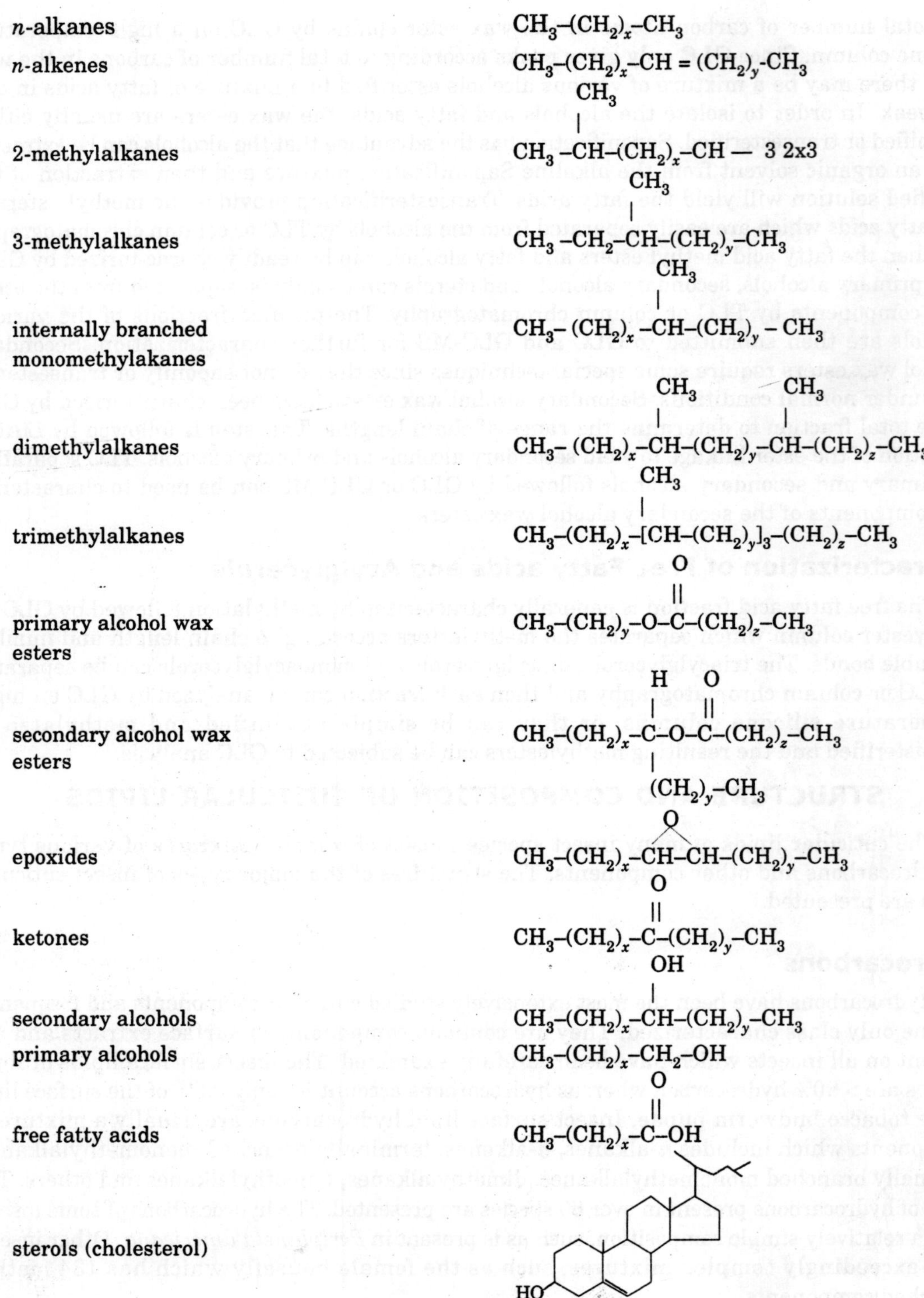

Fig. 4.1. Structures of common insect cuticular components.

The complexity of insect cuticular hydrocarbons has led to the suggestion that hydrocarbon composition might be used as a taxonomic character. However, only a relatively small number of species have been examined and from the data presented there do not appear to be obvious correlations of hydrocarbon compositions within insect groups. In addition, the composition of hydrocarbons often varies tremendously among different developmental forms, between sexes, and at different times of the year. In general, most investigators have concentrated on one developmental stage. These observations indicate that the distribution pattern of hydrocarbon components is unlikely to be a useful discriminating factor in the near future.

n-Alkanes

n-Alkanes occur in all the insect surface lipids so far investigated. The *n*-alkanes of insect surface lipids are generally in the range C_{21}–C_{36} with alkanes having an odd number of carbon atoms predominating. The complexity of the *n*-alkane mixtures varies considerably. For example, *n*-pentacosane comprises 85% of the *n*-alkane fraction of *P. americana*, but *n*-alkanes from C_{17} to C_{34} with both odd and even numbers of carbon atoms are major components of the hydrocarbons of pupal tobacco budworms, *Heliothis virescens*. *n*-Alkanes are not usually found alone, but only *n*-alkanes were reported from the hydrocarbon fraction from cast skins of the beetle, *Tenebrio molitor*. Analysis of the hydrocarbons of seven scale insects showed that they were entirely *n*-alkanes of C_{25}-C_{35}, with odd-numbered carbon chain-lengths predominating.

TERMINALLY BRANCHED MONOMETHYL ALKANES

2-Methylalkanes and/or 3-methylalkanes are prevalent in insect surface lipids. Like the *n*-alkanes, the terminally branched monomethylalkanes range from simple compositions such as in *P. americana* where only 3-methylpentacosane is present to complex mixtures of 3-methylalkanes, 2-methylalkanes or mixtures of 2-and 3-methylalkanes. The 2-methylalkanes are somewhat unique in that components with both odd- and even numbered carbon chains are present in substantial amounts. This appears to reflect their biosynthesis, which involves the carbon skeleton of either valine or leucine. Many of the insects studied to date that have internally branched monomethylalkanes also have 3-methylalkanes, which may reflect a similar biosynthetic origin. The majority of the 3-methylalkanes have an odd-numbered carbon chain (even number of total carbons).

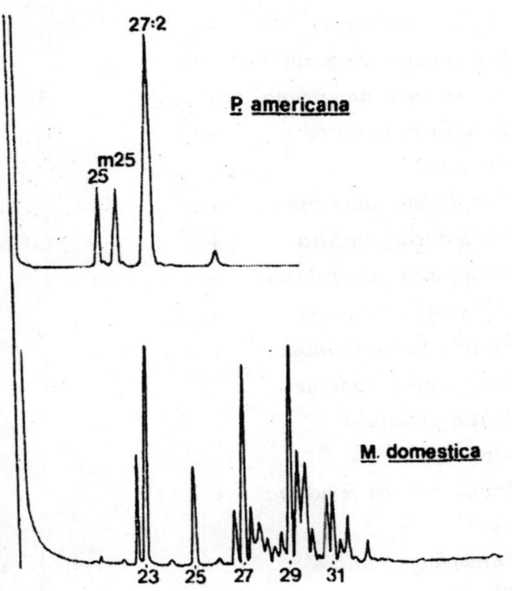

Fig. 4.2. GLC traces of the cuticular hydrocarbons of *Periplaneta americana* and *Musca domestic*. The components in the top trace are 25 (*n*-pentacosane), 25(3-methyl-pentacosane), and 27: 2 ([Z, Z]-6, 9-heptacosadiene). The numbers below the bottom trace refer to the carbon numbers of *n*-alkanes.

Table 4.2. Occurrence of hydrocarbons in insect surface lipids

Species	n-Alkanes	Unsaturated Alkenes	Unsaturated Alkadienes	Branched alkanes: Monomethylalkanes: Terminally branched 2-Me	Branched alkanes: Monomethylalkanes: Terminally branched 3-Me	Branched alkanes: Monomethylalkanes: Internally branched	Branched alkanes: Dimethyl-Alkanes	Branched alkanes: Trimethyl-Alkanes
Plecoptera								
Pteronarcys californica	+	+			+	+		
Orthoptera								
Anabrus simplex	+				+	+	+	
Acheta domesticus	+	+	+	+		+	+	
Gryllus pennsylvanicus	+	+	+	+		+	+	
Nemobius fascialus	+		+	+				
Melanoplus sanguinipes	+				+	+	+	
Melanoplus packardii	+				+	+	+	
Melanoplus bivittatus	+			+		+	+	
Melanoplus femurrubrum	+			+		+	+	
Melanoplus dawsoni	+			+		+	+	
Schistocerca vaga	+					+	+	+
Schistocerca gregaria	+					+	+	
Schistocerca americana	+					+	+	+
Locusla migratoria	+				+	+	+	
Dictyoptera								
Periplaneta americana	+	+	+		+			
Periplaneta japonica	+	+	+		+	+		
Periplaneta australasiae	+	+		+	+			
Periplaneta brunnea	+			+	+			
Periplaneta fuliginosa	+	+		+	+			
Leucophaea maderae				+	+			
Blatta orientalis	+			+	+			
Phasmatodea								
Diapheromera femorata	+			+	+	+		
Diptera								
Sarcophaga bullata	+	+	+		+	+		
Phormia regina	+	+	+	+				
Cochliomyia hominivorax	+	+	+	+				
Calliphora vicina	+	+	+	+				
Musca domestica	+	+		+	+	+	+	
Musca autumnalis	+	+						
Stomoxys calcitrans	+	+				+	+	
Fannia canicularis	+	+						
Fannia pusio	+	+						

Fannia femoralis	+	+						
Glossina morsitans	+						+	+
Lycoriella mali	+							
Drosophila melanogaster	+	+	+	+				
Hymenoptera								
Atta colombica	+	+						+
Atta sexdens	+							+
Atta ceptalotes isthmicola	+							+
Myremecia gulosa	+			+	+	+		
Iridomyrex humilis	+	+			+	+		
Camponotus intrepidus	+	+			+	+		
Formica nigricans	+	+	+			+		
Formica rufa	+	+	+			+		
Formica polyctena	+	+	+			+		
Solenopsis invicta	+				+	+	+	
Solenopsis richteri	+				+	+	+	
Pogonomyrmex rugosus	+				+	+	+	
Pogonomyrmex barbatus	+				+	+	+	
Apis mellifera	+	+	+			+		
Bombus appositus	+	+				+		
Bombus occidentalis	+	+						
Nomia bakeri	+				+			
Bembix pruinosa	+	+			+	+	+	
Coleoptera								
Curculio caryae	+	+	+			+		
Phyllobius maculicornis	+			+		+	+	
Chrysomela sp.	+			+				
Cassida sp.	+			+				
Donacia sp.	+			+				
Anthonomus grandis	+	+				+		
Tribolium confusum	+				+	+	+	
Tribolium castaneum	+			+	+	+	+	
Eleodes armata	+			+	+	+	+	
Cryptoglossa verrucosa	+				+	+		
Centrioptera muricata	+				+	+		
Centrioptera variolosa	+				+	+		
Pelecyphorus adversus	+				+	+		

Tenebrio molitor	+	+			+	+	+	
Tenebrio obscurus	+	+			+	+	+	
Attagenus megatoma	+	+			+	+	+	
Lasioderma serricorne	+				+	+	+	
Cylindrinotus laevioctos triatus	+				+	+	+	+
Phylan gibbus	+				+	+	+	+
Ceutorrhynchus assimilis	+			+		+	+	
Blaps mucronta	+			+		+	+	
Alphitophagus bifasciatus	+	+		+	+	+	+	
Alphitobius diaperinus	+			+	+	+	+	
Rhagonycha fulva	+	+	+					
Cantharis livida	+	+						
Canthans livida	+	+						
Popillia japonica	+			+	+	+		
Trichopsenius frosti	+	+	+	+	+	+	+	
Trichopsenius depressus	+	+	+	+	+	+	+	
Philotermes howardi	+	+	+	+	+	+	+	
Xenitusa hexagonalis	+	+	+	+	+	+	+	
Callosobruchus maculatus	+				+	+	+	
Isoptera								
Zootermopsis angusticollis	+				+	+	+	
Reticulitermes flavipes	+	+	+	+	+	+	+	
Reticulitermes virginicus	+	+	+	+	+	+	+	
Lepidoptera								
Heliothis virescens	+			+	+	+	+	
Heliothis zea	+					+		
Manduca sexta	+	+	+			+	+	
Homoptera								
Acyrthosiphon pisum	+							
Diceroprocta apache	+				+	+		

Internally Branched Mono-, di- and Trimethylalkanes

A very comprehensive review covers the occurrence and analysis of internally methyl-branched hydrocarbons. The majority of the internally branched monomethylalkanes have the methyl branch located on an odd-numbered carbon atom which often is carbon number 5, 7, 9, 11, 13, 15 or 17. The general range is from 20 to 40 carbons. Monomethylalkanes with the methyl branch on an even-numbered carbon atom are usually reported mixed with odd-numbered carbon corners and as minor components.

Most of the long-chain internally branched dimethylalkanes have an isoprenoid type spacing, although they are not derived from isoprenoid units. Common isomers are 9,13-, 11,15-, 13,17-, 15,19- and 17,21-. Methyl branches on positions 11,21- have been reported and an unusual symmetrical component, 5,17-dimethylheneicosane, has been reported from the termite *Zooptermopsis angusticollis*. Other types of dimethylalkanes include a 3,X- and 4,X-dimethylalkane series from the housefly and cigarette beetle, 11,19-dimethylalkanes in the tobacco budworm, 11,12-dimethylalkanes in the cigarette beetle and 13,15-dimethylalkane in the fire ant. As careful analyses of dimethylalkanes are made on more organisms, it appears that the methyl groups can be positioned almost anywhere on the chain. Isomers with methyl groups on odd-numbered carbons are more prevalent than isomers with methyl branches on even-numbered carbons. Internally branched trimethylalkanes with isoprenoid spacing of the methyl branches have been reported from *Manduca sexta*, *Schistocerca vaga* and the tsetse fly *Glossina morsitans*.

Two homologous series of trimethylalkanes have been observed as major constituents of the saturated hydrocarbons from ants. 3,7,11-Trimethylalkanes of C_{34}, C_{36} and C_{38} and 4,8,12-trimethylalkanes of C_{35}, C_{37} and C_{39} are the major constituents of the surface lipid-saturated hydrocarbons of the ant, *Atta Columbia*. In *Alia sexdens*, both 3,7,11- and 4,8,12-trimethylalkanes are present as major components, but in *Atta cephalotes isthmicola,* only the 3,7,11-trimethyl series is present and accounts for only about 15% of the alkanes.

Alkenes

n-Alkenes, with one, two, or three double bonds, have been characterized from about one-half of the insects examined to date. The chain length of cuticular alkenes usually ranges from C_{20} to C_{37}, with odd-numbered chain lengths predominating. The positions of the double bonds can be almost anywhere in the chain. Alkenes with the double bond in the 1-position are present in the red flour beetle, *Tribolium castaneum*. Two and 3-alkenes, as well as a series of alkenes with double bonds in unspecified positions, ranging from C_{20} to C_{31}, have been reported in the pecan weevil *Curculio caryae*. A number of Dipteran species possess *n*-alkenes of 23 to 31 carbons which function in chemical communication . The double bonds are usually either in the 7 or 9 positions. In three cockroaches, *Periplaneta australiasia, Periplaneta fuliginosa* and *Periplaneta japonica*, (Z)-9-alkenes from C_{27} to C_{31} were observed. All castes of the termites *Reticulitermes ftavipes* and *Reticulitermes virginicus* possess (Z)-9-pentacosene as a major cuticular component as do four species of termitophilous beetles associated with them (Howard, R. *et al.*, unpublished data). Alkenes with double bond positions at 8 and 10 have been reported from the honeybee, *Apis mellifera*. The face-fly, *Musca autumnalis,* has a complex series of 6-, 8-, 10-, 11- and 12-pentacosene, 10-, 11-, 12- and 13-heptacosene, and 10-, 11-, 12-, 13-, and 14-nonacosene.

Polyunsaturated hydrocarbons are not as common insect surface lipid components and are often minor components. However, the major hydrocarbon component of the surface lipids of the American cockroach, *Periplaneta americana,* is (Z,Z)-6,9-heptacosadiene. Although most of the other cockroaches studied have methylalkanes as the major components, *P. japonica* hydrocarbons are 17% (Z,Z)-6,9-nonacosadiene and lesser quantities of (Z,Z)-6,9-heptacosadiene, -octacosadiene, -tricontadiene and hentriacontadiene. Cuticular dienes having chain lengths of C_{29}-C_{37} with double bonds in the (Z,Z)-9,19-positions are sex pheromone components of the female yellowheaded spruce sawfly, *Pikonema alaskensis.*

OXYGENATED COMPONENTS

The non-hydrocarbon components of insect surface lipids have not been extensively studied in many insects, but it appears that wax esters seldom predominate except in beeswax. Free alcohols, free fatty acids and lesser quantities of other components are common. Some insects have appreciable quantities of unusual lipids and some of those will be discussed below.

Wax Esters and Sterol Esters

Wax esters are not as common in insect cuticular lipids, nor have they been as widely studied, as hydrocarbons. Wax esters from different insects vary in composition, although most of them are composed of long-chain normal primary alcohols and acids. The cuticular lipids of both the adult and naiad *Pteronarcys californica* contain saturated and unsaturated C_{30} to C_{40} wax esters but the adults have more of the longer-chain wax esters. The pea aphid, *Acyrthosiphon pisum,* contains C_{32} to C_{54} aliphatic wax esters composed of C_{12} to C_{34} normal saturated primary alcohols and acids. Sterol esters are also reported in the cuticular lipids of some insects with the principal sterol being cholesterol.

Novel wax esters were identified in the cuticular lipids of the grasshoppers *Melanoplus sanguinipes* and *Melanoplus packardii.* These wax esters, comprising 18 and 28% respectively of the cuticular lipids of these two insects, consist of saturated fatty acids esterified to secondary alcohols. The fatty acids range from C_{12} to C_{22}. The secondary alcohols range from C_{21} to C_{27} with the hydroxyl group near the center of the carbon chain. In *M. sanguinipes,* 59% of the secondary alcohols consist of the C_{23} compounds in which the major isomer is tricosan-11-ol, with smaller amounts of tricosan-12-ol and tricosan-10-ol. A number of other species of *Melanoplus* also contain these unusual wax esters, and to date they have not been reported from other insects.

The primary alcohols found in the cuticular lipids of insects are usually of even-numbered carbon atoms and range from C_{22} to C_{34}. Seventy percent of the cuticular lipid of the larva of the eri silkworm, *Samia cynthia ricini* is *n*-triacontanol and *n*-octacosanol. The alcohols from *Eleodes armata* (Hadley, N., 1977) are primarily saturated C_{20}-C_{34} alcohols with the suggestion of branched and/or unsaturated alcohols containing 30-40 carbon atoms. Over 90% of the wax on *Eriocampa ovata* larvae is primary alcohol of C_{20}–C_{32}, with hexacosanol comprising about 73% of the fraction (Percy, and Blomquist, unpublished data). In *M. sanguinipes* and *M. packardii* the primary alcohols are C_{24}–C_{30} and comprise only a small percentage of the surface lipids. Trace amounts of free secondary alcohols are also present from *M. sanguinipes* and *M. packardii.* Half of the surface lipid extract of larvae of *Tenebrio molitor* is pentacosan-8,9-diol with a melting point of 115°. A distinction between insects with " soft" surface lipids in which the proportion of hydrocarbons is high, and those with "hard" surface lipids in which long-chain alcohols (and diols) appear to predominate, is suggested.

Fatty Acids

The C_{14}–C_{20} fatty acids with an even number of carbon atoms predominate. In most cases saturated and unsaturated fatty acids have been observed. However, the surface lipid fatty acids of *Lucilia cuprina* are all saturated. Branched fatty acids are not usually present in insect waxes. Free a-hydroxy acids have been observed from the pea aphid, *Acyrthosiphon pisum*.

Other non-hydrocarbon Cuticular Components

The three major non-hydrocarbon components of the surface lipids of female houseflies are (Z)-14-tricosene-10-one, (Z)-9,10-epoxytricosane and 9-bexadecenyl-9-octadecenoate. The first two compounds may be involved in the sex pheromone complex. The surface lipid of male *Drosophila melanogaster* contain (Z)-vaccenyl acetate as a major non-hydrocarbon component. The (Z)-vaccenyl acetate is found in the ejaculatory bulb of adult males and is transferred to females during mating. Heneicosan-8-ol acetate was identified as a major (27%) non-hydrocarbon constituent of 5-day-old male little bouseflies, *Fannia canicularis*. Triacylglycerols are present in some cuticular extracts, but usually as minor amounts. Their presence in extracts of cast skins would suggest that the triacylglycerols are true cuticular components, and not simply contaminants from internal lipids.

Beewax—Surface Lpids and Comb Wax

The composition of the surface lipid of the honeybee, *Apis mellifera,* is quantitatively different from that of the comb wax. The major component of the surface lipid is hydrocarbon (58%), followed by monoesters (23%), diesters (9%), triesters (2%), free fatty acids and more polar lipids (8%). The comb wax composition is hydrocarbon (13–17%), wax monoester (31–35%), diesters (10–14%), triesters (3%), hydroxymonoesters (3–6%) and hydroxypolyesters, free primary alcohols, diols, and acid monoesters (7–10%). The structures of the major types of esters found in beeswax.

BIOSYNTHESIS OF CUTICULAR LIPIDS

Most of the work on the biosynthesis of insect cuticular lipids has focused on the hydrocarbons, with fewer studies on the biosynthesis of the oxygenated components reported. The interest in hydrocarbons probably reflects the large amount of this fraction in surface lipid extracts, the ease in which they can be extracted and quantified and the predominance of unique methyl branched components.

HYDROCARBONS

Considerable progress has been made in the past few years on understanding the biosynthesis of insect hydrocarbons. *In vivo* studies in the 1960s established that labeled acetate was readily incorporated into insect hydrocarbons. Later studies with specific radiolabeled precursors and careful analysis of metabolic products have established the biosynthetic pathways for the most common hydrocarbon components. Experiments using carbon-13 labeled precursors have confirmed some of the conclusions based on radiochemical data. Most experiments to date have used either *in vivo* or tissue slice systems, and only recently have studies with cell-free preparations been initiated. This section will review the information available on the site of hydrocarbon synthesis and the biosynthetic pathways for normal, unsaturated and methyl branched hydrocarbons.

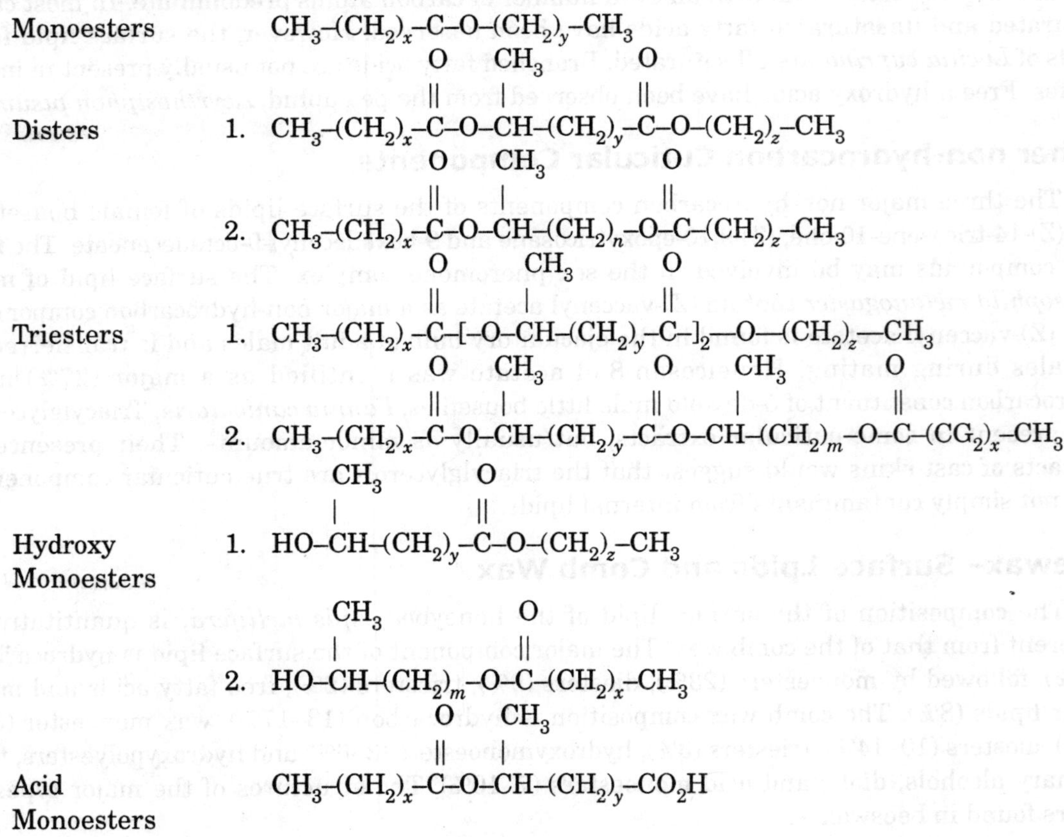

Fig. 4.3. Structures of the major ester components of beeswax.

Dietary Contribution

In a study comparing the cuticular hydrocarbons of the tobacco hornworm to its dietary hydrocarbons, Nelson, D. *et al.* (1971) presented data indicating that a portion of the dietary alkanes was incorporated into the cuticular lipids. A similar conclusion was reached by Blomquist, G. and Jackson, L. (1973*a*) using radiolabeled alkanes fed to a grasshopper. However, in both of these cases the contribution of dietary alkanes to the insect's complement of hydrocarbon appeared small, and the insect biosynthesized the majority of its hydrocarbon components. In a study on the source of the hydrocarbons which serve as kairomones of the corn earworm, Hendry, L. *et al.* (1976) concluded that the diet was the major source of the kairomone (*n*-tricosane). However, they did not examine the ability of the corn earworm to produce *n*-tricosane *de novo*. In as much as many other insects are known to do so readily, it would seem wise to check this point before excluding the possibility.

Site of Synthesis

The most probable site(s) for cuticular hydrocarbon biosynthesis are cells associated with the epidermal layer or the peripheral fat body. Nelson, D. (1969) demonstrated that acetate and palmitate were incorporated *in vitro* into hydrocarbons by the integument of *Periplaneta americana* and *Manduca sexta,* whereas the fat body did not efficiently incorporate either of these precursors into hydrocarbon. Similarly, studies have demonstrated that hydrocarbon production in *Periplaneta americana, Sarcophaga bullata* and *Musca domestica* occurs in epidermal related cells. Diehl, P. (1973, 1975) separated the oenocyte-rich peripheral fat body from the central yellow fat body tissue in the locust, *Schistocerca gregaria,* and observed the highest rate of hydrocarbon synthesis in the oenocyte-rich peripheral fat body. In *Tenebrio molitor,* Romer, F. (1980) demonstrated that isolated oenocytes efficiently and specifically incorporated [1-^{14}C]acetate into hydrocarbon.

Transport

Chino and co-workers have recently shown that the diacylglycerol-carrying lipoprotein (DGLP) (or lipophorin) from the American cockroach and a locust contain large amounts of hydrocarbon. Approximately one-fourth of the total lipids associated with lipophorin were hydrocarbon which were similar in composition to those found on the surface of the insects. They postulate that one function of lipophorin is to carry hydrocarbon from the site of synthesis to the site of deposition.

The two most considered metabolic routes for hydrocarbon biosynthesis involve condensation-reduction and elongation-decarboxylation path-ways. In the bacterium, *Sarcinia lutea,* evidence has been presented favoring a modification of the condensation-reduction pathway, in which two acyl groups condense head-to-head with the loss of a carboxyl group, and the molecule is subsequently reduced. In contrast, Kolattukudy and co-workers have presented convincing evidence that in plants hydrocarbon biosynthesis occurs by the elongation of fatty acids followed by a reductive decarboxylation. Both indirect and direct evidence favors the elongation-decarboxylation pathway in insects.

***(a) Elongation-decarboxylation pathway: indirect evidence* :** Long-chain fatty acids of the same carbon range as the common insect hydrocarbons are not found in most insect cuticular lipid extracts. However, a comparison of the structures of the primary and secondary alcohols with *n*-alkanes lends circumstantial evidence favoring the elongation-decarboxylation pathway for hydrocarbon biosynthesis in insects. The structural relationships between the primary alcohols and *n*-alkanes in several insects suggest the possibility of a similar precursor. The primary alcohols and *n*-alkanes from insects are often of a similar chain-length range (C_{22}–C_{34}). The *n*-alkanes consist primarily of odd-numbered carbon chain components and the primary alcohols are of even-numbered carbon chains. Lambremont, E. (1972) has demonstrated the reduction of palmitic and stearic acid to the corresponding fatty alcohols and the conversion of these fatty alcohols to fatty acids in the tobacco hornworm. It is likely that a similar reduction of very long-chain fatty acids to fatty alcohols is involved in the biosynthesis of cuticular primary alcohols in insects. Similarly, one can envisage the decarboxylation of very long-chain acids to *n*-alkanes one carbon unit shorter. Since very long-chain fatty acids are not commonly found in insect cuticular lipids, it appears that they must be efficiently decarboxylated to alkanes or reduced to the corresponding fatty alcohol. Lack of substantial amounts of very long-chain *n*- or branched fatty acids appears characteristic of most insect cuticular lipids.

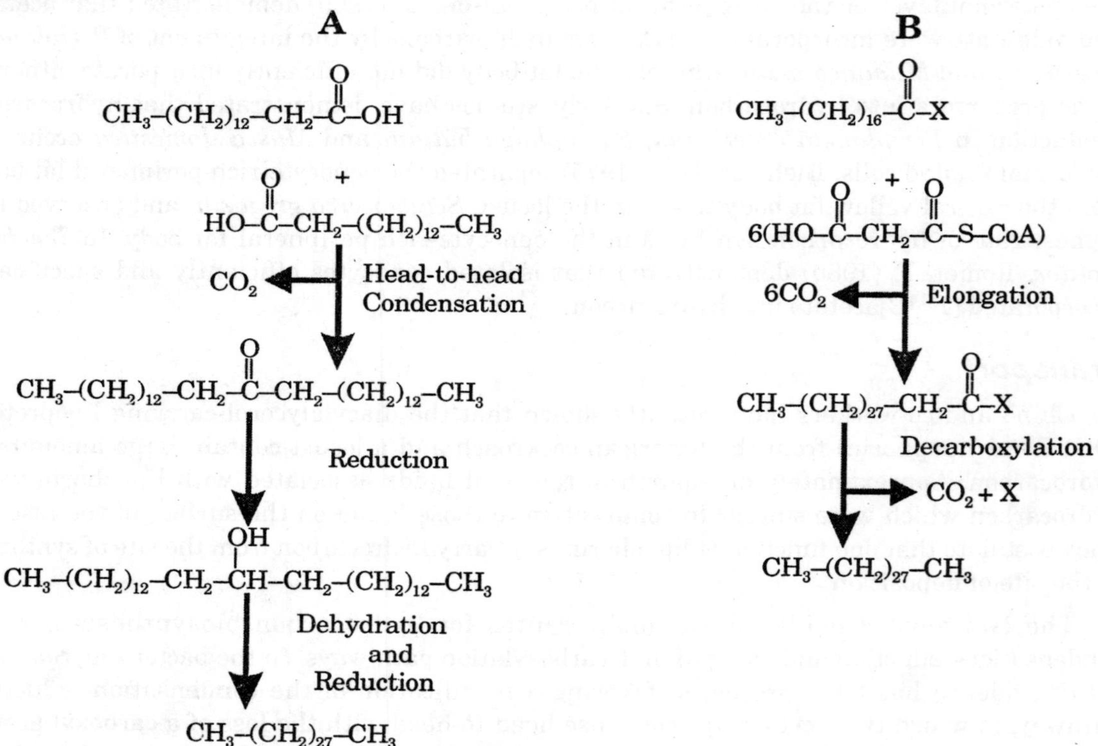

Fig. 4.4. The two most considered pathways for hydrocarbon biosynthesis. A: Head-to-head condensation pathway; B: elongation-decarboxylation pathway.

Over one-half of the secondary alcohols in the grasshopper *Melanoplus sanguinipes* would have to arise from odd-chain fatty acids for a direct condensation mechanism to be operative in their biosynthesis. Tricosan-11-ol is the major isomer comprising over a third of the secondary alcohols in this insect. The fatty acids 11: 0 and 13: 0 would be required for a direct condensation pathway of biosynthesis to occur. Odd-chain fatty acids have not been extensively studied in insects, but unless insects possess an unusually active odd-chain fatty acid synthesizing system, the structures of the secondary alcohols make it appear unlikely that they arise directly from a condensation pathway.

Several lines of evidence from radioisotope tracer studies also argue against a condensation-reduction pathway for hydrocarbon biosynthesis in insects. In the migratory grasshopper, *M. sanguinipes,* the proposed intermediates of the condensation pathway, symmetrical ketones and secondary alcohols of 23, 27 and 31 carbons, were not converted to hydrocarbons. Furthermore, the unidirectional metabolism of alkanes to secondary alcohols in this insect does not support the condensation pathway for hydrocarbon biosynthesis. Similar studies in the cockroach, *Periplaneta fuliginosa,* showed that a symmetrical C_{23} secondary alcohol and ketone were not reduced to n-tricosane, the major n-alkane in this insect. Since the possibility of a non-symmetrical condensation of a C_{10} with a C_{14} or a C_8 with a C_{16} fatty acid was conceivable for the biosynthesis of n-tricosane, the incorporation of 7-, 8-, 10-, 11 - and 12-tricosanols into hydrocarbon was also studied. None of these possible intermediates were incorporated into hydrocarbon in *P. fuliginosa*.

(b) Decarboxylation of long-chain fatty acids to n-alkanes : Direct evidence in favor of the elongation-decarboxylation pathway in insects was obtained by demonstrating the direct decarboxylation of long-chain fatty acids to n-alkanes in several cockroaches and in the termite *Zootermopsis angusticollis*. *In vivo* studies showed that [R-^{3}H] hexacosanoic acid was converted to n-pentacosane in the American cockroach, and that [R-^{3}H] tetracosanoic acid was converted to n-tricosane in *Periplaneta fuliginosa*. The decarboxylation of high specific activity [15, 16-^{3} H] tetracosanoic acid in the termite *Z. angusticollis* was demonstrated both *in vivo* and *in vitro*. An example of the type of data obtained is presented in. Whereas [1-^{14}C] acetate labeled all the cuticular hydrocarbon components [R-^{3}H] tetracosanoic acid labeled n-tricosane exclusively both *in vivo* and *in vitro* α-Hydroxytetracosanoic acid was incorporated into hydrocarbon more efficiently than was tetracosanoic acid, suggesting that the first step in the reductive decarboxylation involves α-hydroxylation. Most of the decarboxylase activity was located in the microsomal fraction, and was stimulated two-fold by the addition of ascorbic acid.

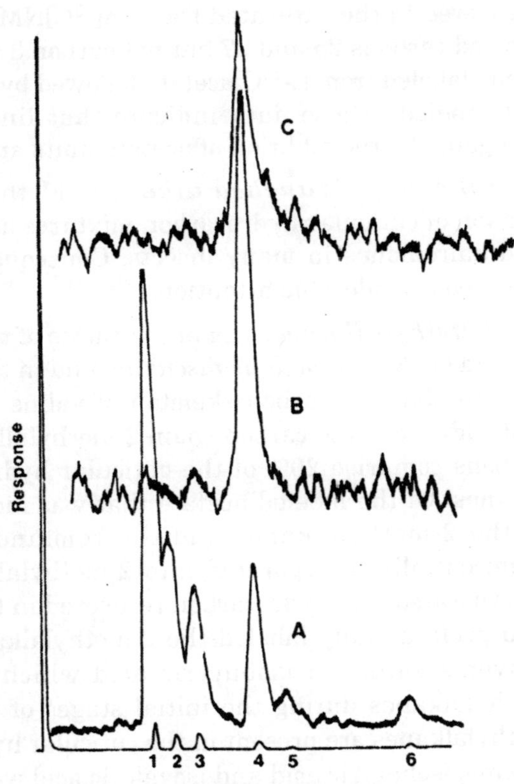

Fig. 4.5. Radio-GLC traces of the distribution of [l-^{14}C] acetate *in vivo* (A) and [15, l6-^{3}H]tetracosanoic acid *in vivo* (B) and *in vitro* (C) into the major cuticular hydrocarbons of *Zootermopsis angusticollis*. The bottom trace is a mass trace. Components are identified as (*I*) *n*-heneicosane, (2) 5-methylheneicosane, (3) 5,17-dimethylheneicosane, (4) *n*-tricosane, (5) 5-methyltricosane, and (6) *n*-pentacosane.

Although all the evidence to date favors an elongation-decarboxylation pathway, alternative pathways cannot be ruled out. There exists no direct evidence demonstrating the elongation of 16 or 18 carbon fatty acids to the putative precursors of hydrocarbons, and very long-chain fatty acids and methyl branched fatty acids are not usually reported in insects. This suggests that if the elongation-decarboxylation pathway does occur, the elongation and decarboxylation reactions must be tightly coupled, as significant amounts of intermediates are not released.

The decarboxylation of long-chain acids cannot be demonstrated in all insects. For example, the female housefly, which synthesizes n-tricosane, did not decarboxylate [15,16-^{3}H]tetracosanoic acid *in vivo*. Rather, most of the radioactivity from this acid was recovered in 16 and 18 carbon fatty acids after a 2h incubation, suggesting a very efficient chain shortening mechanism (Dillwith and Blomquist, unpublished observations).

(c) Biosynthesis of alkenes : A comparison of the data on the incorporation of [1-^{14}C]acetate, [1-^{14}C]stearate, [1-^{14}C]oleate, and [9,10-^{3}H]oleate into hydrocarbon by the housefly, *Musca domestica,* suggests a pathway in which oleic acid is elongated and decarboxylated to form the alkenes, including the sex pheromone component, (Z)-9-tricosene.

The major cuticular hydrocarbon component of the American cockroach is (Z,Z)-6,9-heptacosadiene, comprising 70% of the total hydrocarbons. Labeled acetate was incorporated about equally into the saturated and diene hydrocarbons, whereas [l-^{14}C]linoleic acid was preferentially incorporated into the diene fraction. Recent work showed that [9,10-^{3}H]oleate was incorporated almost exclusively into the diene, with less than 6% of the radiolabel being recovered in the saturated fraction[^{13}C]NMR experiments demonstrated that [2- ^{13}C]acetate labeled carbons 25 and 27 but not carbon 3 of the C_{21} alkadiene. In addition, ozonolysis of the diene labeled from [l-^{14}C]acetate followed by radio-GLC analysis showed that carbons 1-6 were not labeled. These data indicate that linoleate from the diet or synthesized *de novo* is elongated by the addition of acetate units and is then decarboxylated.

(d) Methyl branched alkanes : Methyl branched alkanes often comprise a significant portion of cuticular hydrocarbon mixtures and have been shown to serve as both pheromones and kairomones in many insects. Consequently, the origin of the methyl branch units has received considerable attention.

2-Methylalkanes. The biosynthesis of the 2-methyl-alkanes has been investigated in the ground cricket, *Nemobius fasciatus,* and in the field cricket, *Gryllus pennsylvanicus*. The data suggest that the carbon skeleton of valine and leucine serve as the precursors to the even and odd-numbered carbon chain 2-methylalkanes, respectively. 2-Methylalkanes of 23 and 25 carbons comprise 20% of the cuticular hydrocarbons of *N. fasciatus* with the other 80% *n*-alkanes. Of the labeled acetate that was incorporated into hydrocarbon, 28% was recovered in the 2-methylalkanes and the remainder in the *n*-alkanes. In contrast, valine was preferentially incorporated into 2-methylalkanes, with 98% of the radio-activity that was incorporated into hydrocarbon recovered in the 2-methylalkane fraction. Likewise, isobutyric acid preferentially labeled the 2-methylalkanes. This observation suggests that the cricket converts valine to isobutyric acid which is then incorporated into the even-chain 2-methylalkanes during the initial stages of chain elongation. Both odd- and even-chain 2-methylalkanes are present in the cuticular hydrocarbons of *G. pennsylvanicus.* Labeled valine, leucine, isobutyric acid and isovaleric acid were preferentially incorporated into the branched alkanes, suggesting that leucine was converted to isovaleric acid which served as the precursor to the odd-chain 2 methylalkanes.

3-Methylalkanes. Two pathways have been suggested for the biosynthesis of 3-methylalkanes. In plants and micro-organisms isoleucine is converted to 2-methylbutyric acid which is then incorporated into 3-methylalkanes during the initial stages of chain synthesis. In contrast, considerable evidence has accumulated that in several cockroaches a methylmalonyl derivative serves as the precursor to the branching methyl group of 3-methylalkanes. Isoleucine was not efficiently incorporated into 3-methylpentacosane in *Periplaneta americana.*

A remarkable specificity occurs in the biosynthesis of the branched alkanes of *P. americana,* in that 3-methylpentacosane is the only component present. The lack of methyl branched fatty acids in this insect had led to the suggestion that the methylmalonyl-CoA branching unit was added as the penultimate unit [^{13}C]NMR experiments were used to determine if the branching unit was added as the second unit or near the end of the biosynthetic process. A knowledge cf the position of the labeled carbon from [1-^{13}C]propionate in 3-methylpentacosane would allow a determination of whether propionate was incorporated in early or late stages of the elongation process. If carbon 2 was enriched, the methyl group would have been incorporated during the penultimate step of chain elongation. If carbon 4 was enriched, the methyl group would have been inserted as the second unit in chain elongation. The data from [^{13}C]NMR experiments showed that the labeled carbon from [1-^{13}C] propionate was incorporated exclusively in the 4 position demonstrating that the methyl branch unit is added as the second unit in the elongation process.

Internally branched alkanes. Methyl groups on the internal carbon atoms of alkane chains could originate either during the elongation process by substitution of a methylmalonyl-CoA in place of a malony –CoA, or by the methylation of a preformed chain. In algae, the methylation of a preformed chain by the methyl group of *S*-adenosylmethionine is involved in the biosynthesis of 7- and 8-methylheptadecanes. The variety of positions substituted (5, 1, 9, 11, 13 or 15) and the occurrence of dimethylalkanes with methyl branches separated by 1, 3, 7, 9 or 11 methylene groups favors the methylmalonyl-CoA hypothesis in insects.

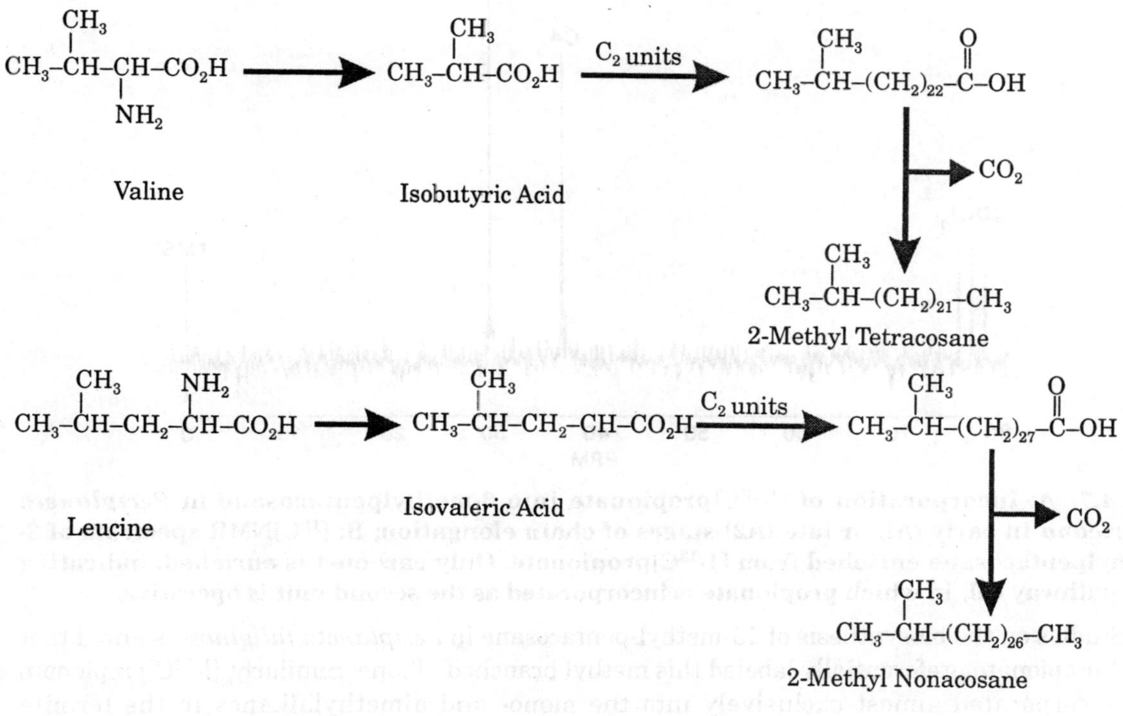

$$\underset{\text{Valine}}{CH_3\text{–}CH(CH_3)\text{–}CH(NH_2)\text{–}CO_2H} \longrightarrow \underset{\text{Isobutyric Acid}}{CH_3\text{–}CH(CH_3)\text{–}CO_2H} \xrightarrow{C_2\ \text{units}} CH_3\text{–}CH(CH_3)\text{–}(CH_2)_{22}\text{–}C(=O)\text{–}OH$$

$$\xrightarrow{-CO_2} \underset{\text{2-Methyl Tetracosane}}{CH_3\text{–}CH(CH_3)\text{–}(CH_2)_{21}\text{–}CH_3}$$

$$\underset{\text{Leucine}}{CH_3\text{–}CH(CH_3)\text{–}CH_2\text{–}CH(NH_2)\text{–}CO_2H} \longrightarrow \underset{\text{Isovaleric Acid}}{CH_3\text{–}CH(CH_3)\text{–}CH_2\text{–}CH\text{–}CO_2H} \xrightarrow{C_2\ \text{units}} CH_3\text{–}CH(CH_3)\text{–}(CH_2)_{27}\text{–}C(=O)\text{–}OH$$

$$\xrightarrow{-CO_2} \underset{\text{2-Methyl Nonacosane}}{CH_3\text{–}CH(CH_3)\text{–}(CH_2)_{26}\text{–}CH_3}$$

Fig. 4.6. Proposed biosynthetic pathways for 2-methylakanes in insects.

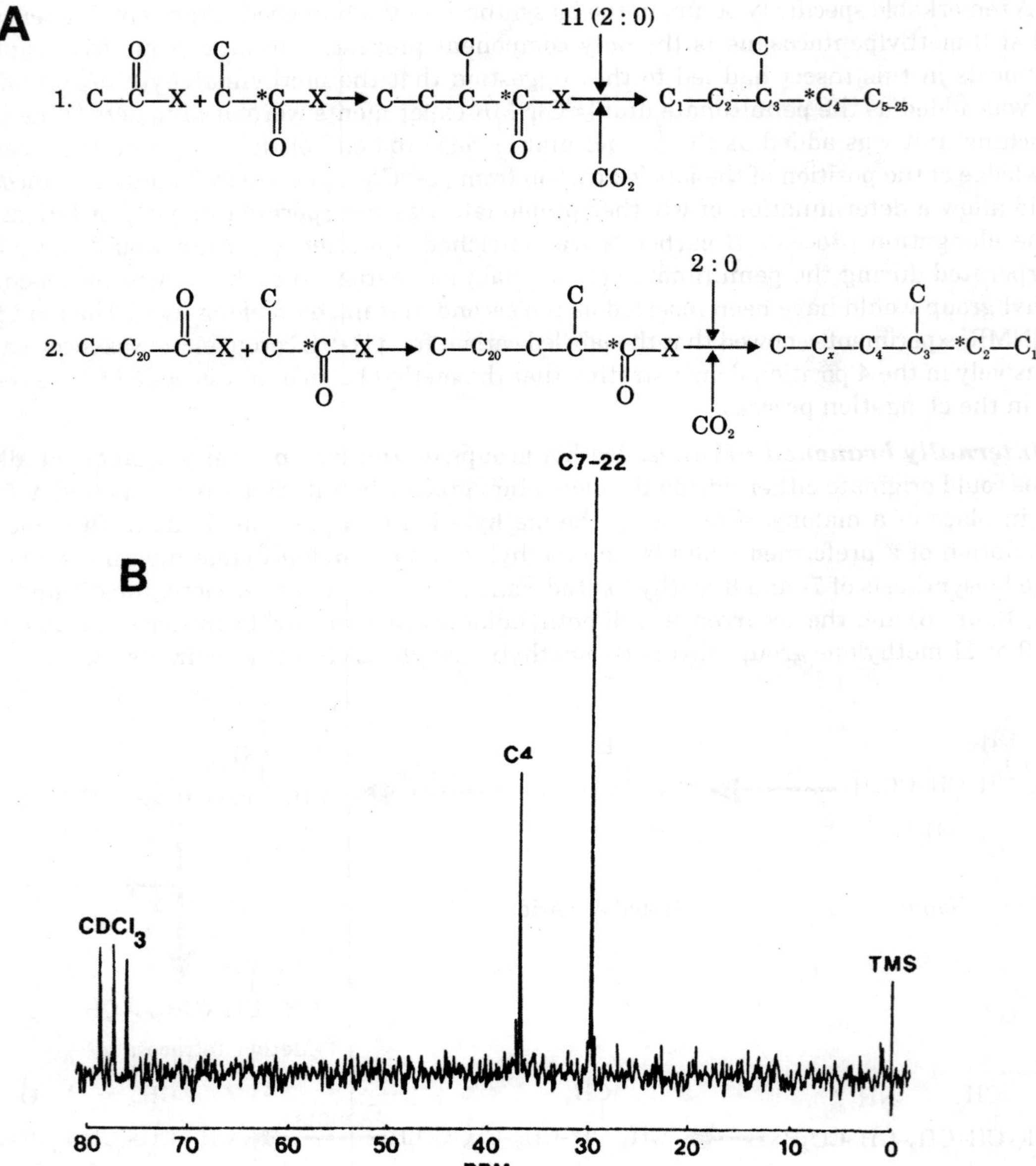

Fig. 4.7. A: Incorporation of [1-^{13}C]propionate into 3-methylpentacosane in *Periplaneta americana* in early (Al) or late (A2) stages of chain elongation; B: [^{13}C]NMR spectrum of 3-methylpentacosane enriched from [1-^{13}C]propionate. Only carbon-4 is enriched, indicating that pathway (Al) in which propionate is incorporated as the second unit is operative.

Studies on the biosynthesis of 13-methyl-pentacosane in *Periplaneta fuliginosa* showed that [1-^{14}C]propionate preferentially labeled this methyl branched alkane. Similarly, [1-^{14}C]propionate was incorporated almost exclusively into the mono- and dimethylalkanes in the termite *Zootermopsis angusticollis* and the housefly, *Musca domestica*. [Methyl^{14} C]Methionine did not serve as an efficient precursor to the branched alkanes in any of these insects. These data

suggest that propionate, as a methylmalonyl derivative, is incorporated in place of malonyl-CoA during chain elongation.

Although propionate is readily incorporated into the methylalkanes in a variety of insects, it may not be the endogenous precursor to the methylmalonyl-CoA used in branched hydrocarbon synthesis. Studies on the biosynthesis of branched alkanes in the termite, *Z. angusticollis*, suggested that succinate is a precursor to the methylmalonyl derivative which is the methyl branched precursor.

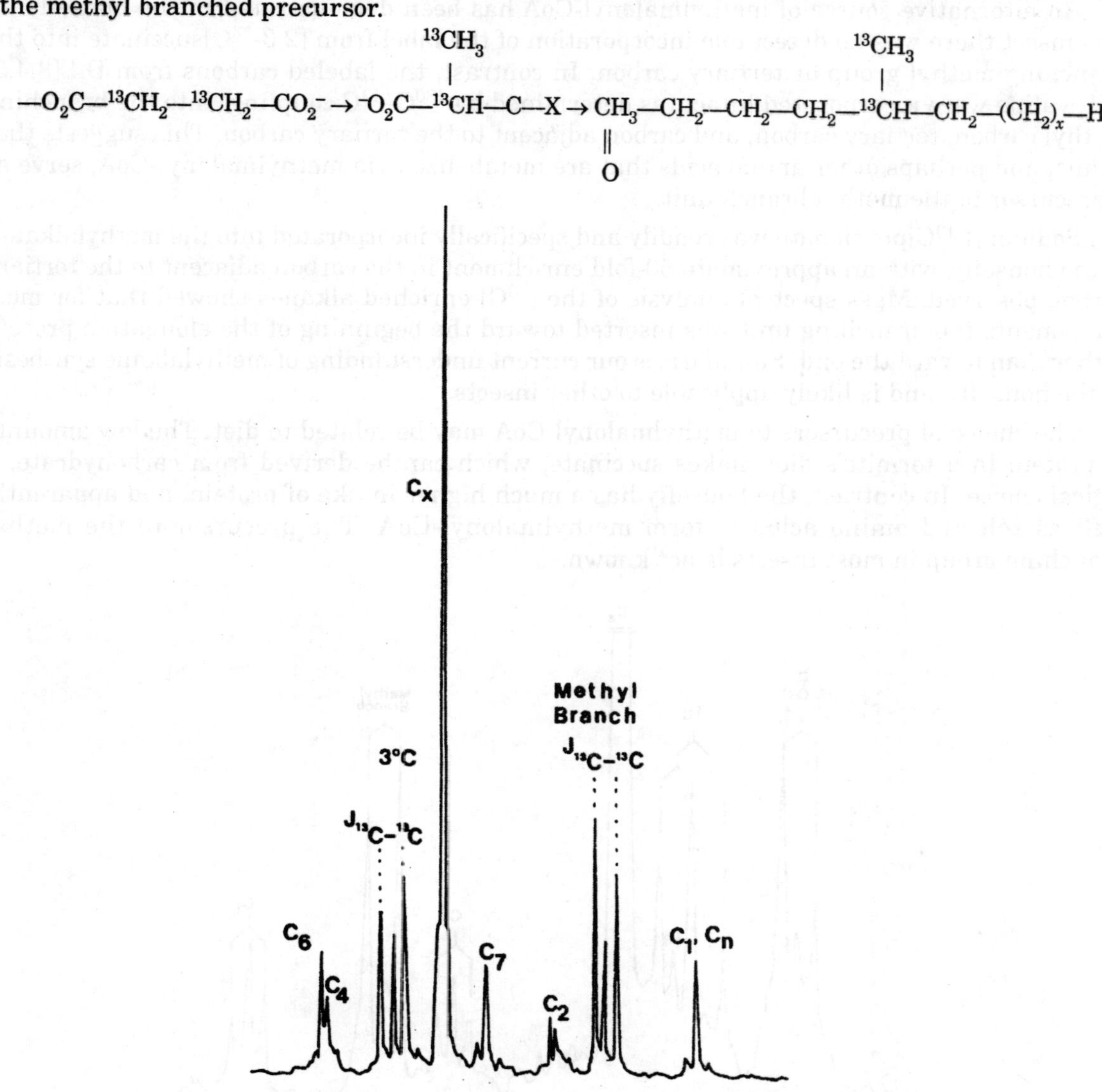

Fig. 4.8. [^{13}C]NMR spectrum of the methylalkanes of *Zootermopsis angusticollis* enriched from [2, 3-^{13}C]succinate. The carbon-carbon coupling shows that the labeled carbons are incorporated intact into the branching methyl carbon and the tertiary carbon.

Definitive evidence that succinate is a source of the branching methyl unit of methylalkanes was obtained by using carbon-13 NMR to examine the incorporation of [2, 3-^{13}C]succinate into methylalkanes in Z. *angusticollis*. The data showed that carbons 2 and 3 of succinate became the methyl branch and tertiary carbon in branched alkanes. Furthermore, the carbon-carbon coupling of adjacent labeled carbons showed that the bond between carbons 2 and 3 of succinate remains intact.

An alternative source of methylmalonyl-CoA has been demonstrated in the housefly. In this insect there was no detectable incorporation of the label from [2,3-^{13}C]succinate into the branching methyl group or tertiary carbon. In contrast, the labeled carbons from D,L[3,4,5-$^{13}C_3$]valine were incorporated intact (as determined by ^{130}C–^{13}C coupling) into the branching methyl carbon, tertiary carbon, and carbon adjacent to the tertiary carbon. This suggests that valine, and perhaps other amino acids that are metabolized via methylmalonyl-CoA, serve as a precursor to the methyl branch unit.

Sodium [l-^{13}C]propionate was readily and specifically incorporated into the methylalkanes of the housefly, with an approximate 50-fold enrichment in the carbon adjacent to the tertiary carbon observed. Mass spectral analysis of the [^{13}C]-enriched alkanes showed that for most components the branching unit was inserted toward the beginning of the elongation process rather than toward the end. Summarizes our current understanding of methylalkane synthesis in the housefly, and is likely applicable to other insects.

The choice of precursors to methylmalonyl-CoA may be related to diet. The low amounts of protein in a termite's diet makes succinate, which can be derived from carbohydrate, a logical choice. In contrast, the housefly has a much higher intake of protein, and apparently utilizes selected amino acids to form methylmalonyl-CoA. The precursor to the methyl branching group in most insects is not known.

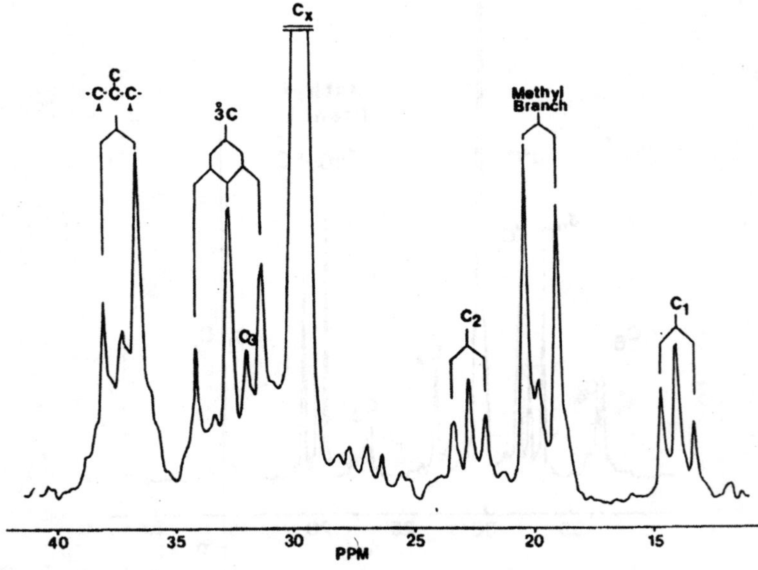

Fig. 4.9. [^{13}C]NMR of the methylalkanes of the female housefly, *Musca domestica*, enriched from D,L-[3,4,5-$^{13}C_3$] valine. The carbon-carbon coupling shows that the three labeled carbons are incorporated intact into the methyl branching carbon, the tertiary carbon, and carbon adjacent to the tertiary carbon.

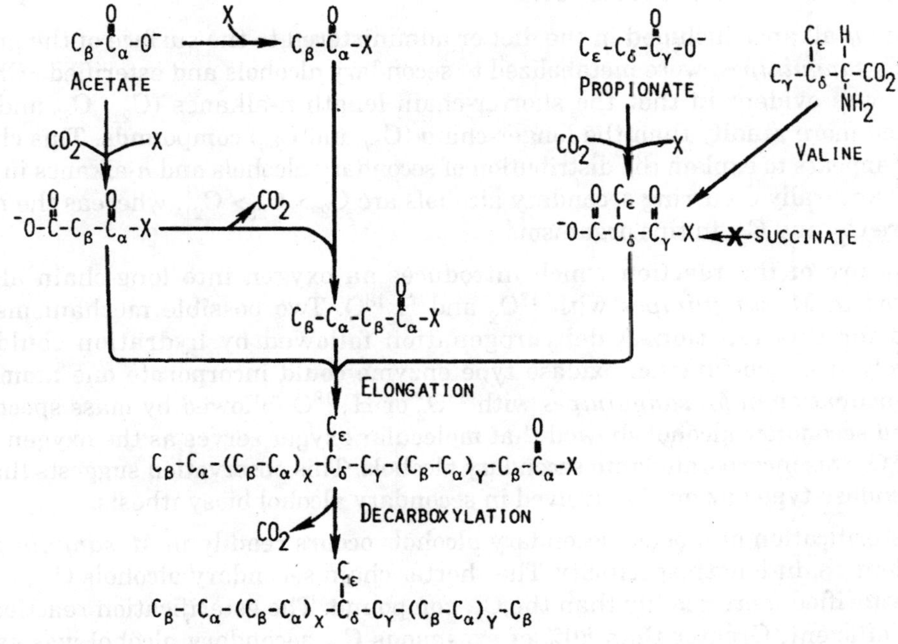

Fig. 4.10. Proposed biosynthetic pathway of the methylalkanes of the female housefly, *Musca domestica.* The origin of each carbon has been determined by [^{13}C]NMR spectroscopy.

In addition to being incorporated as the methyl branching carbon, the labeled carbons from [3-^{13}C]propionate and [methyl- ^{13}C]methyl-malonate also labeled the even-numbered carbons in the alkanes and alkenes. Likewise, [2-^{13}C] propionate labeled the odd-numbered carbons of both the alkanes and alkenes. These data suggest that propionate is converted to an acetyl derivative, with carbon 3 of propionate converted to the carboxyl carbon of acetate, and carbon 2 of propionate converted to the methyl carbon of acetate. The occurrence of this pathway for propionate metabolism in the housefly, which does not utilize the methylmalonyl-CoA mutase reaction, prompted an examination of the housefly and other insects for vitamin B_{12} levels. The housefly contained undetectable levels of vitamin B_{12} which was consistent with its inability to use succinate as a precursor to methylmalonyl–CoA and the presence of enzymes which directly convert propionate to an acetate derivative. In contrast, termite *Z. angusticollis* contained extremely high levels of vitamin B_{12}, which was consistent with its ability to convert succinate to methylmalonate.

BIOSYNTHESIS OF NON-HYDROCARBON CUTICULAR COMPONENTS

The biosynthesis of the non-hydrocarbon components of insect cuticular wax has not been extensively studied. It is postulated that the long-chain fatty acids found in cuticular extracts arise from the elongation of 16 and 18 carbon fatty acids by the addition of two carbon units. Likewise, it is assumed that very long-chain primary alcohols arise from the reduction of the corresponding fatty acid, but direct evidence on these processes is not available.

Secondary Alcohol Wax Esters

Labeled *n*-alkanes included in the diet or administered to the surface of the grasshopper, *Melanoplus sanguinipes,* were metabolized to secondary alcohols and esterified . Chain-length specificity was evident in that the shorter-chain length *n*-alkanes (C_{21}, C_{23} and C_{25}) were metabolized more readily than the longer-chain (C_{27} and C_{29}) compounds. This chain-length specificity appears to explain the distribution of secondary alcohols and *n*-alkanes in this insect. The major naturally occurring secondary alcohols are $C_{23} > C_{25} > C_{21}$, whereas the principal *n*-alkanes are C_{27} and C_{29} in this organism.

The nature of the reaction which introduces an oxygen into long-chain alkanes was investigated in *M. sanguinipes* with $^{18}O_2$ and $H_2{}^{18}O$. Two possible mechanisms had been suggested for this reaction. A dehydrogenation followed by hydration could occur, or alternatively, a mixed-function oxidase type enzyme could incorporate one atom of oxygen from O_2. Incubation of *M. sanguinipes* with $^{18}O_2$ or $H_2{}^{18}O$ followed by mass spectrometry of the isolated secondary alcohol showed that molecular oxyger serves as the oxygen donor. The ^{18}O from $^{18}O_2$ was incorporated into secondary alcohols. This observation suggests that a mixed-function oxidase type enzyme is involved in secondary alcohol biosynthesis.

The esterification of labeled secondary alcohols occurs readily in *M. sanguinipes,* with a fairly evident chain-length specificity. The shorter chain secondary alcohols C_{21}, C_{23}, C_{25} and C_{27} were esterified more readily than the C_{31} compound. The esterification reaction was very rapid and efficient. Greater than 80% of exogenous C_{23} secondary alcohol was esterified in 18 h, and almost all the administered *n*-alkane which was hydroxylated appeared in the form of a secondary alcohol wax ester. In accord with this observation, only trace amounts of free secondary alcohol were present in the cuticular extract of this insect.

Primary Alcohol Wax Esters

Labeled acetate, palmitate and tetracosanol injected beneath the cuticle of the honeybee were incorporated into wax monoester. Acetate and palmitate labeled both the alcohol and acid moieties of the monoester, whereas tetracosanol was incorporated almost exclusively into the alcohol moiety.

The esterification of primary alcohols was studied with a microsomal preparation obtained from the abdomens of insects not actively producing comb wax. Microsomal preparations (105,000 ***g*** pellet) gave about a 2-fold higher specific activity than did a 10,000 ***g*** pellet or a 105,000 ***g*** supernatant. When a microsomal preparation was incubated in the presence of labeled tetracosanol and palmitoyl-CoA, the distribution of products showed that most of the primary alcohol incorporated into the esters was recovered in the monoester fraction, with lesser amounts in the di- and tri-ester fractions.

The rate of wax monoester synthesis from [l-^{14}C]palmitate was stimulated by CoA, ATP and $MgCl_2$ and the addition of palmitoyl-CoA resulted in a 5-fold increase in monoester synthesis from labeled tetracosanol, indicating that the acyl donor group is activated in the form of an acyl-CoA. Palmitoyl-CoA stimulated monoester formation from labeled tetracosanol and increased amounts of wax ester were formed with increasing amounts of palmitoyl-CoA up to 10^{-4}M, after which it inhibited, presumably due to a detergent effect.

The chain-length specificity of the primary alcohols incorporated into monoester was examined, and the results showed that labeled alcohols of 16, 18, 24 and 28 carbons were readily incorporated into monoester with the shorter-chain 16 and 18 carbon alcohols incorporated at higher rates. The naturally occurring primary alcohols are of chain lengths C_{24}–C_{32}. This suggests that the chain-length specificity of the alcohols present in monoester does not reside in the esterification step, but rather in the reduction of fatty acids to alcohols.

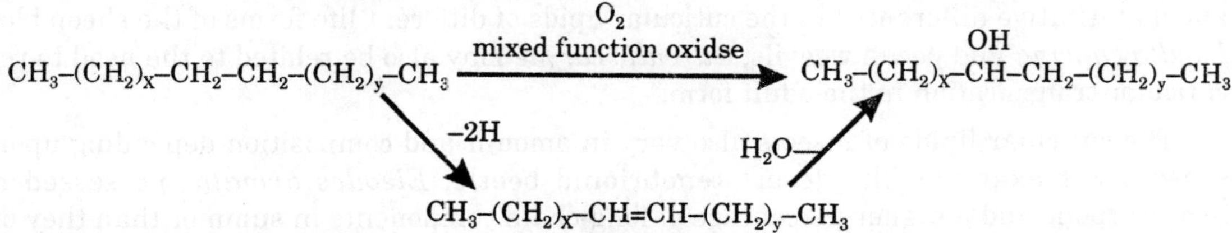

Fig. 4.11. Proposed pathway for the biosynthesis of secondary alcohols in *Melanoplus sanguinipes*. Stable isotope studies have demonstrated that the oxygen in the hydroxyl group arises from molecular oxygen.

The synthesis of wax esters in honeybees actively producing comb wax was examined by Lambremont, E. and Wykle, R. . The post-mitochondrial supernatant fraction readily esterified tetracosanol to palmitoyl-CoA and the characterization of this process suggests that the synthesis of wax esters by the specialized wax glands of the honeybee is similar to that of the epidermal tissue.

Epoxides and Ketones

The biosynthesis of (Z)-9,10-epoxytricosane and 14-tricosen-10-one was examined in the housefly, *Musca domestica.* Both of these components, which are part of the female sex pheromone, were formed from [9,10-^{3}H](Z)-9-tricosene. Both male and female insects were able to convert (Z)-9-tricosene to the corresponding epoxide and ketone with the male possessing higher activity (Blomquist and Dill-with, unpublished data). The male housefly does not possess either of these 2 oxygenated derivatives in its cuticular lipids is apparently due to its inability to synthesize (Z)-9-tricosene.

PHYSIOLOGICAL CONSIDERATIONS

Development and Environment

The variation in insect surface lipids, as affected by developmental and environmental factors, has been studied in a few insect species. The amount and composition of cuticular lipids often changes as an insect goes through developmental stages, particularly if the insect's environment undergoes major changes. For example, the amount and the distribution of lipid classes and the composition of components within a class are quite different in the aquatic naiad form of the big stonefly, *Pteronarcys californica,* compared with the terresterial adult. It appears that in this insect the surface lipid composition varies with life stage, depending upon the need for water conservation. Similarly, the cuticular hydrocarbons vary both quantitatively and qualitatively throughout the life cycle of the flesh-fly, *Sarcophaga bullata,* and the quantity of cuticular hydrocarbon correlates well with the water conservation needs of the insect. Marked differences were noted in the cuticular lipids of immature and adult forms of the cigarette beetle, *Lasioderma serricorne*, the black carpet beetle, *Attagenus megatoma*, two species of imported fire-ants, *Solenopsis invicta* and *Solenopsis richteri* and the desert cicada, *Diceroprocta apache.* There were marked increases in either the percentage of hydrocarbons among the cuticular lipid classes, the total amount of hydrocarbon, the composition of hydrocarbons, or all three. In the desert cicada there was approximately five times more hydrocarbon in the adult than in the nymphal exuvium, and this difference correlated with the increased transpiration potential experienced by the adult. Qualitative and quantitative differences in the cuticular lipids of different life forms of the sheep blowfly, *Lucilia cuprina* and pecan weevils, *Cttrculio caryae* may also be related to the need to reduce cuticular transpiration in the adult form.

The cuticular lipids of insects also vary in amount and composition depending upon the season. For example, the desert tenebrionid beetle, *Eleodes armata,* possessed more hydrocarbons and a higher percentage of long-chain components in summer than they did in winter. Similar results were obtained when winter-collected beetles were acclimated to 35°, suggesting that this change in hydrocarbon is an adaptation to an altered environment. Again, the changes appear to be related to the water conservation needs of the insect.

Diapausing pupae of the tobacco hornworm. *Manduca sexta* secrete three times as much surface wax as do non-diapausing pupae. The extra thickness of the wax layer apparently protects the insect from desiccation and the authors speculate that the deposition of additional wax may result from hormonal changes accompanying entry into diapause.

ENDOCRINE REGULATION

Very little is known about the general area of the control of hydrocarbon biosynthesis in insects. Work by Armold, M. and Regnier, F. however, suggests that 20-hydroxyecdysone stimulates hydrocarbon formation. They found that in the development of *S. bullata* there are two periods of rapid accumulation of hydrocarbons. The most rapid accumulation is during the 4-day period preceding the pupal–adult ecdysis (>20 μg day^{-1} insect^{-1}). An earlier accumulation (5-8 μg day^{-1} insect^{-1}) occurs during pupariation and the 3-day period following pupariation. At pupariation, exogenous 20-hydroxyecdysone was shown to stimulate hydrocarbon biosynthesis from [^{3}H]acetate at a rate of 1 .3 times the rate of controls 10.5 h after hormone administration and at a rate of 3.3 times the rate of controls 24 h after hormone administration. In an *in vitro* study they showed that integuments of 20-hydroxyecdysone-treated insects incorporated acetate into hydrocarbons 9.8 times better than integuments of control insects. This shows that cuticular hydrocarbon biosynthesis in S. *bullata* occurs in the integument, and that the locus of regulation is also present in the integument.

Dramatic changes in the composition of the cuticular lipids of the housefly, *Musca domestica,* occur with age in the adult insect. Newely emerged males and females have almost identical cuticular compositions. Dramatic changes take place in the female between days 2 and 3. (Z)-9-Tricosene, (Z)-9,10-epoxytricosane, and (Z)-14-tricosen-10-one first appear at this time, and increasing amounts of these components and methylalkanes accumulate in the female as the insect ages. These components, which are part of the housefly sex pheromone, are absent, or present in very small amounts, in the male. Because of the apparent similarity in timing of the beginning of pheromone production and ovarian maturation, the production of the pheromone components was correlated to ovarian development.

Production of the cuticular sex pheromone components of the female housefly, including (Z)-9-tricosene (muscalure), (Z)-9,10-epoxytricosane, (Z)-14-tricosen-10-one, and methyl branched alkanes, correlated with ovarian development. No pheromone was produced by females with ovaries in pre-vitellogenesis, production was initiated in insects during early vitellogenesis, and larger amounts of pheromone were found in females as the ovarian follicles matured.

Females ovariectomized shortly after adult emergence did not produce pheromone. Reimplanting pre-vitellogenic ovaries into ovariectomized females and allowing them to reach early vitellogenic stages restored pheromone production. Implanting ovaries into male insects resulted in altering the composition of their cuticular lipids such that they then produced (Z)-9-tricosene (muscalure), the C_{23} epoxide and ketone, and large amounts of methylalkanes. Injection of 20-hydroxyecdysone into ovariectomized females or into newly emerged males caused the same alteration in cuticular lipids as did implanting of ovaries, and production of pheromone components was initiated. The major cuticular component produced by ovariectomized females and control males is (Z)-9-heptacosene. The presence of maturing ovaries or injection of 20-hydroxyecdysone causes a change in the chain length of the alkenes such that (Z)-9-tricosene is produced, and increases the amount of methylalkanes synthesized. Adult male and female insects both possess the enzyme(s) which convert (Z)-9-tricosene to the corresponding epoxide and ketone at all ages, and the production of these oxygenated derivatives occurs when substrate ((Z)-9-tricosene) is present.

BIOSYNTHESIS OF BEESWAX

In addition to producing a thin layer of surface lipids, some insects, such as the honeybee, produce a much larger amount of wax for the honeycomb. The biosynthesis of beeswax has been investigated by Pick, T. (1963) and Young, R. (1963), who show that labled acetate was incorporated into the hydrocarbon, ester and acid fractions of beeswax. the biosynthesis of wax by the honeybee was studied in winter and summer. It was observed that in winter, bees not only incorporated acetate into hexen-eztractable lipids at much levels than in summer, but also demonstrated a 4-fold higher incorporation of acetate into hydrocarbon than into monoester. In contrast, in summer, when many of the bees were actively producing comb wax, acetate was incorporated at about the same rates into monoester and into hydrocarbon. A comparison of the cuticular and comb wax of the honeybee showed that the major component in the cuticular wax was hydrocarbon, which comprised about 58% of this wax. Monoester accounted for 23%. In contrast, hydrocarbon and monoester account for 14 and 35% respectively, of the comb wax of the honeybee. Data from studies on the incorporation of labeled acetate into wax components by insects not producing comb wax showed that it was incorporated into the various wax fractions in about the same proportion as the amount of each component. In bees actively producing comb wax, a lower rate of incorporation into hydrocarbon and a higher rate of incorporation into monoester was observed.

It is generally accepted that the activities which bees engage in are age-related phenomena, although there is great flexibility in the age–activity relationship. The comb wax of bees is secreted by wax glands located on the ventral abdomen and the wax glands are best developed and most productive in bees 12-18 days old. Bees from 13 to 17 days old incorporated acetate about equally into both the hydrocarbon and monoester fractions. From age 19 to 31 days there was an increase in the incorporation of acetate into hydrocarbon and a decreased incorporation into monoester, so that by day 31 over 71% of the label that was incorporated into wax was present in the hydrocarbon fraction, and less than 10% in monoester. Thus, it appears that from day 13 to 17, both cuticular and comb wax are synthesized, whereas in older bees, only cuticular wax is being formed.

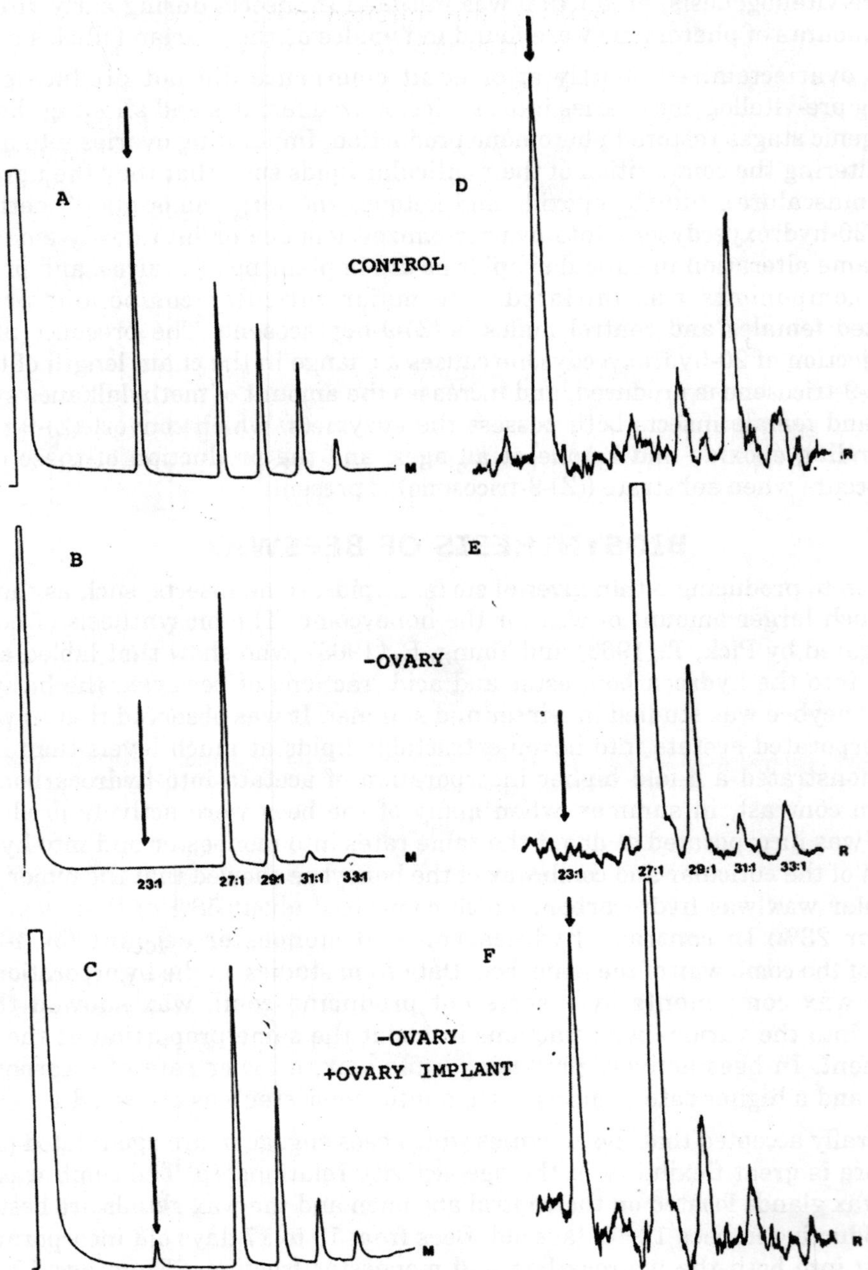

Fig. 4.12. The effect ovariectomy and re-implanting ovaries on alkene synthesis in the female housefly, *Musca domestica*. A-C are mass traces and D-F are radioactivity traces. Ovaries were removed wihin 12 h after adult emergence. Ovaries from 24 hold donor insects were implanted into 4-day-old ovariectomized female. Six-day- old imsects were injected with [1-^{14} C]acetate, sacrificed after 3h, and the alkenes isolated and analyzed by radio–GLC.

To further explore this phenomenon, the incorporation of labeled acetate into fat body, thorax, and both dorsal and ventral integument tissue from the abdomens of bees was studied. The ventral abdominal integument tissue contained wax glands. An analysis of the data showed that fat body tissue did not efficiently incorporate labeled acetate into either hydrocarbon or monoester. Both the thorax and the dorsal abdominal integument tissue incorporated acetate into wax. These tissues incorporated radioactivity at about a 4- to 5-fold higher rate into hydrocarbon than into monoester. In contrast, the ventral abdominal tissue, which contained the wax glands, incorporated radioactivity about equally into both hydrocarbon and monoester. These data demonstrate that the site of wax synthesis in the honeybee is associated with the integument and points out the difference in the wax produced by the glands (comb wax) compared with the cuticular wax.

CONCLUDING REMARKS

The development of microanalytical techniques, particularly GLC-MS, has allowed rapid advances in our understanding of the chemistry of insect cuticular lipids. However, our knowledge of the chemical composition of insect cuticular lipids is still limited to representative species of only seven orders, and in many cases is restricted to one lipid class—the hydrocarbons. Although it has been known for decades that a major function of insect cuticular lipids is to restrict water loss, a generally accepted model for the observed waterproofing is not available. Indeed, the recent data which argue against the long-held view of an "oriented monolayer" should result in a renewed interest in gaining an understanding of the waterproofing properties of cuticular lipids. The recognition that cuticular lipid components function in chemical communication in certain species has resulted in a great deal of interest to determine how general this phenomenon is, and to examine the types of roles cuticular lipids play. The general metabolic pathways have been determined for some of the cuticular lipid components, and information is becoming available on the endocrine regulation of cuticular lipid formation in a few species. However, these studies have only begun to explore the processes by which cuticular lipid components are formed and regulated and many intriguing questions remain. A strong case can be made for the joining of forces between behavioral ecologists, physiologists, morphologists, biochemists and chemists to better our understanding of insect cuticular lipids.

CHAPTER 5

Biochemistry of Prostaglandins

This year has seen something of a revival in the number of papers dealing with the synthesis of prostaglandins (PG's) and their analogues. From about 70 publications in this area, approximately 50 are concerned with the synthesis of PG analogues and about 20 deal with the synthesis of primary PG's. This distribution is undoubtedly due to the continued search by several pharmaceutical companies for more selective, potent, and longer-acting PG's and also to the inevitable decline in the rate of generation of original primary-PG syntheses, most of which appeared during the feverish activity in the first few years of this decade.

The stage seems to have been reached where the rich store of synthetic methodology already built up is sustaining the activities of most drug firms and, at the same time, is providing a disincentive for the investigations of alternative chemistry. It is therefore perhaps timely that this year has witnessed the emergence of Thromboxane A_2, a biogenetic and structural relative of the PG's which, with its high biological potency and intriguing structure, may well trigger a spate of further original synthetic work in the near future.

Several general reviews on PG's have appeared in 1975 dealing with various aspects of their occurrence, biosynthesis, synthesis, metabolism, production, and use in medicine and a summary of some of the highlights from the International PG Conference in Florence has been published.

The number of commercially available PG's is growing. The pre-1975 list of $PGF_{2\alpha}$ (Upjohn), PGE_2 (Upjohn), and I.C.I. 81008 (I.C.I.) expanded in 1975 to include the 15-mehtyl-4, 5-allene $PGF_{2\alpha}$ analogue (Syntex) and I.C.I 80996 (I.C.I.) I both of which have use in veterinary medicine.

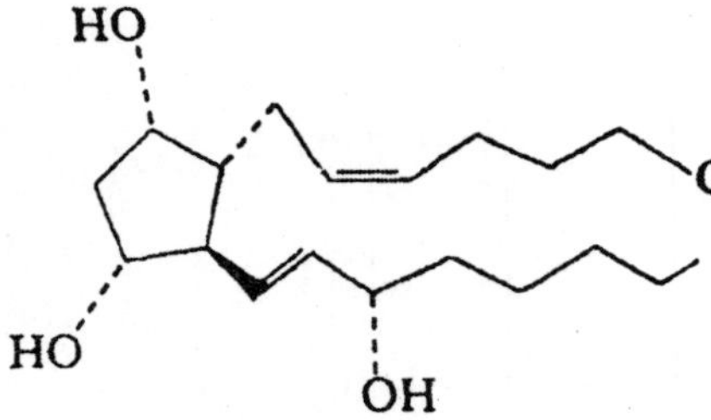

(1) $PGF_{2\alpha}$

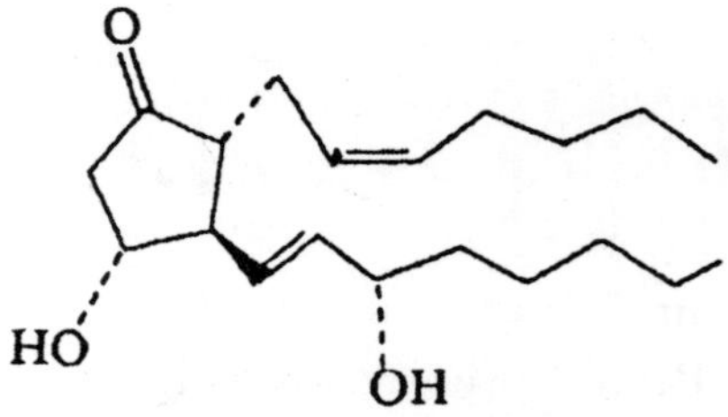

(2) PGE_2

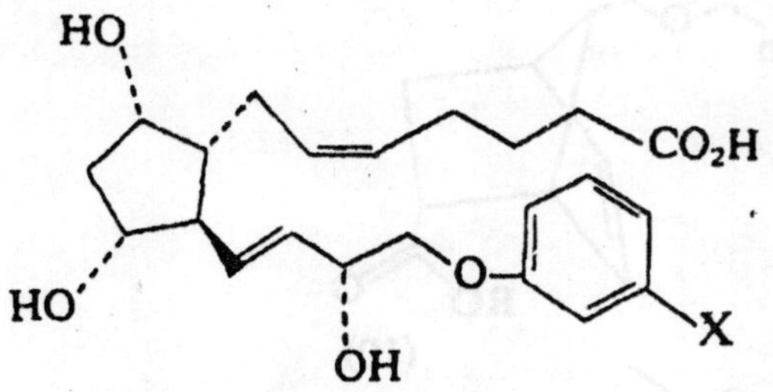

(3) X = CF_3
(5) X = Cl

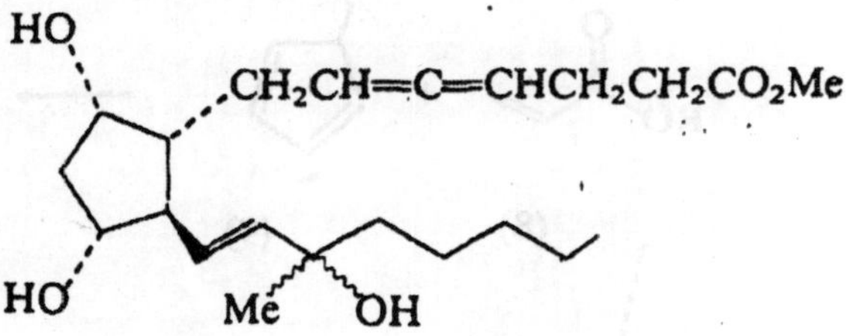

(4)

NOMENCLATURE

Although the structures of the biosynthetic endoperoxide intermediates PGH_2(6) and PGG_2(7), they were not at that time so named. It may therefore be helpful to remind the reader of these structures since they will be referred to later.

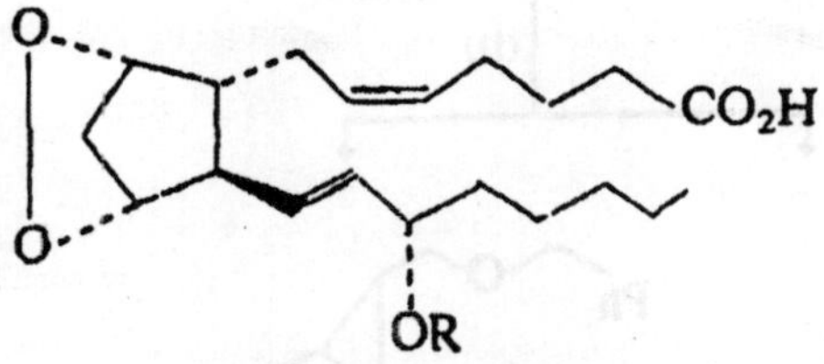

(6) R = H (PGH_2)
(7) R = OH (PGG_2)

SYNTHESIS OF PRIMARY PROSTAGLANDINS

Corey's Bicyclo[2, 2, 1] heptane Route—*Refinements.* Corey and Ensley, in an elegant piece of work, have introduced chirality at the very foundation of the bicyclo[2,2,1]-heptane route by effecting an asymmetric induction in the Diels-Alder reaction between the optically pure acrylate (8) and 5-benzyloxymethylcyclopentadiene (9) to give the *endo*-adducl (10) (Scheme 1). The absolute configuration of (10) was demonstrated in the subsequent conversion into the known iodohydrin (15). Treatment of the enolate from (10) with triethyl phosphite in the presence of oxygen gave a mixture of *endo*- and exo-hydroxy-esters (11) which was reduced to the diol mixture (12) and the optically active alcohol (13), available in pure form for recycling after a simple separation on silica. [The alcohol (13) originally used to prepare (8) was derived from (–)-pulegone, which in turn was produced from (–)-citronellol by oxidation with pyridinium chlorochromate, a reagent introduced this year as a convenient substitute for the Collins reagent]. Periodate cleavage of (12) then afforded the optically active ketone (14), which was transformed by standard methods to (15) having the configuration corresponding to the natural PG's. In this way (15) was obtained in 66% yield from (8) without a resolution step.

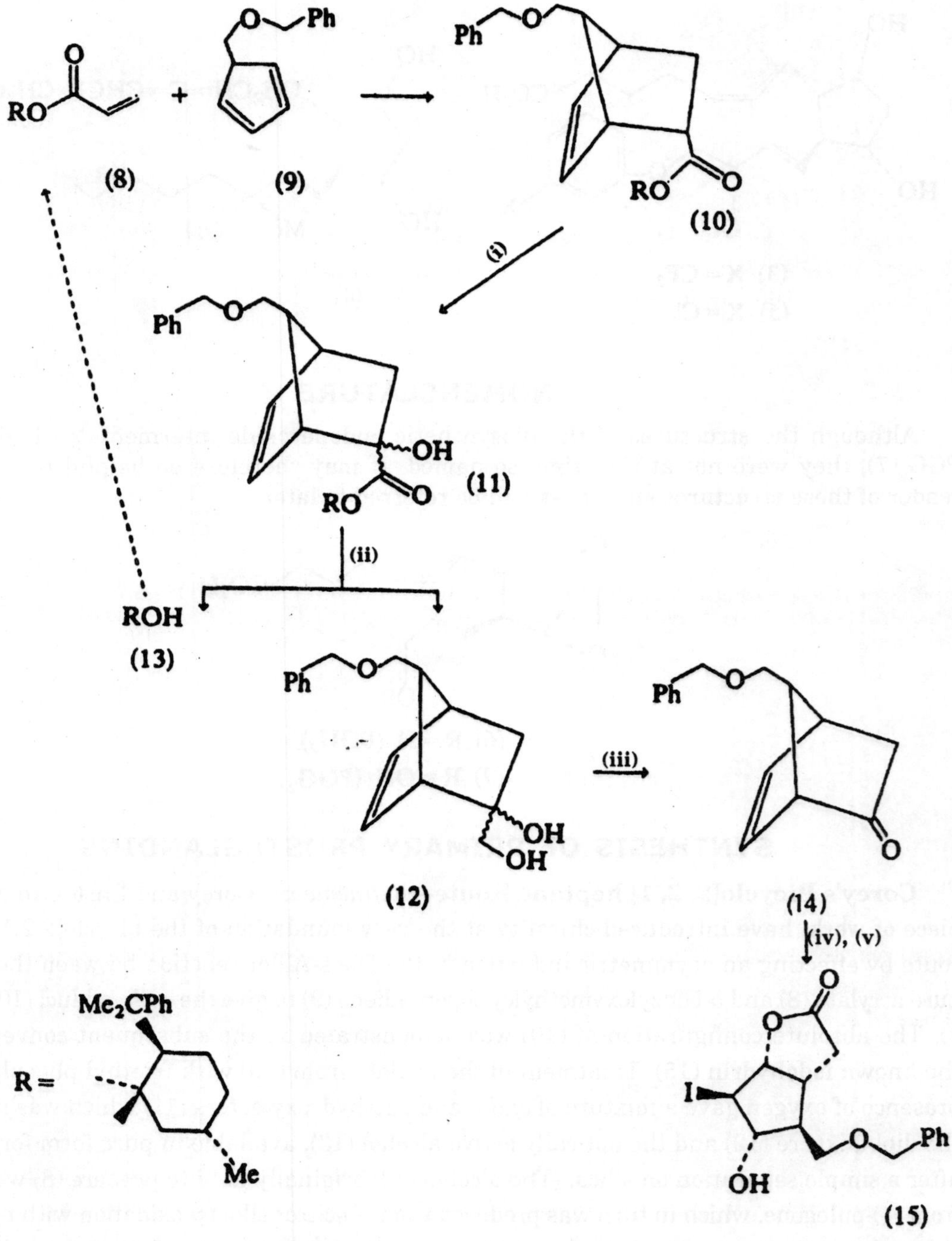

RO, (8), (9), Ph, O, (10), (i), OH, (11), (ii), ROH, (13), (12), (iii), (14), (iv), (v), I, (15), R = Me₂CPh, Me

Reagents : (i), $LiNPr_2$–O_2–$(EtO)_3P$, (ii), $LiAlH_4$; (iii), IO_4^- ; (iv), Baeyer-Villiger; (v), iodolactonization.

Scheme 1

Deiodination of derivatives such as (15) is a key step in the Corey synthesis and is normally carried out with tri-*n*-butyltin hydride (TBTH), which can be generated *in situ* from tri-n-butyltin oxide and polymethylhydrogen siloxane (PMHS). The final product has to be separated from a full equivalent of tri-*n*-butyltin iodide and the PMHS residues. Corey and Suggs have reported an improved procedure in which only 0.1–0.3 equivalents of tri-n-butyltin chloride is reduced in ethanol with sodium borohydride to give diborane (trapped by the solvent) and the required TBTH. The latter effects deiodination to produce tri-n-butyltin iodide which is reduced back to TBTH to complete the cycle.

The general utility of the Meerwein-Ponndorf reduction for the conversion of enones of the type (16) into the corresponding enols free from 1, 4-reduction products has been confirmed by Picker *et al* during work on the 17, 17-dimethylenone (16; R = Me).

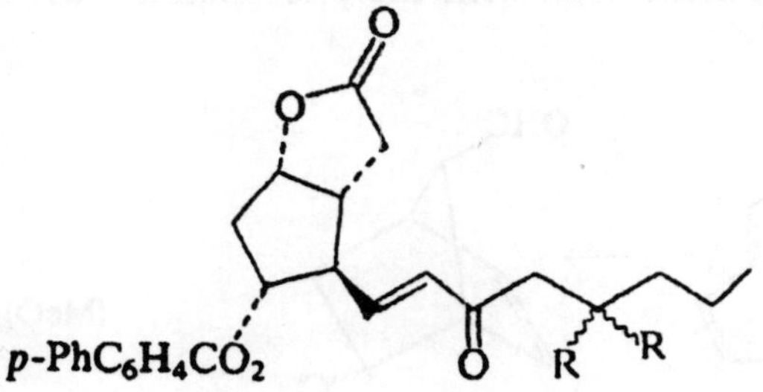

p-$PhC_6H_4CO_2$ O R R

(16)

New Routes to Intermediates. Ranganathan and co-workers have described the simple transformation of their previously derived intermediate (17) to the known Corey intermediate (22) (Scheme 2). Model work showed that, under iodolactonization conditions, the bicyclic lactone (18) gave only (20) and not the expected iodohydrin (19). Thus, (17) similarly on treatment with hypobromite gave a high yield of (21) which, *via* debromination and acetylation, produced (22). Stereo- and regio-selectivity resulted from Br^+ attack from the convex face of the bicyclic lactone (17), followed by HO^- attack in the usual *trans* fashion at the 11-rather than the 10-position (PG numbering) where unfavourable eclipsing of the 9-substituent would occur. (17) is therefore now a source of PGA, E, and F, and of the 11-deoxy-PG's.

The I.C.I. group have produced two variations to their acetoxyfulvene route. Firstly, Brown and Lilley have described a short PG synthesis in which the *syn*-aldehyde (23), available

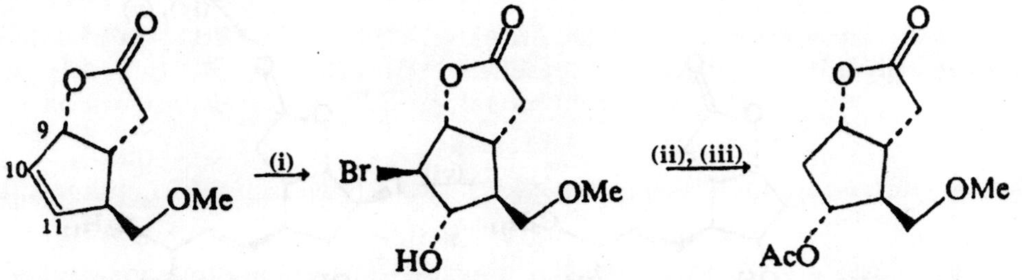

O 9 10 11 OMe (i) Br OMe HO (ii), (iii) OMe AcO

(17) (21) (22)

Reagents: (i), MeCONHBr–H_2O-Me_2CO; (ii), Ni-H_2; (iii), acetylation

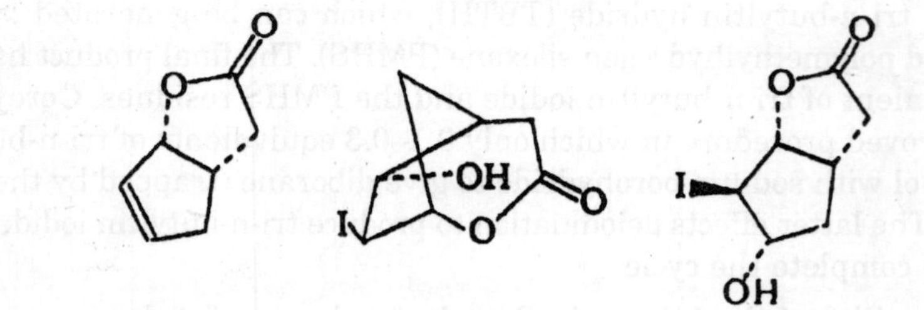

(18) (19) (20)

in three steps from acetoxyfulvene, was converted into $PGF_{2\alpha}$ (1) in seven steps (Scheme 3). The aldehyde (23) was condensed with the phosphonate (24) to give the enone (25), which

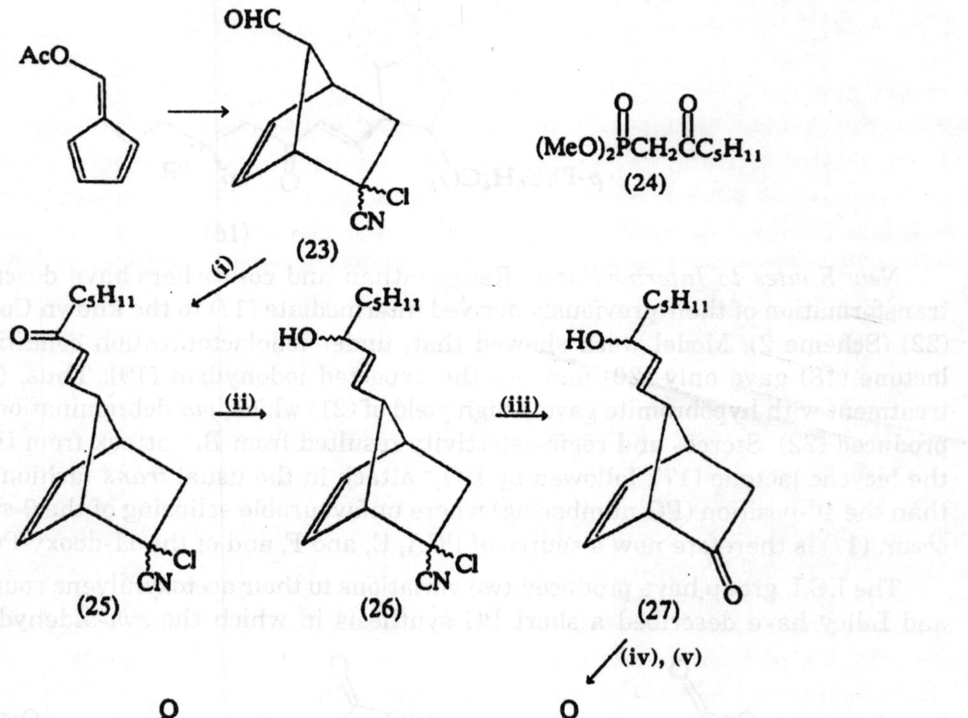

(23) (24) (25) (26) (27) (28) (29)

Reagents : (i) lithio-(24); (ii) $Al(OPr^i)_3$; (iii) KOH-DMSO; (iv) H_2O_2—OH^-; (v) KI-I_2; (vi) $Bu_3{}^nSnH$

Scheme 3

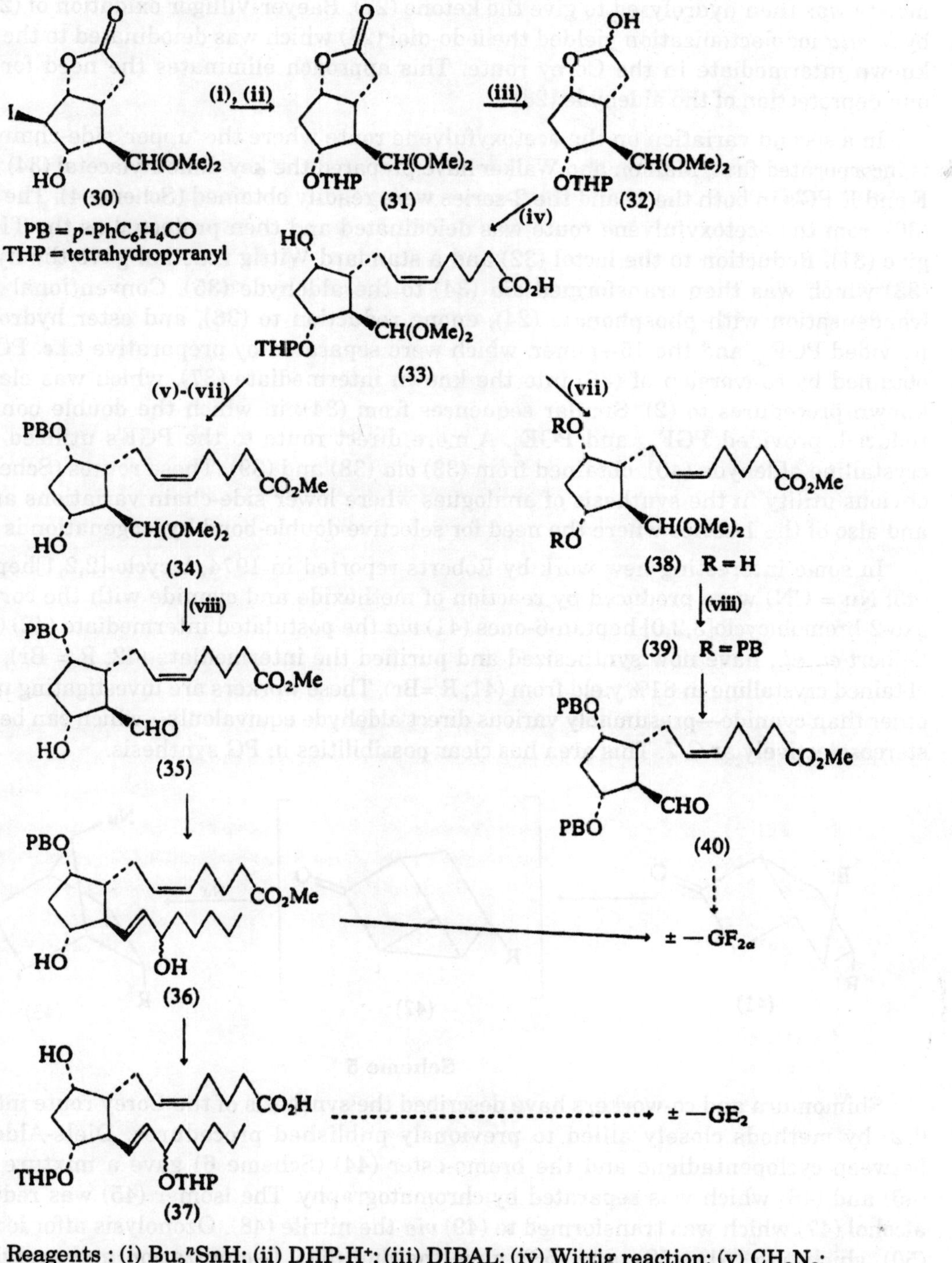

(30) PB = p-PhC_6H_4CO
THP = tetrahydropyranyl

(i), (ii) (31) (iii) (32) (iv) (33) (v)-(vii) (34) (vii) (38) R = H (viii) (35) (viii) (39) R = PB (40) (36) ± — $GF_{2\alpha}$ (37) ± — GE_2

Reagents : (i) $Bu_3{}^nSnH$; (ii) DHP-H^+; (iii) DIBAL; (iv) Wittig reaction; (v) CH_2N_2; (vi) PBCl-py; (vii) TsOH-MeOH; (viii) conc. HCl—$CHCl_3$—Pr^iOH.

Scheme 4

was reduced with aluminium isopropoxide to a 1:1 mixture of enols (26). The chloronitrile moiety was then hydrolysed to give the ketone (27). Baeyer-Villiger oxidation of (27) followed by *in situ* iodolactonization yielded the iodo-diol (28) which was deiodinated to the diol (29), a known intermediate in the Corey route. This approach eliminates the need for protection and deprotection of the aldehyde (23).

In a second variation on the acetoxyfulvene route where the 'upper' side-chain (C-l—C-5) is incorporated first, Mallion and Walker have prepared the key dimethylacetal (34) from which F and E PC's in both the 1- and the 2-series were readily obtained (Scheme 4). The iodohydrin (30) from the acetoxyfulvene route was deiodinated and then protected as the THP ether to give (31). Reduction to the lactol (32) and a standard Wittig reaction gave the hydroxy-acid (33) which was then transformed *via* (34) to the aldehyde (35). Conventional procedures [condensation with phosphonate (24), enone reduction to (36), and ester hydrolysis] then provided $PGF_{2\alpha}$ and the 15-epimer, which were separated by preparative t.l.c. PGE_2 (2) was obtained by conversion of (36) into the known intermediate (37), which was elaborated by known procedures to (2). Similar sequences from (34), in which the double bond was first reduced, provided $PGF_{1\alpha}$ and PGE_1. A more direct route to the PGF's utilized the stable, crystalline aldehyde (40), obtained from (33) *via* (38) and (39). These routes (Scheme 4) have obvious utility in the synthesis of analogues where lower side-chain variations are required and also of the 1-series where the need for selective double-bond hydrogenation is obviated.

In some interesting new work by Roberts reported in 1974, bicyclo-[2,2,1]heptan-2-ones (43; Nu = CN) were produced by reaction of methoxide and cyanide with the corresponding oxo-2-bromobicyclo[3,2,0] heptan-6-ones (41) *via* the postulated intermediate (42) (Scheme 5). Gilbert *et. al.*, have now synthesized and purified the intermediate (42; R = Br), which was obtained crystalline in 81% yield from (41; R =Br). These workers are investigating nucleophiles other than cyanide—presumably various direct aldehyde equivalents—which can be introduced stereoselectively at *C*-7. This area has clear possibilities in PG synthesis.

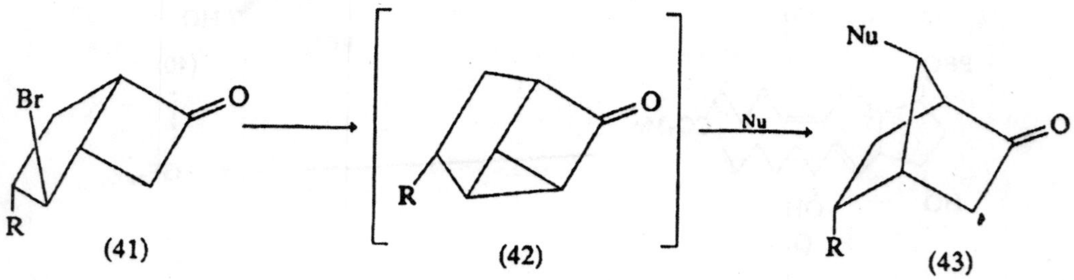

Br O R (41) → [O R (42)] —Nu→ Nu O R (43)

Scheme 5

Shimomura and co-workers have described the synthesis of the Corey route inter-mediate (22) by methods closely allied to previously published procedures. Diels-Alder reaction between cyclopentadiene and the bromo-ester (44) (Scheme 6) gave a mixture of isomers (45) and (46) which was separated by chromatography. The isomer (45) was reduced to the alcohol (47) which was transformed to (49) *via* the.nitrile (48). Ozonolysis afforded the diacid (50) which, *via* (51), (52), and (53), gave the diol (54). Lactonization and acetylation then gave (22).

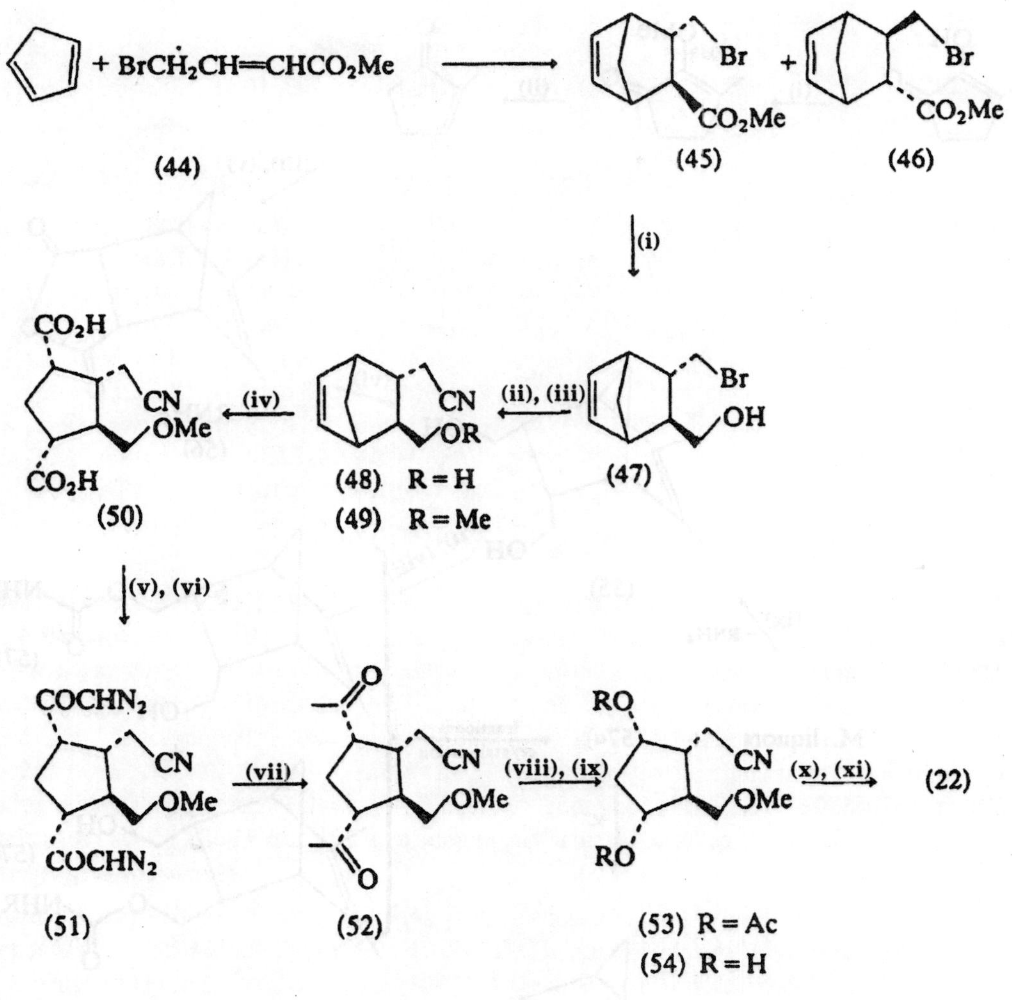

Reagents: (i) $LiAlH_4$; (ii) NaCN; (iii) NaH-MeI; (iv) O_3; (v) $(COCl)_2$; CH_2N_2; (vii) HI; (viii) CF_3CO_3H–CH_2Cl_2–NA_2HPO_4; (ix) OH^-; (x) H^+—lactonization; (xi) acetylation.

Scheme 6

In a very appealing synthesis, ingenious if perhaps rather long, a Swiss group from Hoffman-La Roche have generated a chiral PG intermediate (57) by combining the optically inactive meso-compound (55) with a chiral moiety (56) to give a mixture of diastereomers (57*a* and (*b*) from which that required, (57*a*), could be obtained by fractional crystallization (Scheme 7). Simple hydrolysis of the material in the mother liquors returned the starting *meso*-compound (55) and the chiral amine (56) which were then both recycled *(cf.* Corey and Ensley). In this way high chiral efficiency was obtained. The *meso-dio*] (55) was synthesized as outlined in Scheme 7. The absolute configuration of (57*a*) was determined by *X*-ray analysis on the bromo-ether (58), which was prepared directly by the action of bromine on (57*a*) itself. (57*a*), which has the absolute configuration corresponding to natural $PGF_{2\alpha}$, was converted

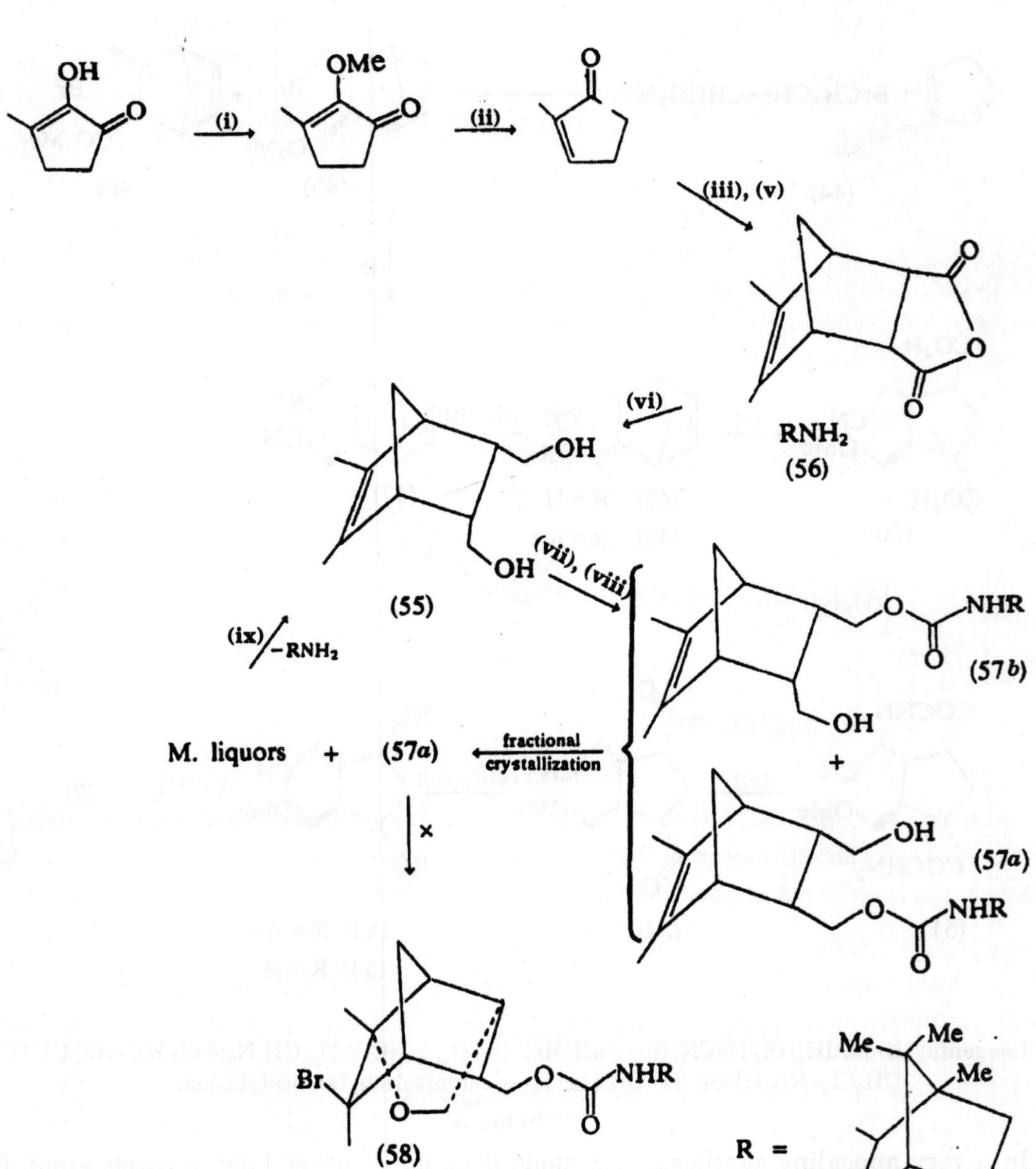

OH OMe O (i) (ii) (iii), (v) O O O RNH$_2$ (56) (vi) OH OH (55) (vii), (viii) O NHR O OH (57b) + OH (57a) O NHR O (ix) –RNH$_2$ M. liquors + (57a) fractional crystallization x Br O O NHR O (58) R = Me Me Me

Reagents: (i) MeOH-HCl; (ii) DIBAL; (iii) MeMgI; (iv) p-TsOH; (v) maleic anhydride; (vi) REDAL; (vii) $COC1_2$; (viii) (56); (ix) KOH; x, Br_2.

Scheme 7

into the known Corey route intermediate (69) as shown in Scheme 8. Transformation to the nitrile (60) *via* the mesylate (59) followed by hydrolysis gave the lactone (61) which in turn afforded the hydroxy-amide (62) on treatment with pyrrolidine. Oxidation then furnished the *endo*-aldehyde (63) which, after epimerization to the *exo*-isomer (64), was reduced to the alcohol (65). Protection of (65) as the benzyl ether (66) and ozonolysis gave an unstable bis-methyl ketone (67) which was transformed in a Baeyer–Villiger step to the diacetate (68). Hydrolysis, lactonization, and acylation then afforded the known (69). Similar operations on (57*a*) performed in a different order produced *ent*-(69) in 13 steps *via* the key intermediates (70) – (73).

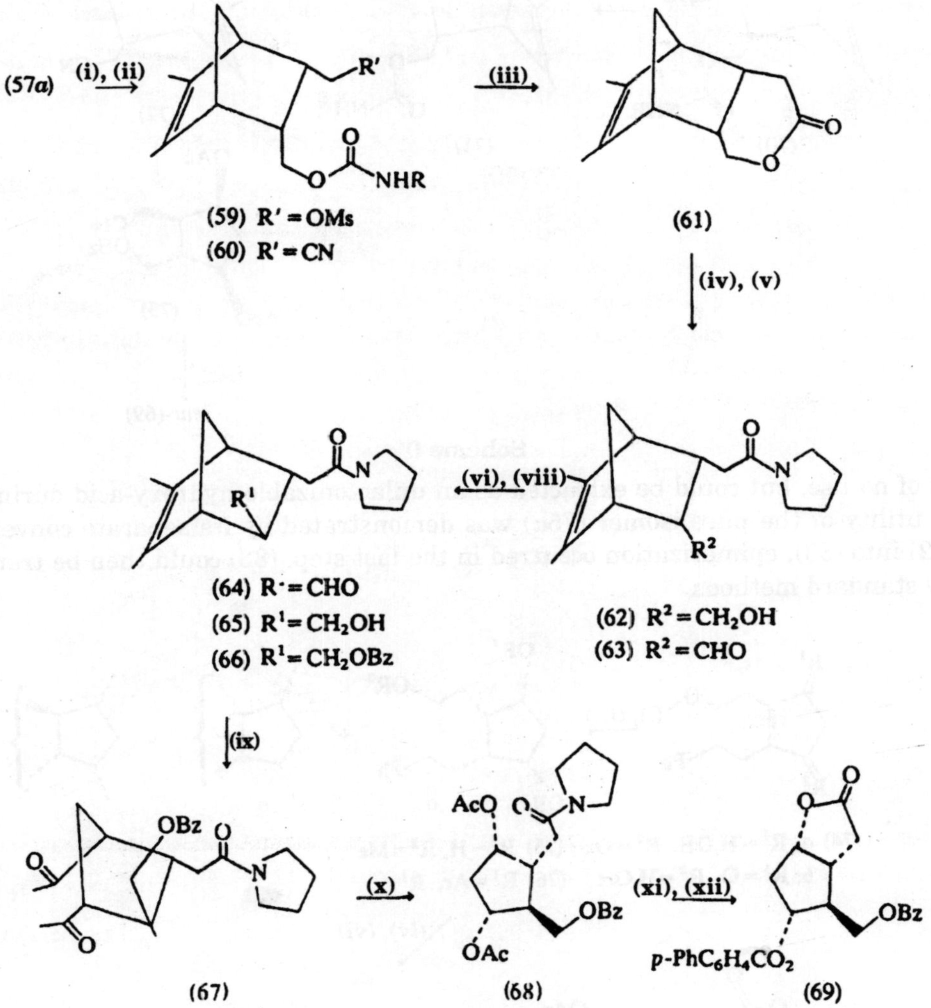

(57*a*) (i), (ii) → (59) R' = OMs; (60) R' = CN → (iii) → (61) → (iv), (v) → (62) $R^2 = CH_2OH$; (63) $R^2 = CHO$ → (vi), (viii) → (64) $R^1 = CHO$; (65) $R^1 = CH_2OH$; (66) $R^1 = CH_2OBz$ → (ix) → (67) → (x) → (68) → (xi), (xii) → (69)

Reagents: (i) MsCl-py; (ii) NaCN-DMSO; (iii) OH^-/HT^+; (iv) HN◻; (v) SO_3– py–DMSO; (vi) pyridinium acetate-benzene; (vii) $NaBH_4$; (viii) BzCI-NaH; (ix) O_3; (x) TCF_3CO_3H; (xi) OH^-; (xii) lactonization; (xiii) p-PhC_6H_4COCl–py.

Scheme 8

An alternative entry into the Corey route has been published by Coen *et al.*, who have described a new method for the reduction of intermediates of the type (74) to (75) together with an alternative procedure for conversion of (75) into the Corey-type aldehyde (82) (Scheme 10). Hydrogenation of the mixture (74a and b) over W_2-Raney nickel gave a mixture of isomers

(75*a*, *b*, and *c*). (75*b*) was the required isomer, though (75*a*) could also be used because of the ease with which it could be epimerized at C-12 later on, *i.e.* during (82) → (83).

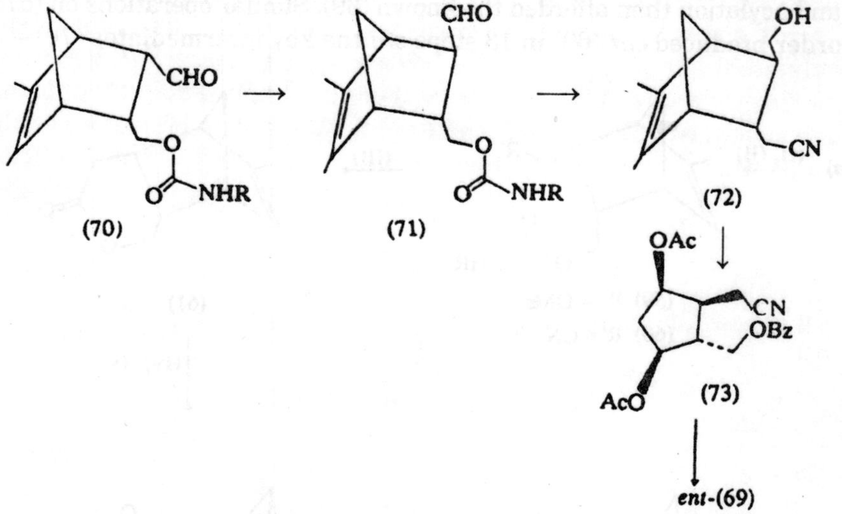

CHO; CHO; OH; CHO; O; NHR; O; NHR; CN; (70); (71); (72); OAc; CN; OBz; AcO; (73); *ent*-(69)

Scheme 9

(75*c*) was of no use, but could be extracted as an unlactonizable hydroxy-acid during (79) → (80). The utility of the pure isomer (75*a*) was demonstrated by its separate conversion *via* (76) → (82) into (83); epimerization occurred in the last step. (83) could then be transformed to PC's by standard methods.

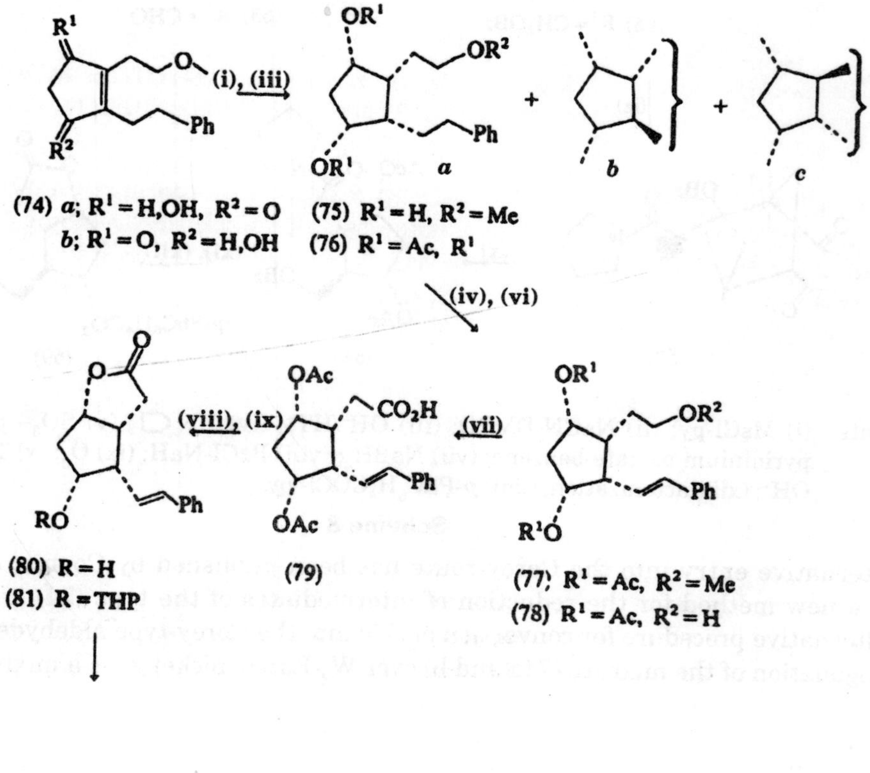

R^1; O; (i), (iii); OR^1; OR^2; Ph; R^2; Ph; OR^1; *a*; *b*; *c*

(74) *a*; R^1 = H,OH, R^2 = O (75) R^1 = H, R^2 = Me
b; R^1 = O, R^2 = H,OH (76) R^1 = Ac, R^1

(iv), (vi)

O; O; OAc; CO_2H; (viii), (ix); (vii); OR^1; OR^2; Ph; Ph; Ph; RO; OAc; R^1O

(80) R = H (79) (77) R^1 = Ac, R^2 = Me
(81) R = THP (78) R^1 = Ac, R^2 = H

↓

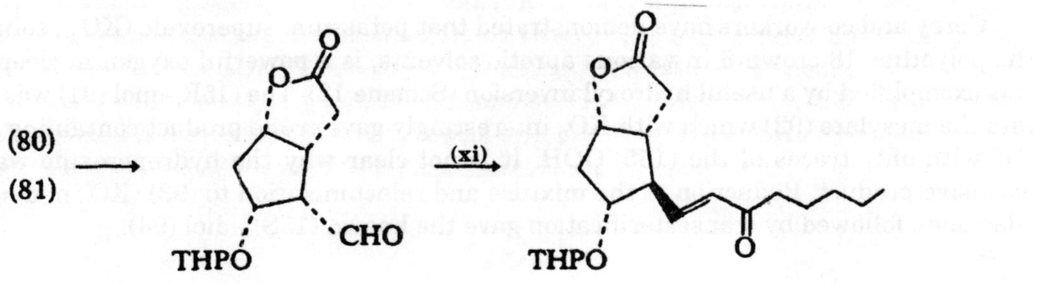

(80) (81) ⟶ (82) ⟶(xi) (83)

Reagents: (i) W_2-Raney nickel-H_2; (ii) chromatography; (iii) acetylation; (iv) NBS-CCU; (v) BBr_3; (vii) Jones oxidation; (viii) hydrolysis; (ix) DHP-H^+; (x) OsO_4^-—IO_4; (xi) $(MeO)_2P(O)C^-H$–COC_5H_{11}

Scheme 10

A group from the same laboratories has published another route to the Corey aldehyde (90) *via* two series, (*a*) and (*b*) (Scheme 11), which cuts across and combines several features of other routes. Reduction of (84*a* and *b*) with lithium-ammonia gave *all-trans*-(85); this was converted by classical steps *via* (86) into the lactone (87*a* and *b*), from which could be derived the tosylates (88*a*, *b*, and *c*) [*i.e.* (87*b*) → (87*c*)]. These were then elaborated by Sutherland's procedure, ring-opened with hydroxylamine, and recyclized to the γ-lactones (89*a*, *b*, and *c*), which were converted by known procedures into (90).

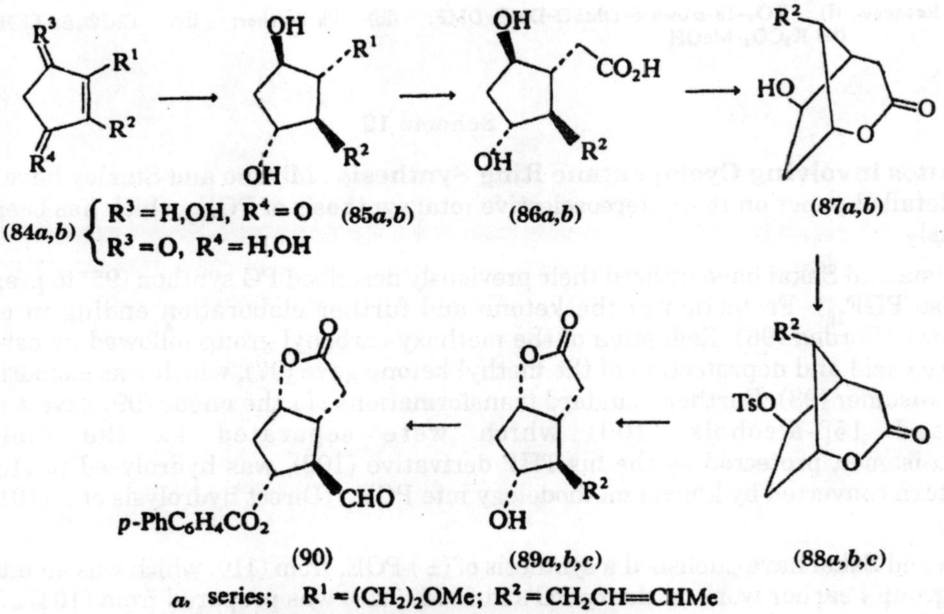

(84*a,b*) { R^3 = H,OH, R^4 = O; R^3 = O, R^4 = H,OH } ⟶ (85*a,b*) ⟶ (86*a,b*) ⟶ (87*a,b*) ⟶ (88*a,b,c*) ⟶ (89*a,b,c*) ⟶ (90)

a, series; R^1 = $(CH_2)_2OMe$; R^2 = $CH_2CH{=}CHMe$

b, series; R^1 = $CH_2CH{=}CMe_2$; R^2 = $(CH_2)_2Ph$

c, series; R^2 = $CH{=}CHPh$

Scheme 11

Corey and co-workers have demonstrated that potassium superoxide (KO_2), solubilized by the polyether 18-crown-6 in various aprotic solvents, is a powerful oxygen nucleophile. This was exemplified by a useful hydroxyl inversion (Scheme 12). The (15R)-enol (91) was converted into the mesylate (92) which with KO_2 interestingly gave crude product containing the (15S)-OH with only traces of the (155)-OOH. It is not clear why the hydroperoxide was not the exclusive product. Reduction of the mixture and relactonization to (93) (KO_2 had opened the γ-lactone), followed by transesterification gave the known (15S) - diol (94).

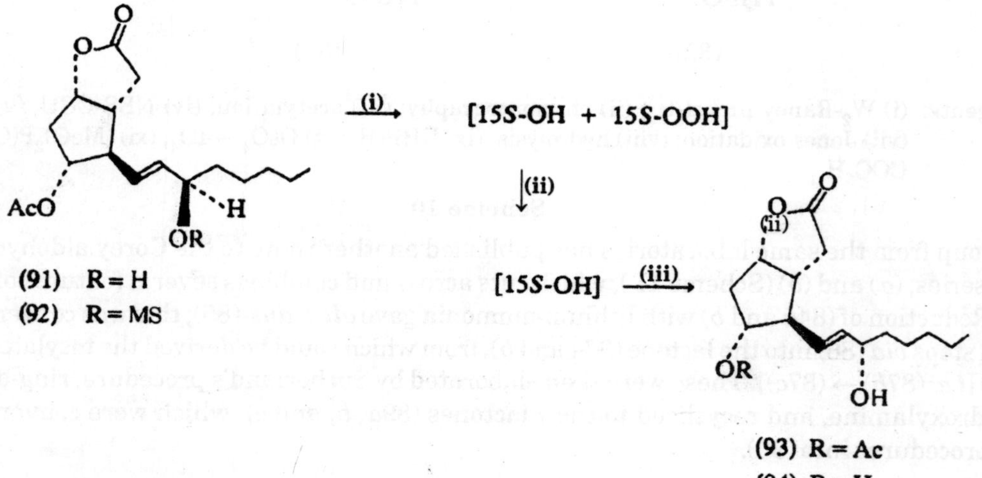

AcO H OR

(91) R = H

(92) R = MS

(i) → [15*S*-OH + 15*S*-OOH]

↓(ii)

[15*S*-OH] (iii) →

(ii)

OR OH

(93) R = Ac

(94) R = H

Reagents: (i) KO_2–18-crown-6–DMSO–DME–DMF; (iii) Ph_3P–ether; (iv) $ClCO_2Et$–LiOH; (iv) K_2CO_3–MeOH

Scheme 12

Routes involving Cyclopentane Ring Synthesis : Miyano and Stealey have published a full, detailed paper on their stereoselective total synthesis of PGE_1 which has been reported previously.

Kojima and Sakai have utilized their previously described PG synthon (95) to prepare PGE_1 (and also $PGF_{1\beta}$). Protection of the ketone and further elaboration ending in a selective hydrolysis afforded (96). Reduction of the methoxy-carbonyl group followed by esterification of the free acid and deprotection of the methyl ketone gave (97), which was isomerized to the all-*trans*-isomer (98). Further standard transformations *via* the enone (99) gave a mixture of 15α- and 15β-alcohols (100) which were separated as the diols (101). The 15α-isomer, protected as the bis-THP derivative (102), was hydrolysed to (103), which was in turn converted by known methodology into PGE_1. [Direct hydrolysis of α-(101) afforded $PGF_{1\beta}$.]

Oda and Sakai have published a synthesis of (±)-PGE_1 from (110) which was an intermediate in this group's earlier work on deoxy-PG's in 1975. (110) was prepared from (104) *via* (105) — (109) (Scheme 14); of note in this sequence were the stereospecific alkylation (107) → (108) and the selective hydrolysis (108) → (109). (110) was transformed *via* (111) — (115) to (116), which was in turn converted into PGE_1 by methods standard in the PG field.

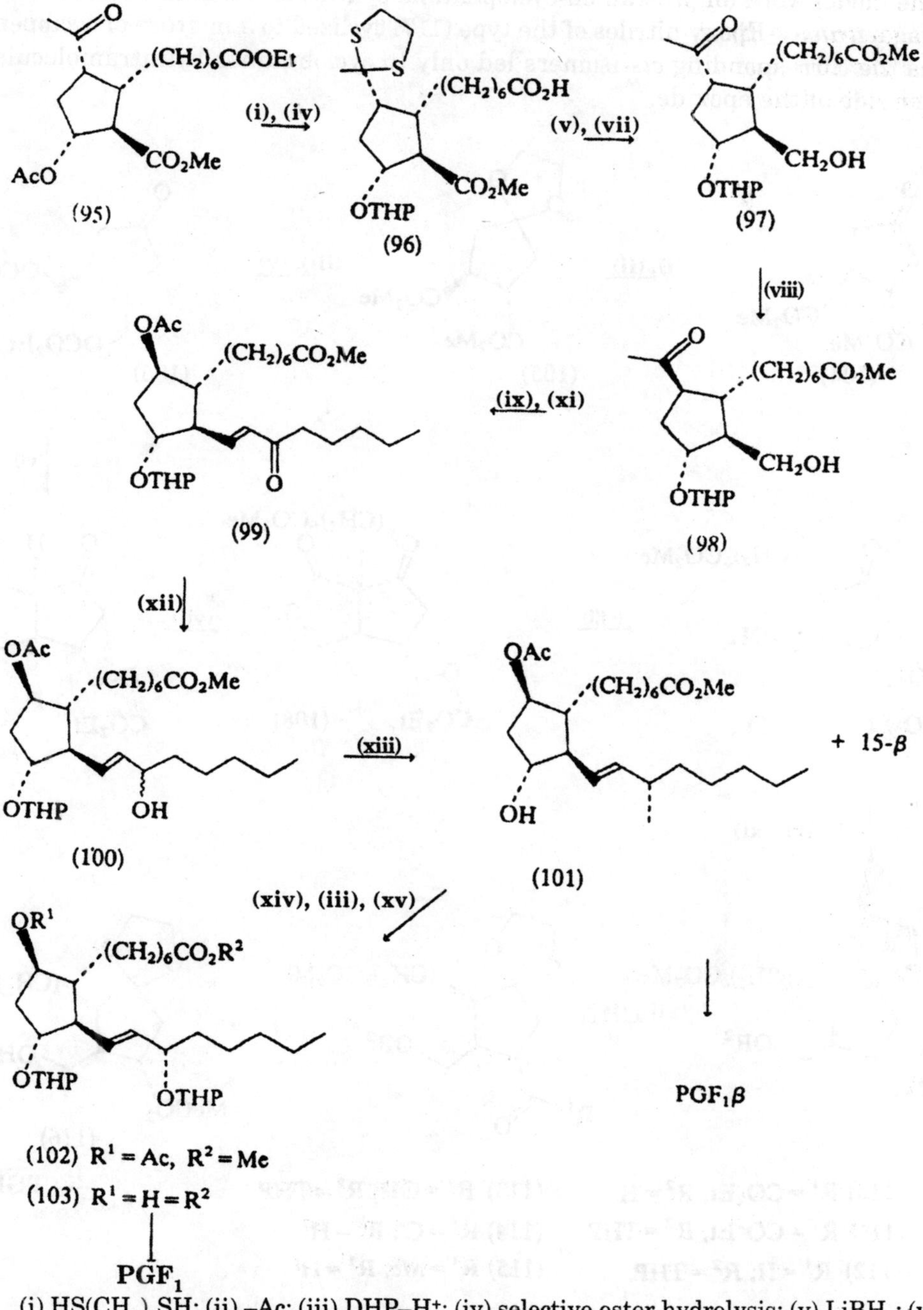

(95) (96) (97) (98) (99) (100) (101)

(102) $R^1 = Ac$, $R^2 = Me$

(103) $R^1 = H = R^2$

PGF$_1$

PGF$_1\beta$

Reagents: (i) $HS(CH_2)_2SH$; (ii) –Ac; (iii) DHP–H^+; (iv) selective ester hydrolysis; (v) $LiBH_4$; (vi) CH_2N_2; (vii) Loss of [dithiolane], (viii) K_2CO_3 (epimerization); (ix) Baeyer-Villiger; (x) oxidation; (xi) $\geqslant P = CHCO_5H_{11}$; (xii) $NaBH_4$; (xiii) H^+; (xiv) chromatography; (xv) OH^-

Scheme 13

Some model work on prostanoid-cyclopentane synthesis has been described by Lallemand and Onaga. *trans-γ*-Epoxy-nitriles of the type (117) cyclized to a mixture of cyctopentanes (118), whereas the corresponding *cis*-isomers led only to cyclobutanes by intramolecular attack on the other side of the epoxide.

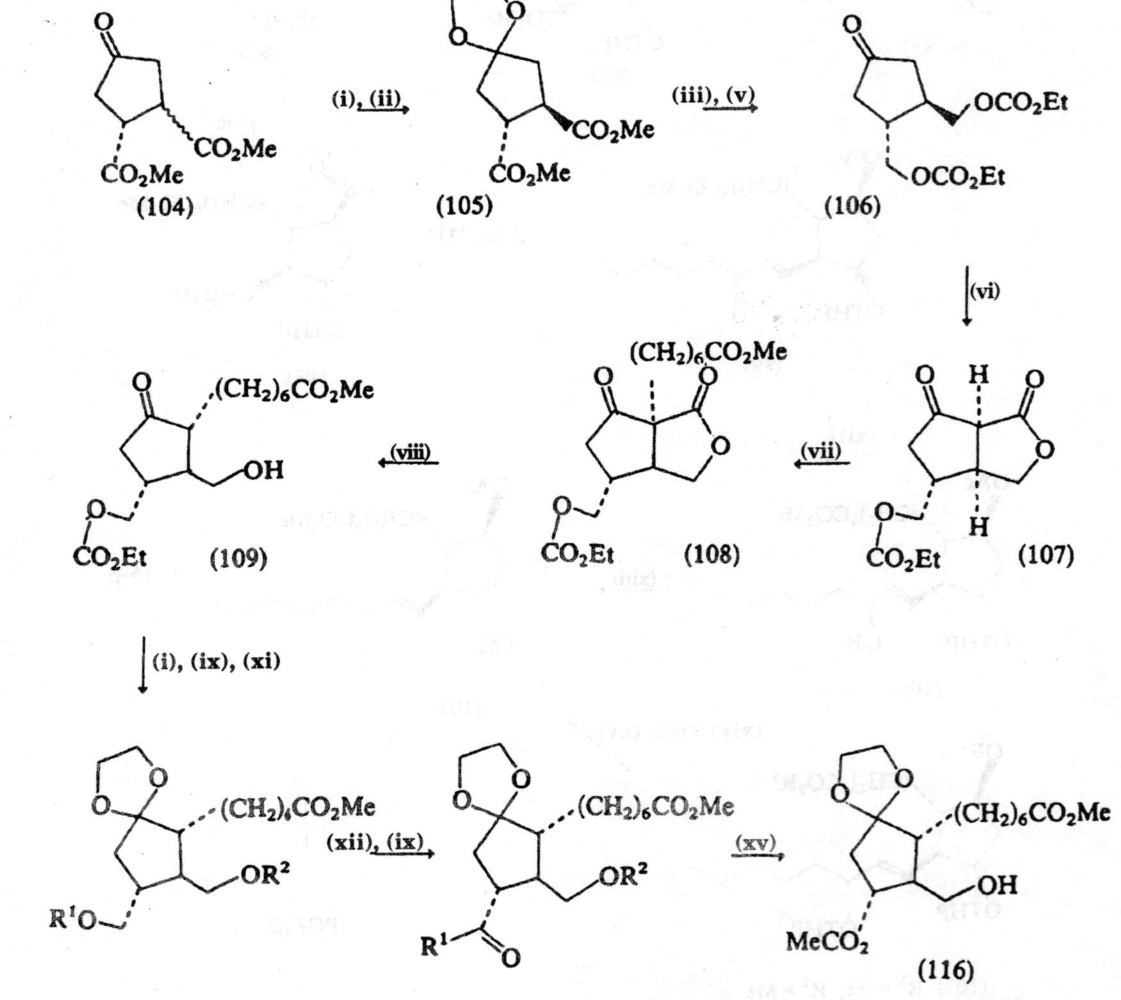

(104) (i), (ii) → (105) (iii), (v) → (106) (vi) → (107) (vii) → (108) (viii) → (109) (i), (ix), (xi) → (110)–(112) (xii), (ix) → (113)–(115) (xv) → (116) ⟶ PGE_1

(110) $R^1 = CO_2Et$, $R^2 = H$
(111) $R^1 = CO^2Et$, $R^2 = THP$
(112) $R^1 = H$, $R^2 = THP$.
(113) $R^1 = OH$, $R^2 = THP$
(114) $R^1 = Cl$, $R^2 - H$
(115) $R^1 = Me$, $R^1 = H$

Reagents: (i) $HO(CH_2)_2OH–H^*$; (ii) NaOMe; (iii) $LiAlH_4$; (iv) $ClCO_2Et$-py; (v) *p*-$TsOH$-H_2O; (vi) KOBu'- THF; (vii) $I(CH_2)_6CO_2Me$-MOBu^t-DMSO; (viii) Na_2HPO_4-dioxan-H_2O; (ix) DHP-H^+; (x) OH^-; (xi) CH_2N_2; (xii) CrO_3-H_2O-py; (xiii) $(COCl)_2$; (xiv) Me_2CuLi; (xv) Baeyer-Villiger.

Scheme 14

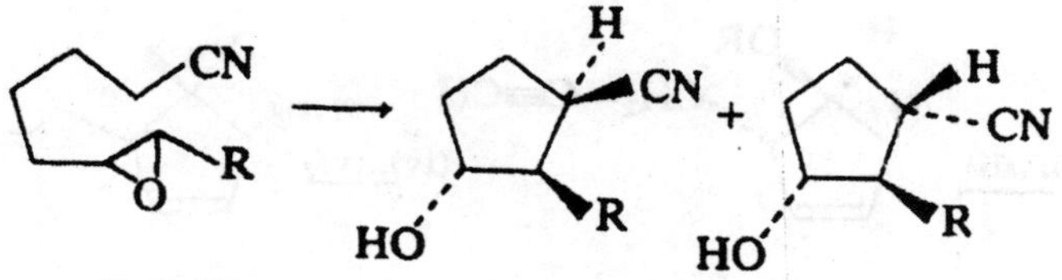

(117) (118)

Routes *via* Conjugate Addition to Cyclopentenones: Two papers have appeared describing full details of the elegant work by Sih and co-workers which culminated in the asymmetric total synthesis of (—)-PGE_1 and (—)-PGE_2. This has been reported previously.

Stork *et al.,* in a beautiful piece of work, have devised a route to PG's *via* a general synthesis of 4-hydroxycyclopentenones. At the heart of this approach was the intra-molecular, chloral-assisted hydration (and dehydration) of the model compound (120) which was the unrequired isomer obtained as the kinetic product when the epoxide (119) was treated with triethylamine. The required isomer (121), which could not be obtained satisfactorily from (120) through normal inter molecular hydration, because of the *trans* configuration of the OH and R groups, was obtained smoothly on treatment with chloral. This was utilized in a synthesis of PGE_2. The epoxide (122) was transformed *via* (123) to (124) which was elaborated further,

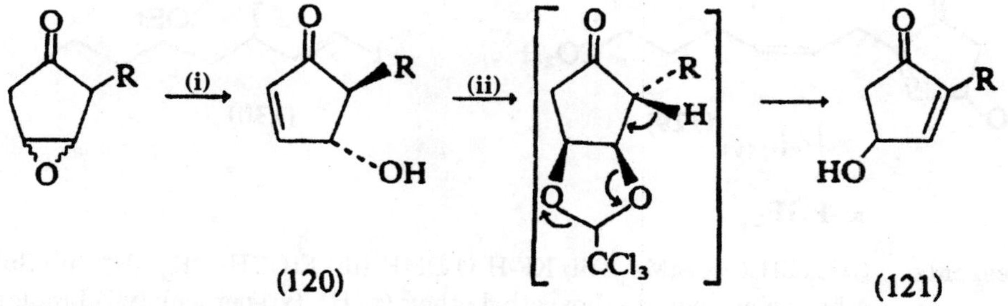

(120) (121)

(119) R = $CH_2C \equiv CBu$

Reagents: (i) Et_3N; (ii) CCl_3CHO

Scheme 15

via (125) — (127), to the known carboxylic acid (128). Treatment of (128) with triethylamine and then directly with chloral afforded a 69% yield of the required (129) without isolation of the inter-mediate corresponding to (120). Conjugate addition of the divinylcuprate reagent derived from (130) and deprotection of the 15-hydroxy-group gave (±)-PGE_2 and the 15-epimer, which were separated by standard methods.

A further elegant contribution to PG synthesis has appeared from Stork's group in a new route to (±)-$PGF_{2\alpha}$ *via* the methylenecyclopentanone (140) which features the efficient trapping of enolate anions with monomeric formaldehyde together with some skilful use of protecting groups. This route also started with cyclopentadiene epoxide (122), which was converted in five steps into (134). A key step was the reaction of (134) with 'HOBr', which regioselectively attacked only the cyclic double bond to furnish (135). This result arose from deactivation of

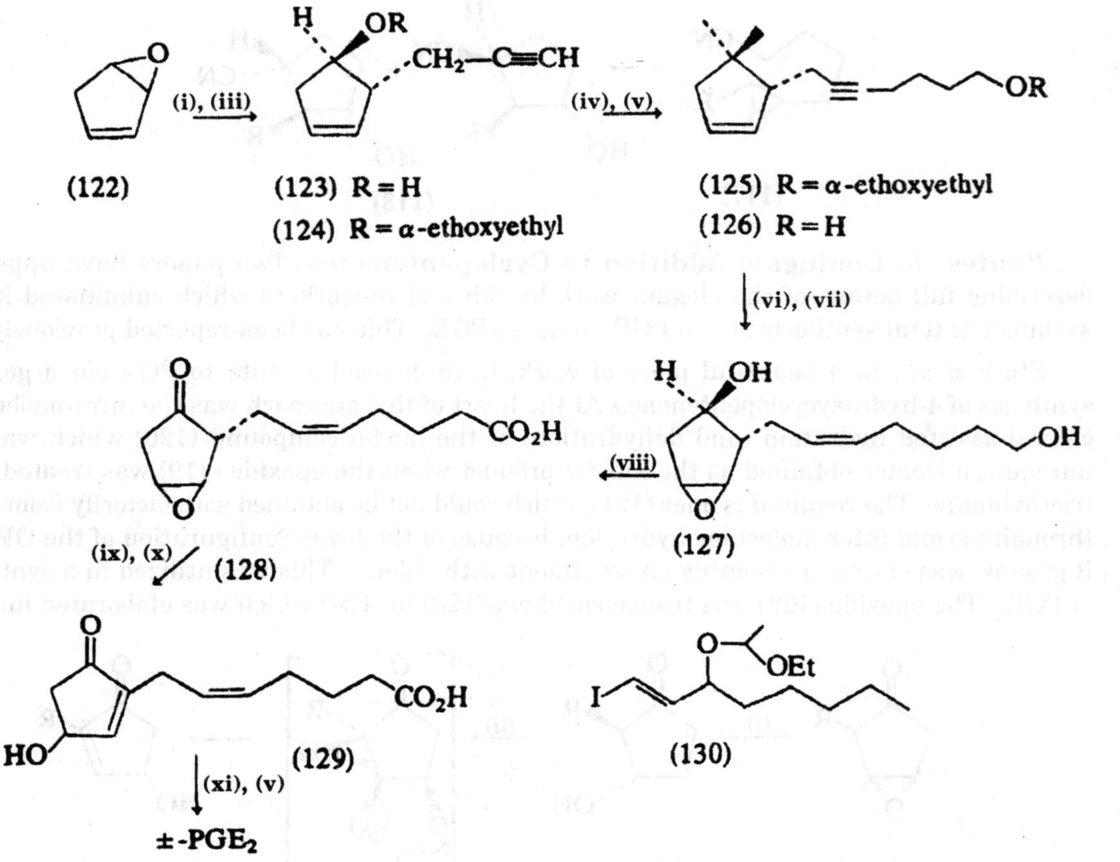

Reagents: (i) $Li(CH_2C{\equiv}CSiMe_3)$; (ii) KF-H_2O-DMF; (iii) $EtOCH{=}CH_2$—H^+; (iv) Bu^nLi-HMPA-THF-4-bromobutanol, α-ethoxyethyl ether; (v) H^+; (vi) peracid; (vii) Lindlar-H_2; (viii) Jones oxidation; (ix) Et_3N-ether-CH_2Cl_2; (x) CCl_3CHO; (xi) $LiCuR_2$ from (130)

Scheme 16

the side-chain double-bond by the judiciously chosen enol-protecting group and from displacement of the kinetic α-bromonium ion from the least hindered face of the cyclopentane ring. Oxidation to (136) followed by reaction with methyl diphenylphosphinite gave the enol phosphinate (137) which, with *n*-butyl-lithium, yielded the lithium enolate (138); this latter intermediate was captured by formaldehyde in the key step to produce (139). Dehydration to (140) and subsequent conjugate addition of the divinylcuprate derived from (141) gave (142). Acid treatment then removed only the α-ethoxyethyl group (leaving the 15-OH still protected) to expose the 1-OH group, which was then oxidized to the acid (143). Stereoselective reduction of (143) to the 9α-alcohol and removal of the 11- and 15-protecting groups gave (±)-$PGF_{2\alpha}$ and the 15-epimer, which were separated as the methyl esters by preparative h.p.l.c. The protecting group at the 11-position was carefully chosen; it is stable to acid and does not undergo elimination during the conversion of (139) into (140), but can be cleaved selectively in the final step, the conditions for which do not affect the allylic alcohol.

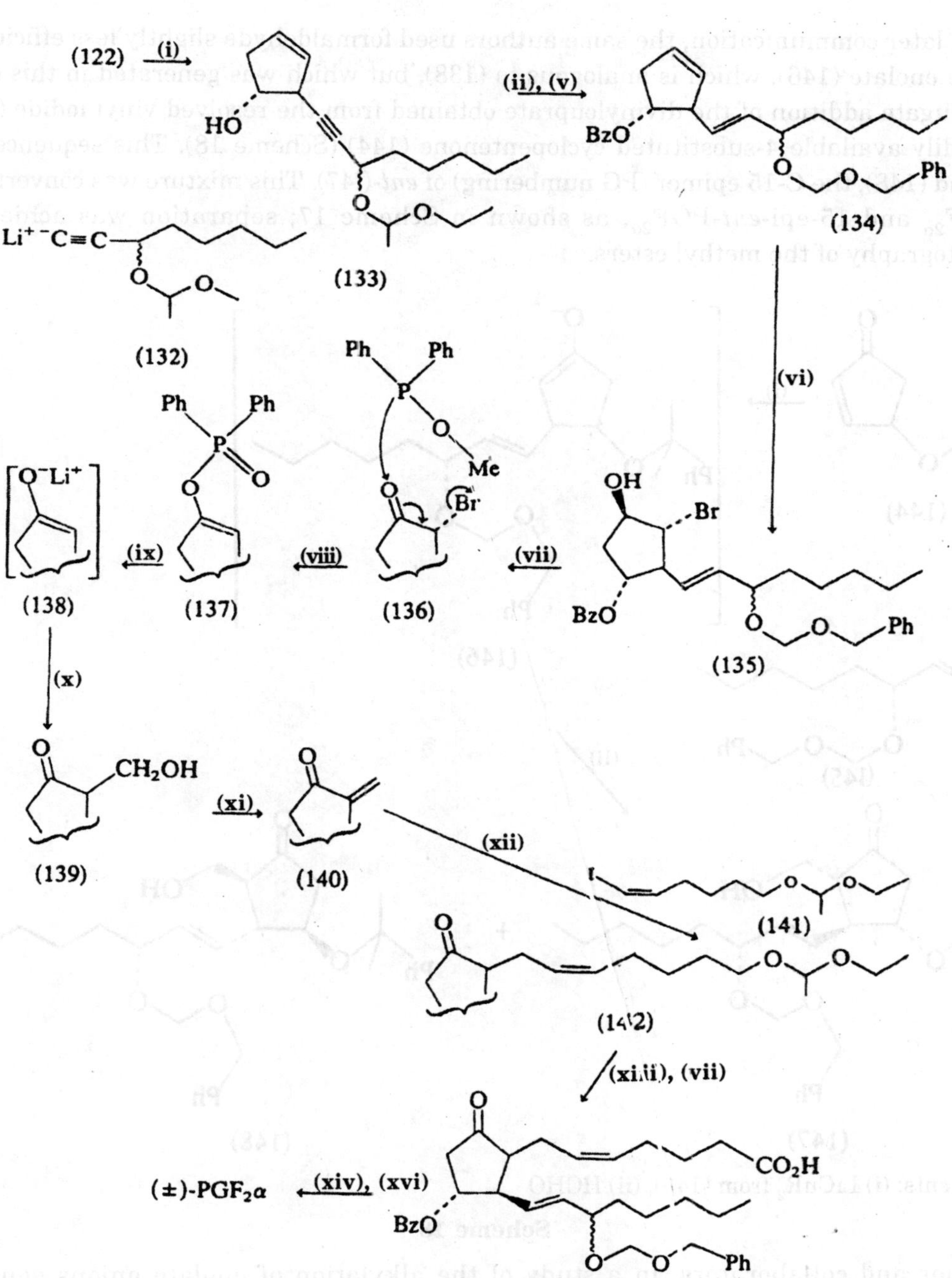

(122) →(i) ... (ii), (v) → (133) ... (134)

$Li^{+}\ ^{-}C{\equiv}C$ (132)

(vi) → (135) →(vii) (136) →(viii) (137) →(ix) (138)

(x) → (139) →(xi) (140) →(xii) (141) → (142)

(xi.ii), (vii) → (143) → (xiv), (xvi) → (±)-$PGF_{2}\alpha$

Reagents: (i) (132); (ii) BzI on Li salt; (iii) H^+; (iv) $LiAlH_4$ ® enol'; (v) Ph Λ O Λ Cl-$EtNPr_2$; (vi) NBS-DMSO-H_2O; (vii) Jones oxidation; (viii) methyl di phenyl phosphinite; (ix) BuLi-$ZnCl_2$; (x) HCHO; (xi) MsCl, then Et_3N; (xii) $LiCuR_2$ from (141); (xiii) 50% AcOH-H_2O, then oxidation; (xiv) $LiBu_3{}^{s}BH$; (xv) Na-liqNH_3-EtOH, (xv.i) chromatography.

Scheme 17

In a later communication, the same authors used formaldehyde slightly less efficiently to trap the enolate (146), which is analogous to (138), but which was generated in this case by the conjugate addition of the divinylcuprate obtained from the resolved vinyl iodide (145) to the readily available 4-substituted cyclopentenone (144) (Scheme 18). This sequence led to (147) and (148), the C-15 epimer (PG numbering) of *ent*-(147). This mixture was converted into (+)-$PGF_{2\alpha}$ and 15-epi-*ent*-$PGF_{2\alpha}$, as shown in Scheme 17; separation was achieved by chromatography of the methyl esters.

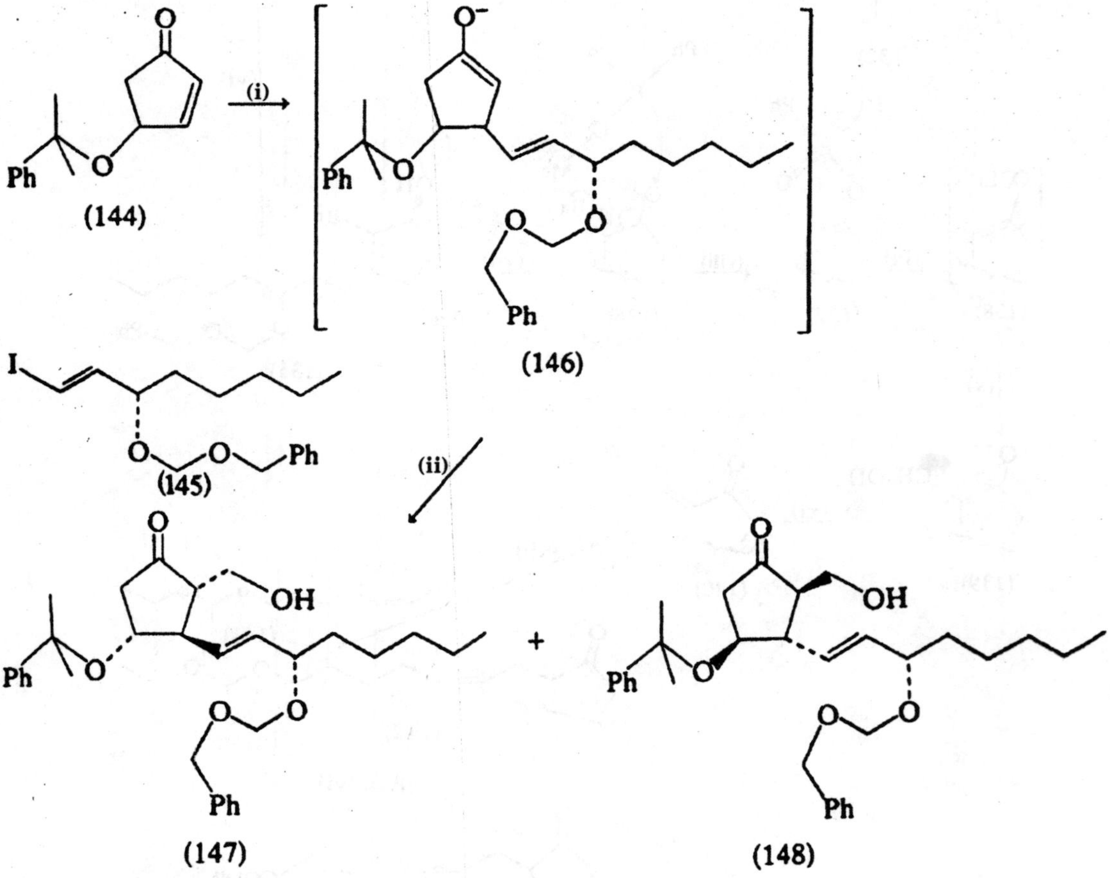

(144) (145) (146) (147) (148)

Reagents: (i) $LiCuR_2$ from (145); (ii) HCHO

Scheme 18

Posner and collaborators, in a study of the alkylation of enolate anions generated regiospecifically by addition of organocuprates to cyclopentenones, have described a new mixed cuprate (149) which transfers the vinyl group very selectively (25:1). The authors point out that this model work, if applied to PG synthesis proper, would considerably increase the efficiency with which valuable side-chain vinyl intermediates [*e.g.* (141), (145)] could be utilized.

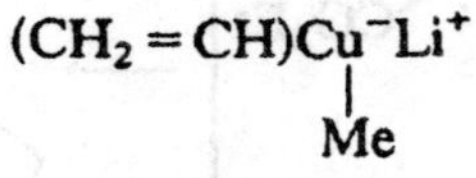

(149)

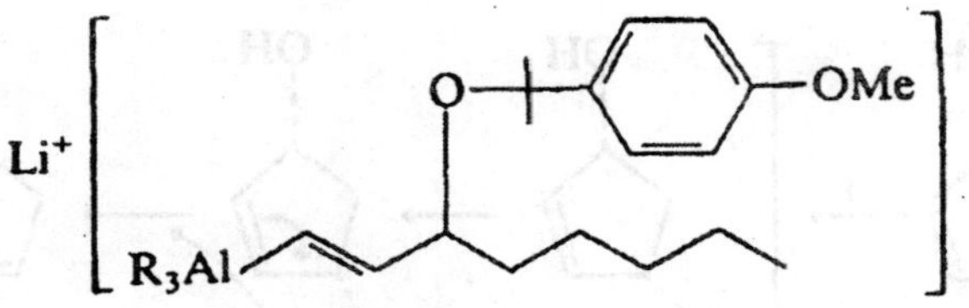

(150)

Bernady and co-workers have reported an improved procedure for the preparation of side-chain synthons as lithium trialkylvinylalanates of the type (150) and their conjugate addition to 2-substituted cyclopentenones. The effect of varying the R group and reaction solvent were investigated.

A synthesis of the PG synthon (156) [which could easily lead to (129)] for use in a conjugate addition sequence has been devised by Ellison and co-workers (Scheme 19). The anion from (151) was alkylated with (152) to give (153), which was selectively deprotected to (154). This latter intermediate then was elaborated in five steps to (156) *via* (155).

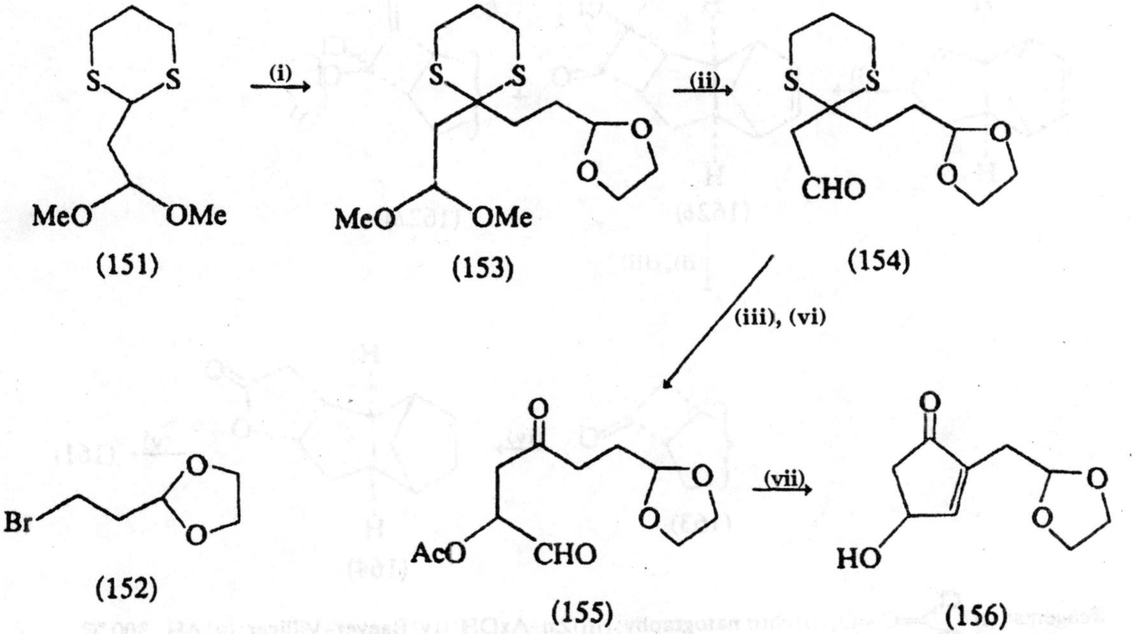

Reagents: (i) Bu^nLi – (152); (ii) 5% CF_3CO_2H– $CHCl_3$; (iii) M MgBr; (iv) $(Ac)_2O$; (v) NCS – $AgNO_3$–$MeCN$–H_2O; (vi) OsO_4– $NaIO_4$; (vii) NaOH – H_2O – dioxan.

Scheme 19

Two syntheses of the intermediate (161) have appeared. In the first of these, Kondo and colleagues treated *cis*-cyclo-pentene-3,5-diol (157) (itself a starting material for other routes to PG's; with triethyl orthoacetate. On distillation, the initially farmed ester (158) eliminated ethanol to give (159) which underwent a Claisen rearrangement to the cyclopentene (160), lactonization of which gave (161).

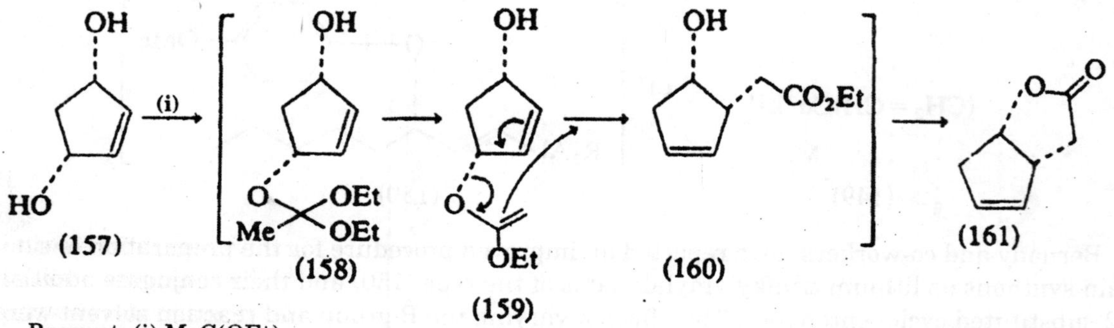

(157) (158) (159) (160) (161)

Reagent: (i) $MeC(OEt)_3$

Scheme 20

In the second route. Weil and Rouessac obtained (161) by a retro=Diels – Alder reaction on the tricyclic lactone (164), which was prepared in four steps from *endo*-dicyclopentadiene *via* (162*a*) and (163).

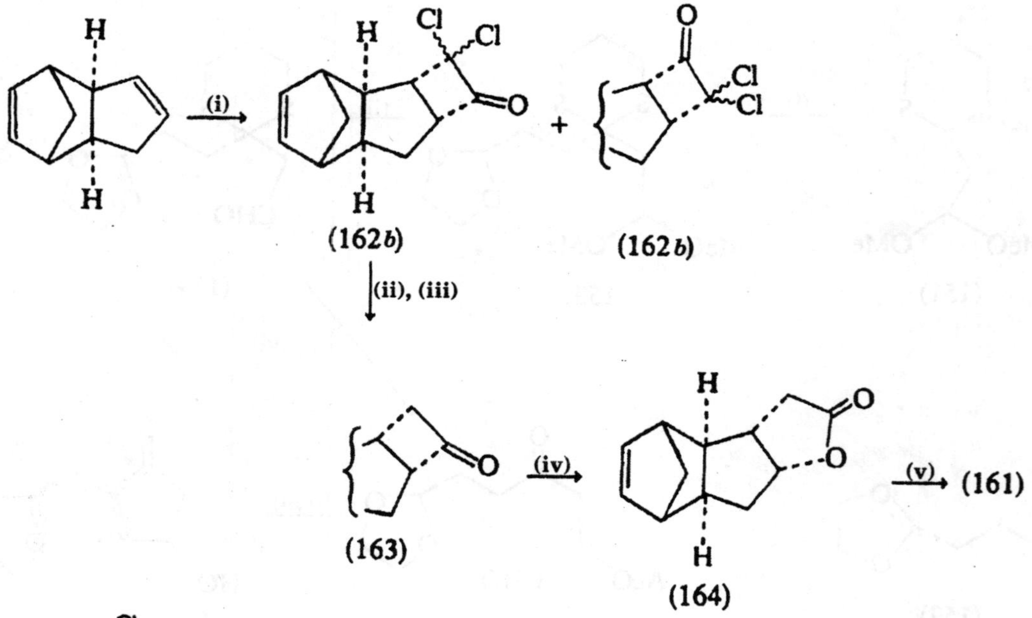

(162*b*) (162*b*) (163) (164) (v) (161)

Reagents: (i) $Cl_2C{=}C{=}O$; (ii) chromatography; (iii) Zn–AxOH; (iv) Baeyer–Villiger; (v) ΔH, 300 °C

Scheme 21

SYNTHESIS OF MODIFIED PROSTAGLANDINS

Deoxyprostaglandins : *11-Deoxyprostaglandins.* Several new syntheses of 11-deoxy-PG's appeared in 1975. Corey and Wollenberg have illustrated how their new nucleophilic vinylating agent (165), which is readily available from propargyl tetrahydropyranyl ether, can be used in a synthesis of (±)-11-deoxy-PGE_2 (173) (Scheme 22). Addition of the mixed cuprate derived from (165) to cyclopentenone gave (166), which was transformed *via* the alcohol (167) and the mesylate (168) to the bicyclic ketone (169). Pyrolysis effected rearrangement to (170) which was reduced stereospecifically to the α-alcohol (171). Oxidative ring cleavage in (171) ultimately gave a mixture of α- and β-aldehydes which, on equilibration, gave the β-isomer (172) which was then transformed by conventional steps to (±)-11-deoxy-PGE_2 (173).

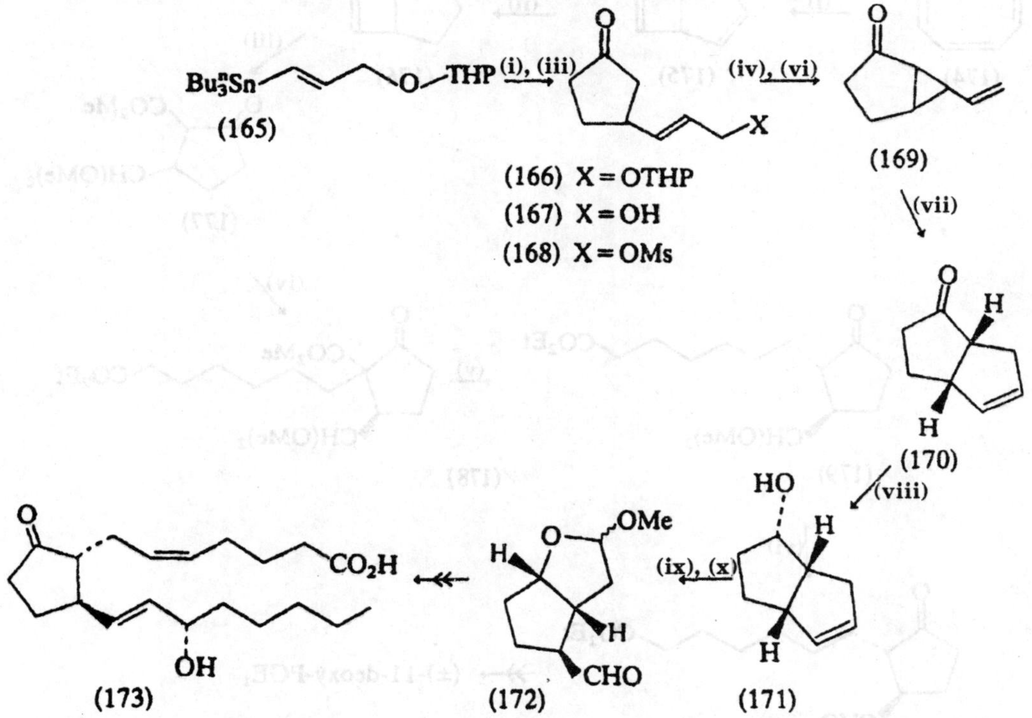

Bu^n_3Sn (165) (166) X=OTHP (167) X=OH (168) X=OMs (169) (170) (171) (172) (173)

Reagents: (i) BuLi–THF; (ii) 1-pentynylcopper; (iii) =O; (iv) H^+; (v) mesylation; (vi) KOBu^t; (vii) ΔH, 300 °C; (viii) $NaBH_4$–MeOH, −40 °C, (ix) OsO_4–$NaIO_4$; (x) BF_3–ether–MeOH

Scheme 22

Greene and Crabbe have published an interesting route to the corresponding 1-series compound, (±)-11-deoxy-PGE_1. The tropolone (174) on irradiation in methanol underwent valence tautomerism and skeletal rearrangement to give (175). Selective reduction then afforded (176) which yielded (177) *via* ozonolysis and acetal formation. Alkylation of (177) furnished (178) which in a key step gave (179). Deprotection of (179) then gave (180) which was converted into (±)-11-deoxy-PGE_1 by standard methods.

The intermediate (181), suitable for conversion into (+)-ll-deoxy-PGE_1 has been prepared by Oda and Sakai from (107).

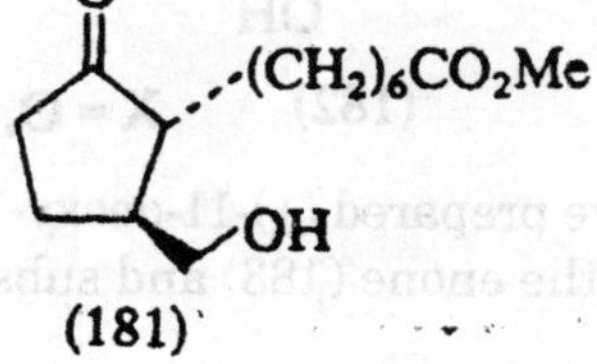

$(CH_2)_6CO_2Me$ OH

(181)

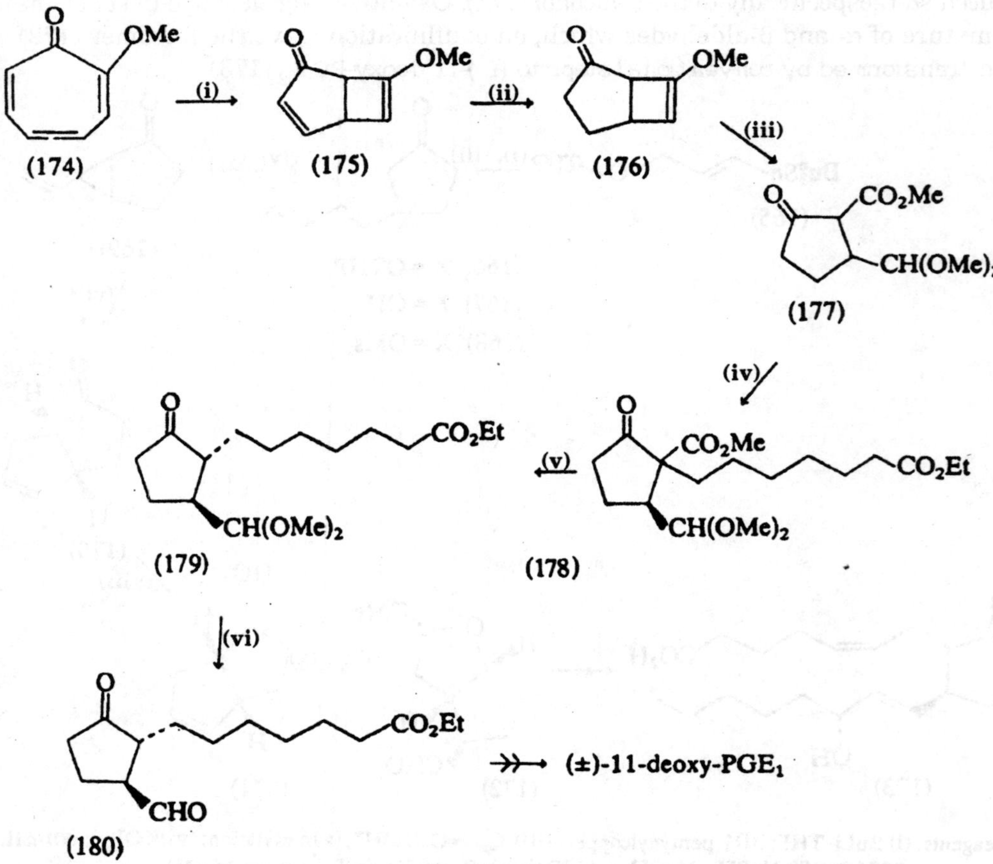

(174) —(i)→ (175) —(ii)→ (176) —(iii)→ (177) —(iv)→ (178) —(v)→ (179) —(vi)→ (180) →→ (±)-11-deoxy-PGE$_1$

Reagents: (i) $h\nu$, MeOH; (ii) Pt–H_2–EtOAc; (iii) O_3, then acetalization; (iv) KH-DMSO-I⌇⌇⌇CO_2; (v) NaCN–HMPA, (vi) *p*-TsOH–acetone

Scheme 23

A variety of (+)-11-deoxy-PGE$_1$ analogues of the type (182) have been reported by Bernady and co-wokers in their study of the addition of various lithium alanates to cyclopentenone.

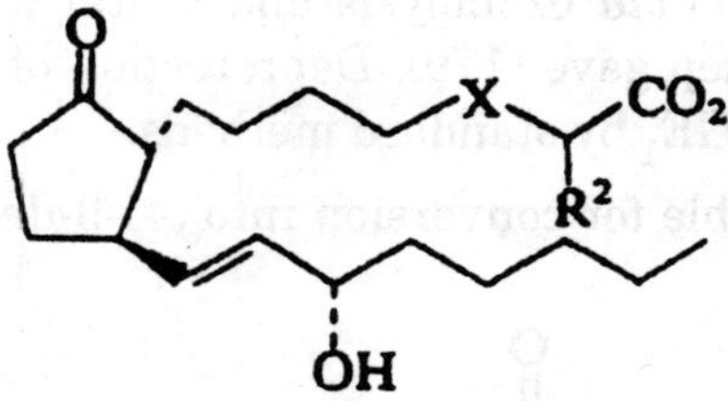

(182) X = O, S, or CH_2

Greene and collaborators have prepared (±)-11-deoxy-15-deoxy-13-hydroxy-13-methyl-PGE$_1$. (185) by ketone transposition in the enone (183) and subsequent standard elaboration of (184).

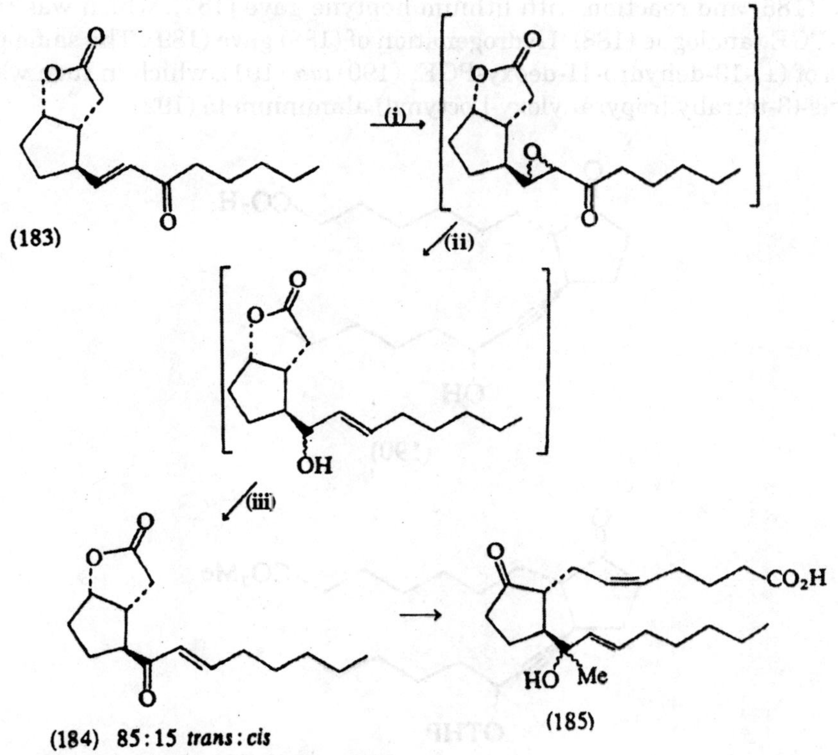

(183)

(i)

(ii)

OH

(iii)

(184) 85 : 15 *trans* : *cis*

(185)

HO Me

Reagents: (i) H_2O_2–OH^-; (ii) NH_2NH_2–AcOH; (iii) MnO_2

Scheme 24

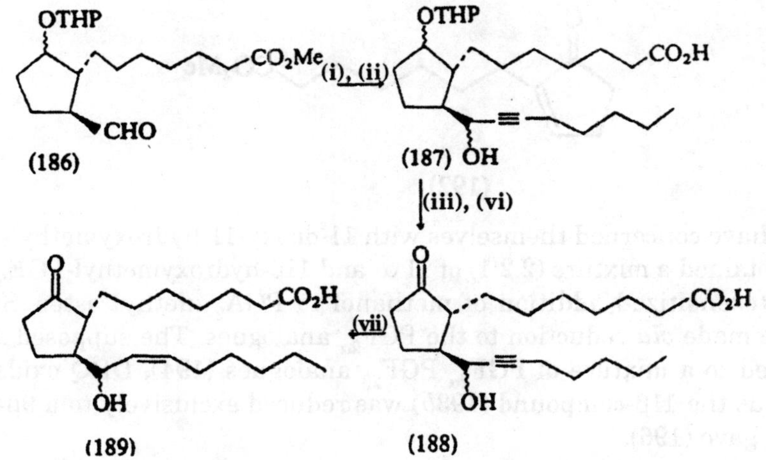

OTHP CO_2Me CHO (186)

(i), (ii)

OTHP CO_2H (187) OH

(iii), (vi)

CO_2H OH (189)

(vii)

CO_2H OH (188)

Reagents: (i) ester hydrolysis; (ii) Li-heptyne; (iii) acetylation; (iv) H^+; (v) Jones oxidation; (vi) C-13 acetate → OH; (vii) Lindlar-H_2

Scheme 25

Similar PG analogues have been synthesized by Bagli and Bogri from (186) (Scheme 25). Hydrolysis of (186) and reaction with lithium heptyne gave (187), which was transformed to the 11-deoxy-PGE_1 analogue (188). Hydrogenation of (188) gave (189). The same paper contains the synthesis of (±)-13-dehydro-11-deoxy-PGE_1 (190) *via* (191), which in turn was obtained by addition of tris-(3-tetrahydropyranyloxy-l-octynyl) aluminium to (192).

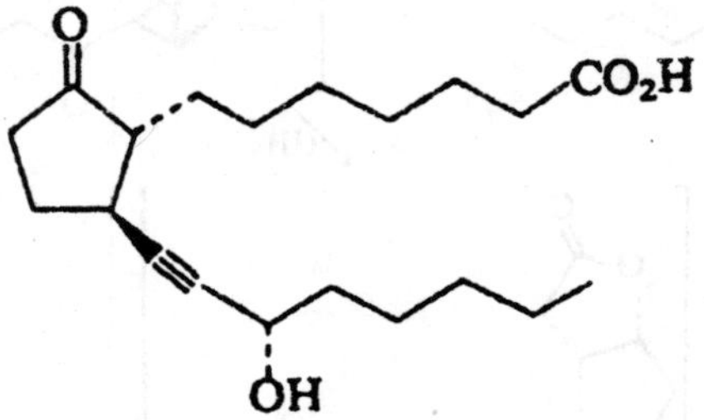

O CO₂H OH

(190)

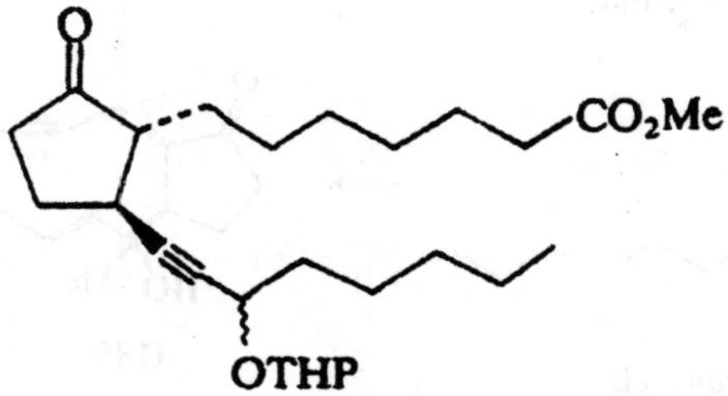

O CO₂Me OTHP

(191)

O CO₂Me

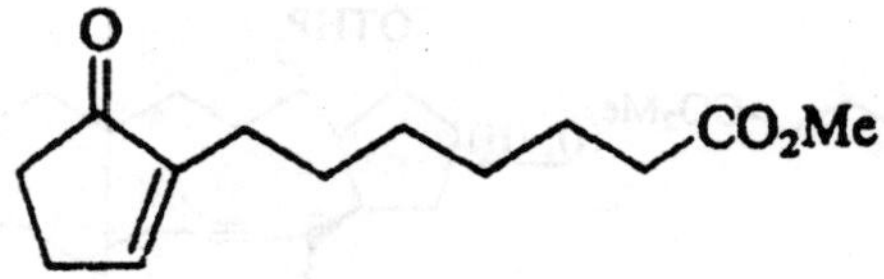

(192)

A few groups have concerned themselves with 11-deoxy-11-hydroxymethyl-PG's. Guzman and Muchowski obtained a mixture (2.2:1) of 11 α- and 11β-hydroxymethyl-PGE_2 methyl esters (193) by the photosensitized addition of methanol to PGA_2 methyl ester. Stereochemical assignments were made *via* reduction to the $PGF_{2\alpha}$ analogues. The supposed 11α-compound (193*a*) was reduced to a mixture of $PGF_{2\alpha}$ $PGF_{2\beta}$ analogues (194), DDQ oxidation of which gave (195), whereas the 11β-compound (193*b*) was reduced exclusively to a 9β-alcohol which on DDQ oxidation gave (196).

The free acids corresponding to (193) have been prepared by Bundy by the same method from PGA_2 during work on endoperoxide analogues. The 4:1 mixture of 11α- and 11β-isomers was separated by chromatography and the α-isomer selectively reduced to the optically active 11-deoxy-11-hydroxymethyl-$PGF_{2\alpha}$ (197) with lithium perhydro-9*b*-boraphenalyl hydride.

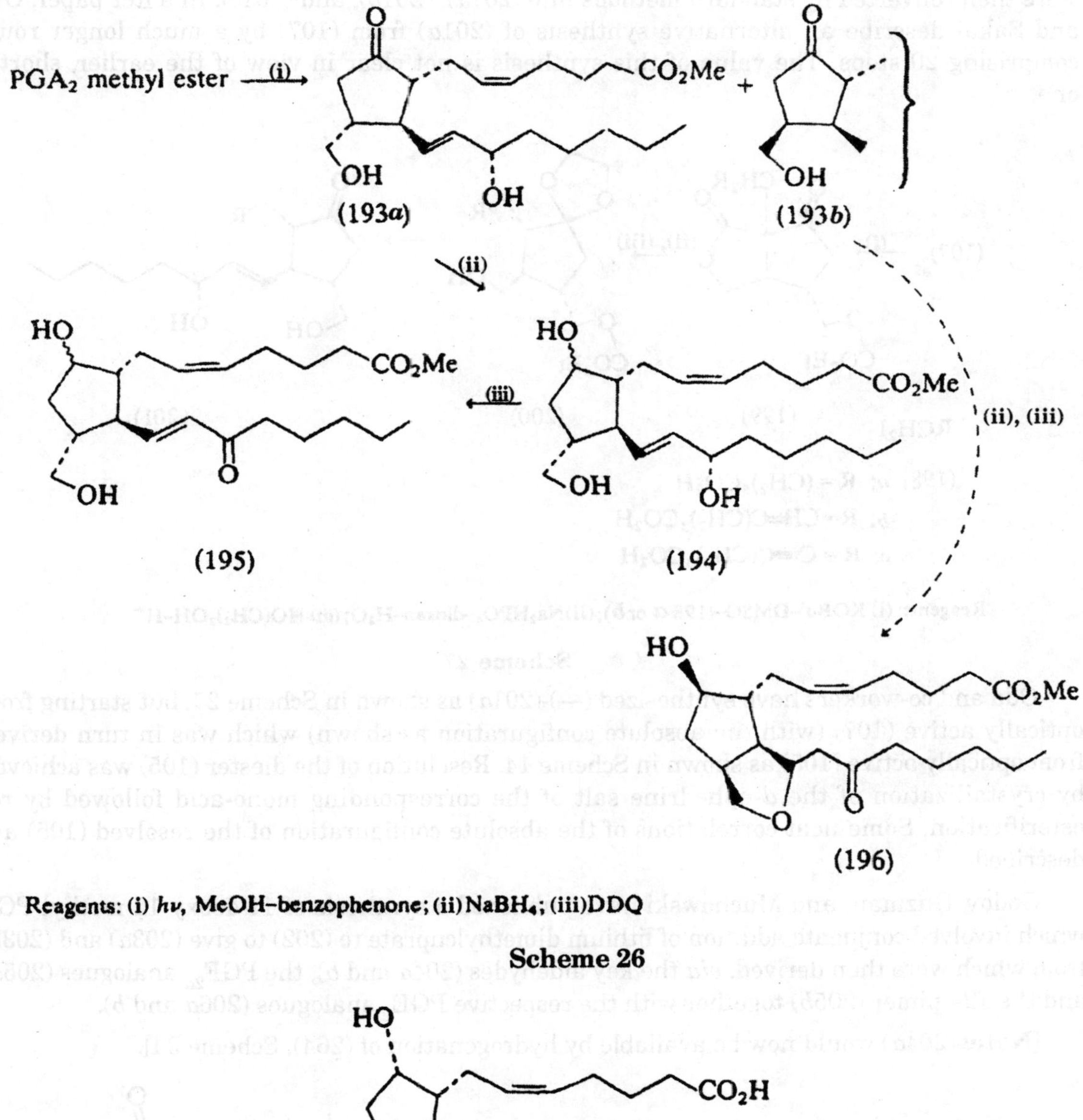

PGA_2 methyl ester —(i)→ (193*a*) + (193*b*)

(193*a*) —(ii)→ (194) —(iii)→ (195)

(193*b*) - - (ii), (iii) - -→ (196)

Reagents: (i) *hν*, MeOH–benzophenone; (ii) $NaBH_4$; (iii) DDQ

Scheme 26

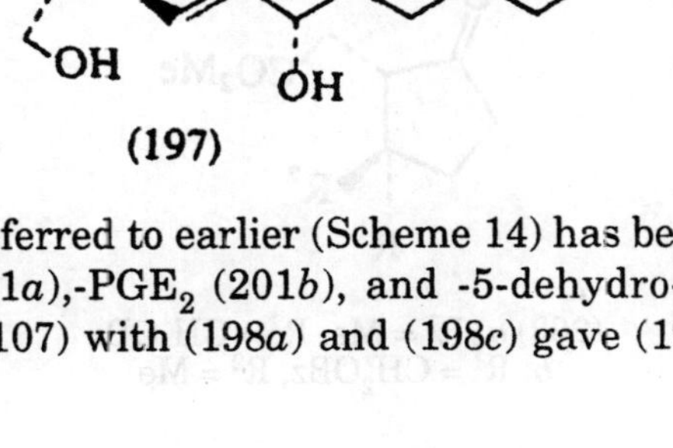

(197)

The intermediate (107) referred to earlier (Scheme 14) has been transformed to 11-deoxy-11α-hydroxymetyl-PGE_1 (201*a*),-PGE_2 (201*b*), and -5-dehydro-PGE_2 (201*c*) (Scheme 27). Stereospecific alkylation of (107) with (198*a*) and (198*c*) gave (199*a*) and (199*c*) which were then converted into (200*a*) and (200*c*). Hydrogenation of (200*c*) gave (200*b*). (200*a*, *b*, and *c*) were then converted by standard methods into (201*a*), (201*b*), and (201*c*). In a lter paper, Oda and Sakai describe an alternative synthesis of (201*a*) from (107) by a much longer route comprising 20 steps. The value of this synthesis is not clear in view of the earlier, shorter one.

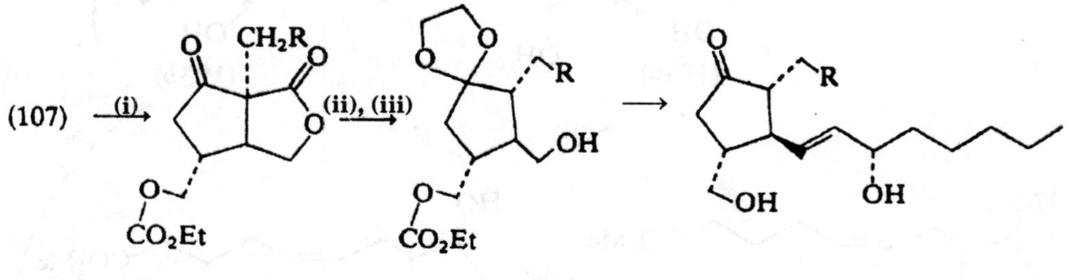

(107) →(i) (199) →(ii), (iii) (200) → (201)

RCH_2I

(198) *a*; $R = (CH_2)_5CO_2H$
b; $R = CH{=}C(CH_2)_3CO_2H$
c; $R = C{\equiv}C(CH_2)_3CO_2H$

Reagents: (i) $KOBu^t$–DMSO–(198 *a* or *b*); (ii) Na_2HPO_4–dioxan–H_2O; (iii) $HO(CH_2)_2OH–H^+$

Scheme 27

Oda and co-workers have synthesized (—)-(201*a*) as shown in Scheme 27, but starting from optically active (107) (with the absolute configuration as shown) which was in turn derived from optically active (105) as shown in Scheme 14. Resolution of the diester (105) was achieved by crystallization of the *d*-ephedrine salt of the corresponding mono-acid followed by re-esterification. Some neat correlations of the absolute configuration of the resolved (105) are described.

Godoy, Guzman, and Muchowski have published a synthesis of 11-deoxy-12-methyl-PG's which involved conjugate addition of lithium dimethylcuprate to (202) to give (203*a*) and (203*b*) from which were then derived, *via* the key aldehydes (204*a* and *b*), the $PGF_{2\alpha}$ analogues (205*a*) and the 12-epimer (205*b*) together with the respective PGE_2 analogues (206*a* and *b*).

[**Note:** (204*a*) would now be available by hydrogenation of (264), Scheme 34].

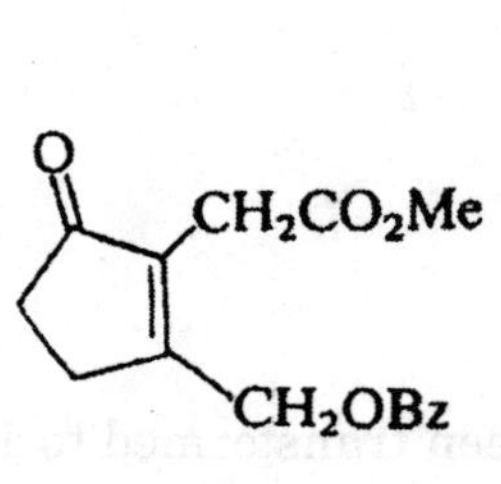

(202)

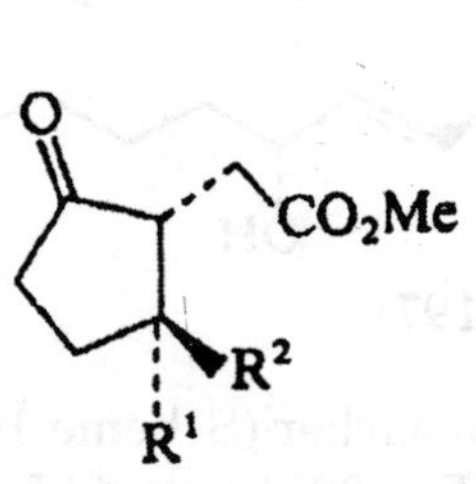

(203) *a*; $R^1 = Me$, $R^2 = CH_2OBz$
b; $R^1 = CH_2OBz$, $R^2 = Me$

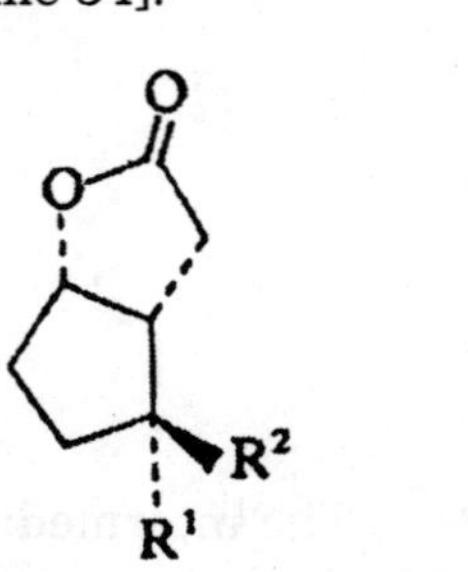

(204) *a*; $R^1 = Me$, $R^2 = CHO$
b; $R^1 = CHO$, $R^2 = Me$

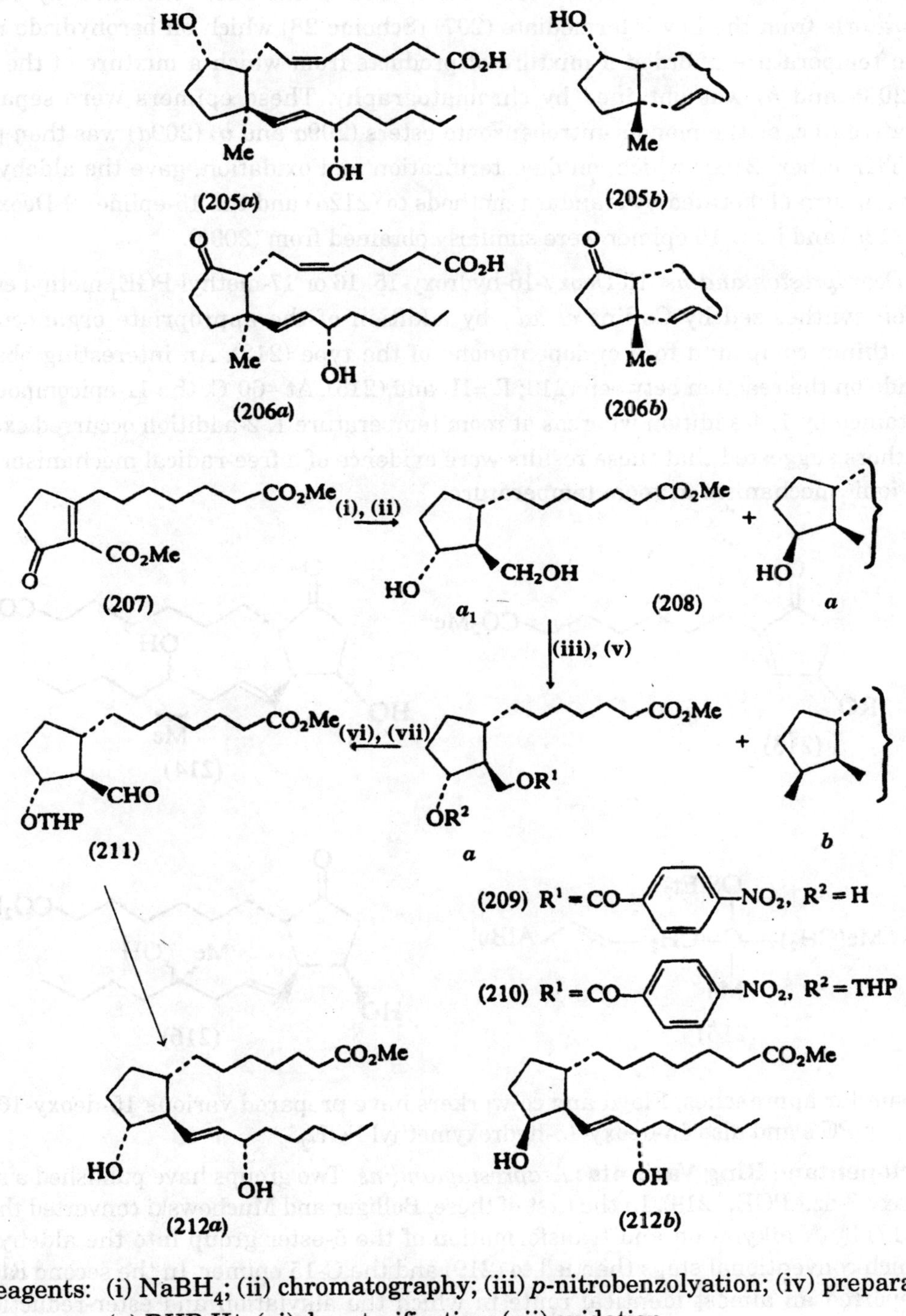

(205*a*) (205*b*)

(206*a*) (206*b*)

(207) (208) (209) $R^1 = CO-C_6H_4-NO_2$, $R^2 = H$

(210) $R^1 = CO-C_6H_4-NO_2$, $R^2 = THP$

(211) (212*a*) (212*b*)

Reagents: (i) $NaBH_4$; (ii) chromatography; (iii) *p*-nitrobenzolyation; (iv) preparative t.l.c.; (v) DHP-H^+, (vi) de-esterification; (vii) oxidation.

Scheme 28

9-Deoxyprostaglandins. (±)-9-Deoxy-PGE_1 (212*a*) has been obtained by Finch and collaboratorls from the key intermediate (207) (Scheme 28) which on borohydride reduction at room temperature afforded a mixture of products from which a mixture of the required diols (208*a* and *b*) was obtained by chromatography. These epimers were separated by preparative t.l.c. as the mono-*n*-nitrobenzoate esters (209*a* and *b*) (209*a*) was then protected as the THP ether (210*a*) which, on de-esterification and oxidation, gave the aldehyde (211); this was in turn elaborated by standard methods to (212*a*) and the 15-epimer 9-Deoxy-11-epi-PGE_1 (212*b*) and its C-15 epimer were similarly obtained from (209*b*).

15-Deoxyprostaglandins. 15-Deoxy-16-hydroxy-15, 16 or 17-methyl-PGE_1 methyl ester (214) has been synthesized by Collins *et. al.*, by addition of the appropriate organocuprate or organolithium compound to a cyclopentenone of the type (213). An interesting observation was made on the reaction between (213; R =H) and (215). At –60°C, the 11-epicompound (216) was obtained by 1, 4-addition whereas at room temperature 1, 2-addition occurred exclusively. The authors suggested that these results were evidence of a free-radical mechanism at –60°C and an ionic mechanism at room temperature.

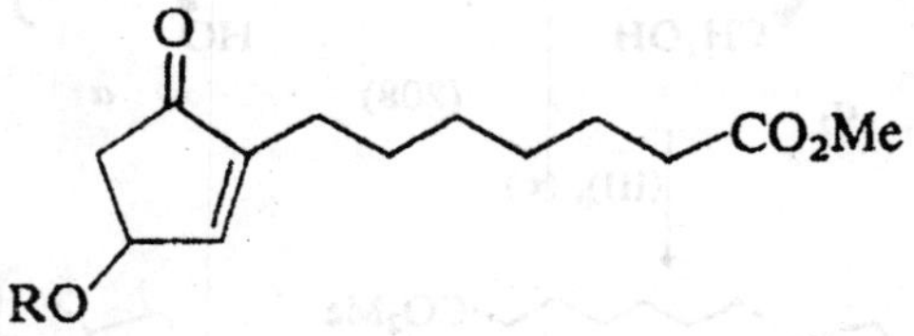

O, CO_2Me, RO

(213)

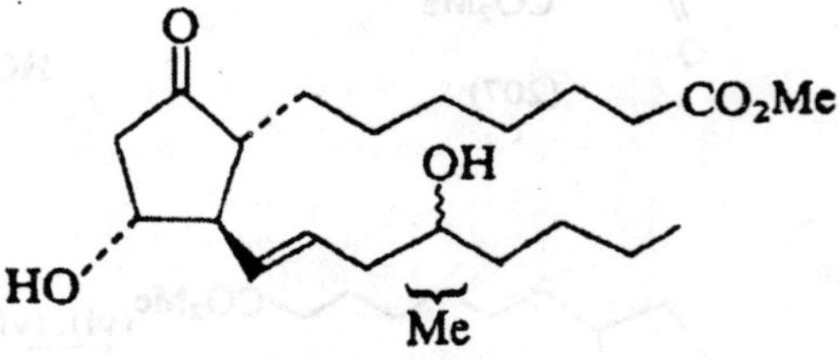

O, CO_2Me, OH, HO, Me

(214)

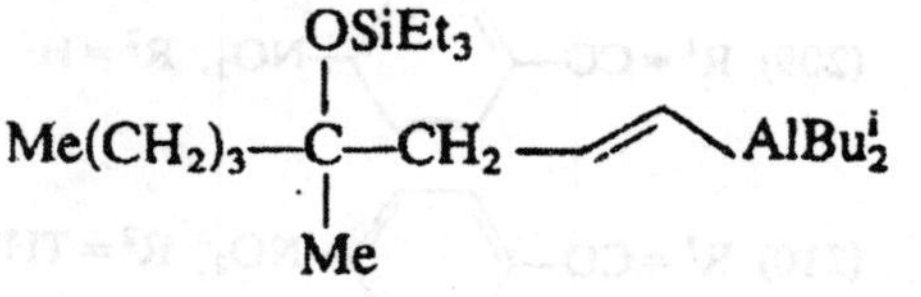

OSiEt$_3$

$Me(CH_2)_3$—C—CH_2—AlBu$^i_2$

Me

(215)

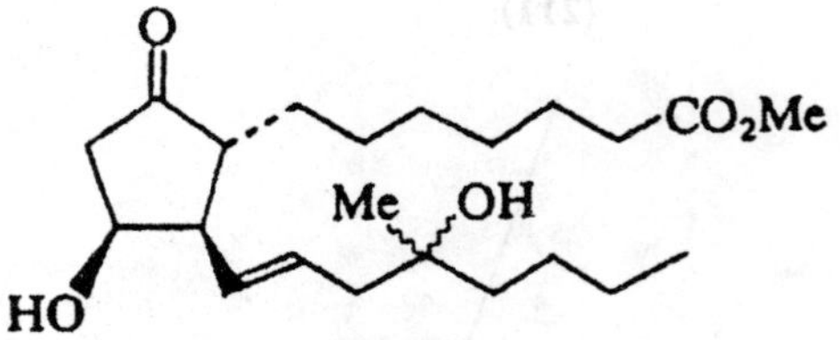

O, CO_2Me, Me, OH, HO

(216)

By similar approaches, Floyd and co-workers have prepared various 15-deoxy-16,17- and 20-hydroxy-PG's and also 15-deoxy-15-hydroxymethyl PGE_2.

Cyclopentane Ring Variants : *Azaprostaglandins.* Two groups have published a synthesis of 11-deoxy-8-aza-PGE_1 (219). In the first of these, Bolliger and Muchowski converted the known ester (217) by *N*-alkylation and transformation of the 5-ester group into the aldehyde (218) from which conventional steps then led to (219) and the C-15 epimer. In the second case Bruin *et al.* reported an almost identical route in which the alkylation and ester-reduction steps were reversed. This paper also described the preparation of 11-deoxy-8-aza-PGE_2 (221) *via* hydrogenation of the acetylenic ester (220).

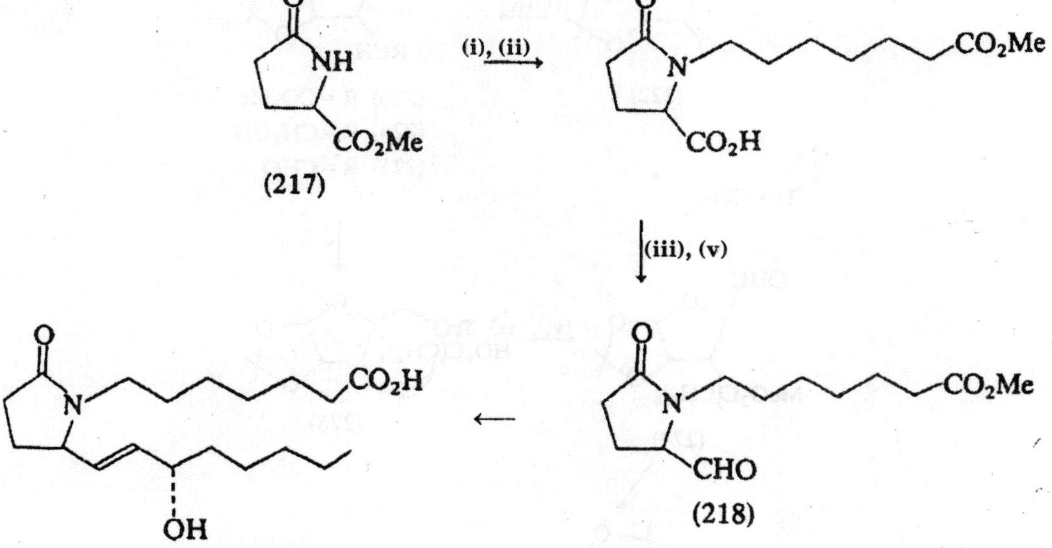

(217)

(i), (ii)

(iii), (v)

(218)

(219)

Reagents: (i) $Br(CH_2)_6CH_2Me$, NaH-DMF-NaI; (ii) selective hydrolysis; (iii) Et_3N-$ClCO_2Me$-THF; (iv) $NaBH_4$; (v) Collins oxidation

Scheme 29

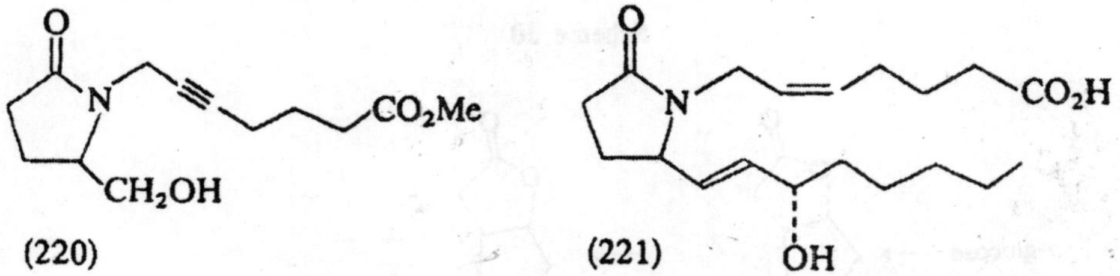

(220) (221)

Oxaprostaglandins. 11-Deoxy-l l-oxa-$PGF_{2\alpha}$(232) was reported last year and its synthesis from 1,4-anhydro-D-glucitol was described. In a similar approach, Lourens and Koekemoer have derived some 11-oxa-analogues, including (232), with the same absolute configuration as the natural PG's (222), obtained from D-xylose, was converted *via* the ester (223), the alcohol (224), and the aldehyde (225) into (226). Further transformations led to the aldehyde (227) which by standard steps gave the oxa-analogue (228).

In a following paper, (232) was obtained *via* two routes from D-glucose and D-xylose to give respectively (229) and (230) which were both precursors of the key aldehyde (231). (232) and its C-15 epimer, which were separated by preparative t.l.c., were obtained routinely from (231).

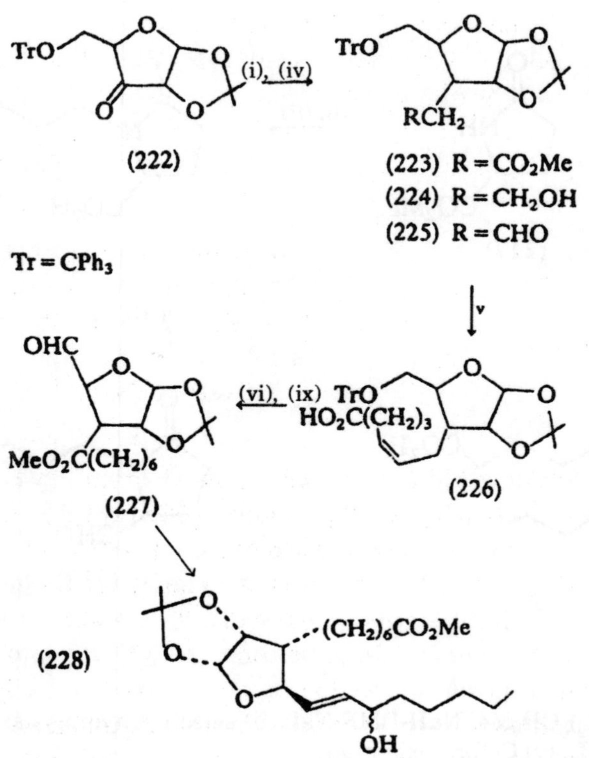
TrO
(i), (iv)
TrO
RCH2
(222)
(223) R = CO2Me
(224) R = CH2OH
(225) R = CHO
Tr = CPh3
v
OHC
(vi), (ix)
TrO
HO2C(CH2)3
MeO2C(CH2)6
(227)
(226)
(CH2)6CO2Me
(228)
OH

Reagents: (i) K salt of trimethyl phosphonoacetate; (ii) Ni–H_2; (iii) $LiAlH_4$; (iv) Collins oxidation; (v) Wittig reaction; (vi) Pt-H_2–AcOH; (vii) –Tr; (viii) CH_2N_2; (ix) Moffat oxidation

Scheme 30

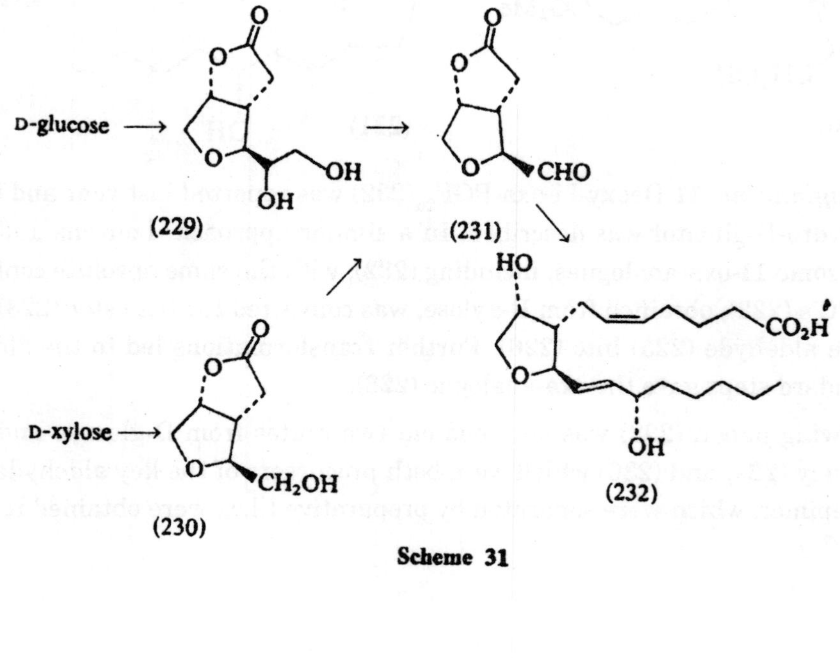
D-glucose
OH
OH
(229)
CHO
(231)
HO
CO2H
D-xylose
CH2OH
(230)
OH
(232)

Scheme 31

Thiaprostaglandins. Harrison and co-workers have synthesized 11-thia-PGE_1(237) and its sulphoxide (238), together with 11-thia $PGF_{1\alpha}$ (239). The hetero-cyclic compound (234) was assembled from (233) and methyl thioglycollate then elaborated *via* (235) to (236). This aldehyde was transformed to (237) and (238) by steps standard in the PG field. (239) was obtained in a similar way through stereospecific reduction of the ketone (235).

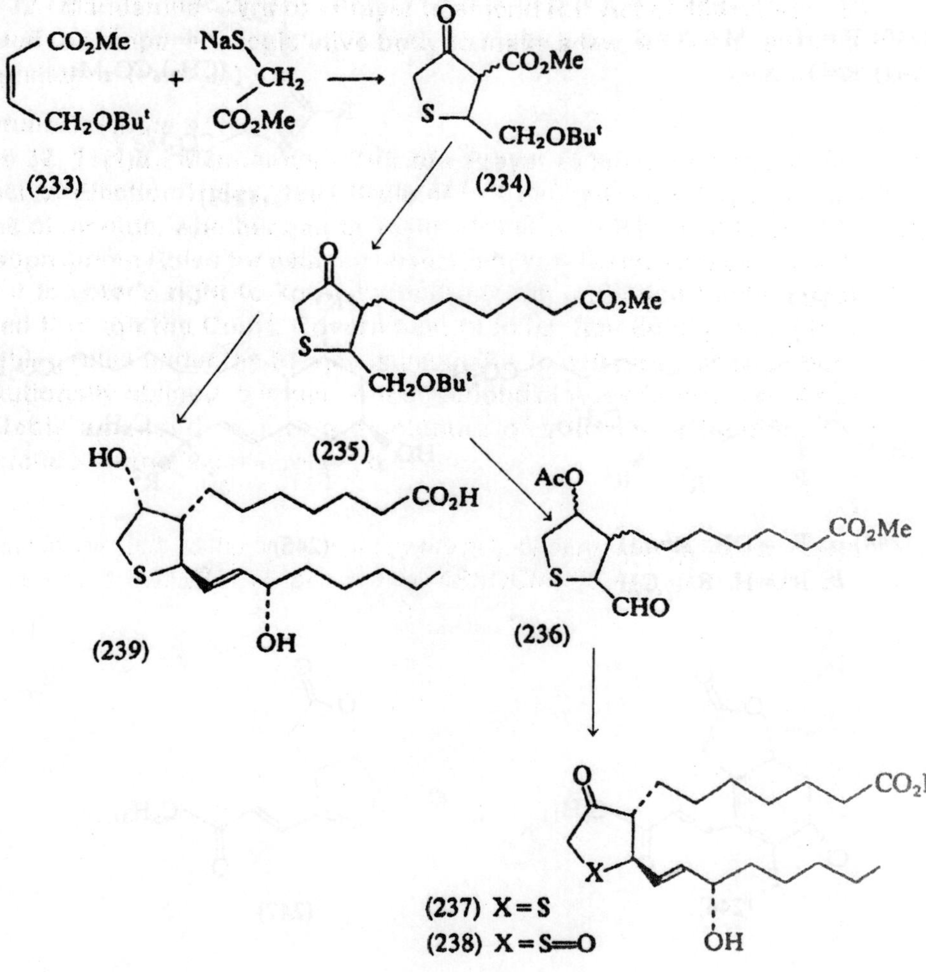

CO_2Me CH_2OBu^t (233) + NaS CH_2 CO_2Me → CO_2Me S CH_2OBu^t (234)

CO_2Me S CH_2OBu^t (235)

HO CO_2H S OH (239)

AcO CO_2Me S CHO (236)

CO_2H X OH

(237) X=S
(238) X=S=O

Scheme 32

Heteroaromatic Prostaglandins. The thiazole and oxazole analogues (240) and (241) have been prepared by Ambrus and Barta. The heterocyclic ring was assembled from (242) and thioformamide, thioacetamide, or formamide to furnish (243), from which (240) and (241) were derived by standard methods.

Cyclohexane Ring Analogues. Muchowski and Verlarde have outlined the preparation of analogues (244) and (245) *via* (246), which was one of several products obtained from the reaction of chlorofluorocarbene on the dienone (247).

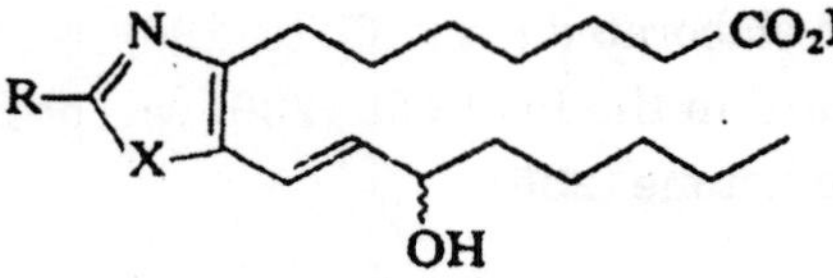

$MeOCO(CH_2)_6COCH(Cl)CO_2Me$

(242)

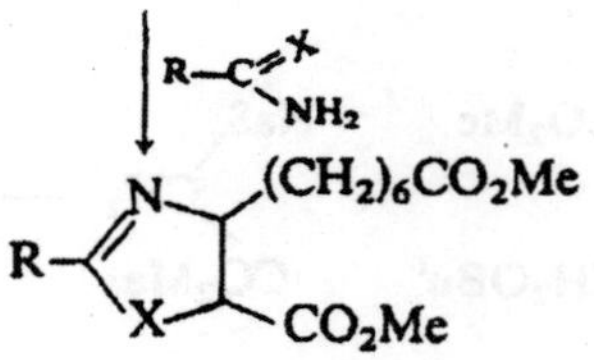

(240) R = H or Me, X = S
(241) R = H, X = O

(243)

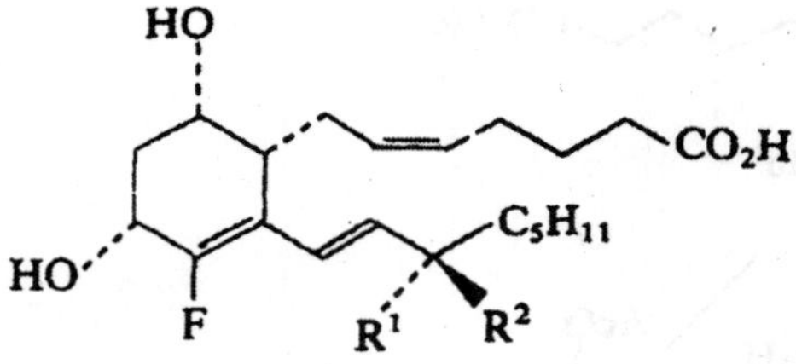

(244) *a*; R^1 = OH, R^2 = H
b; R^1 = H, R^2 = OH

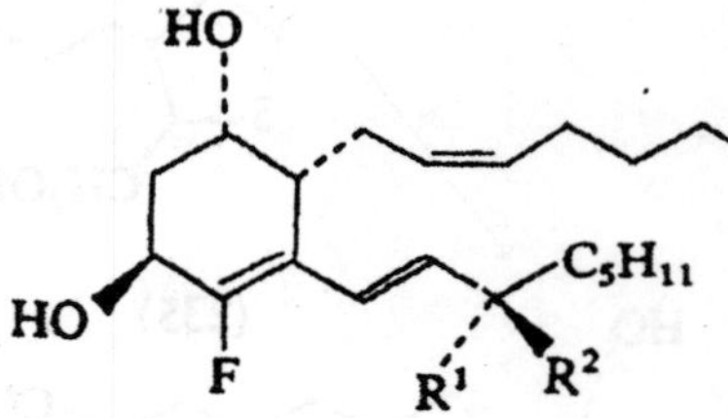

(245)

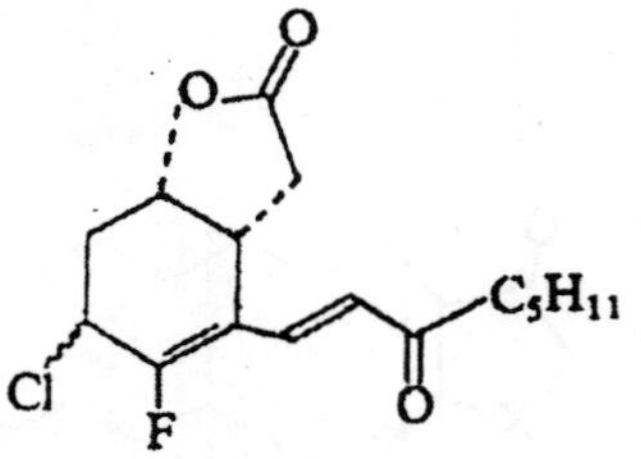

(246)

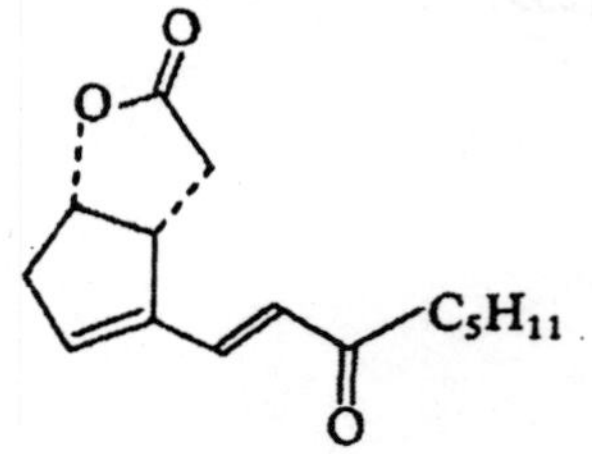

(247)

Cyclobutane Ring Analogues. Guzman and co-workers have published the first synthesis of the cyclobutane PG analogues (252) and (253). The known epoxide (248) was transformed through the azidoketone (249) and the enolic α-diketone (250) to the α-diazo-ketone (251) which gave the epimeric acids (252) and (253) on photolysis. These were separated, and their structures assigned by ^{13}C n.m.r. The α-hydroxymethyl analogues (254) and (255) were obtained by reduction of the mixed anhydrides of (252) and (253).

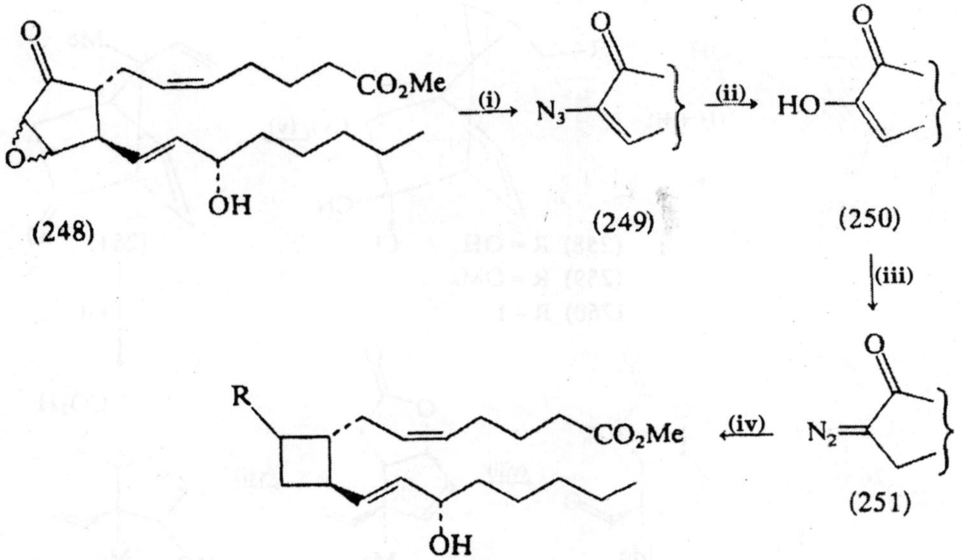

(248) (249) (250) (251)

(252) R = α- CO_2H
(253) R = β- CO_2H
(254) R = α- CH_2OH
(255) R = β- CH_2OH

Reagents: (i) NaN_3–MeOH–H_2O–dioxan; (ii) HjO–MeOH–$(NH_4)_2S$; (iii) toluene-*p*-sulphonyl-hydrazine; (iv) *hv*

Scheme 33

Methyl- and Methylene-prostaglandins : *2-Methylprostaglandins.* In attempts to produce more biologically stable analogues, two groups have generated separate and different syntheses of 12-methyl-PGA_2 (265). To use the original Corey route, the disubstituted cyclopentadiene (256) or equivalent would be required. No syntheses of (256) were available.

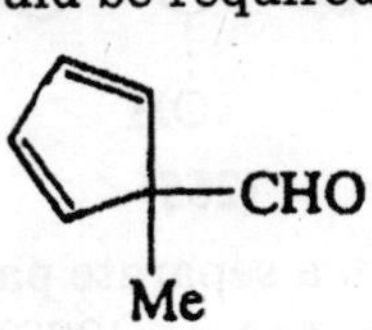

(256)

Corey's group have circumvented this problem in an imaginative way. Reaction of cyclopentadiene with epichlorhydrin afforded the spirocyclic alcohol (257) which gave (258) in the Diels-Alder step. The iodide (260), prepared *via* the mesylate (259), gave the vinyl intermediate (261) following a radical deiodination and subsequent hydrolysis. Baeyer-Villiger oxidation then yielded (262), which was in turn lactonized to (263). Selective cleavage of the vinyl group gave (264) from which (265) and its C-15 epimer were prepared by standard methods.

Grieco and collaborators arrived at the key aldehyde (264) in a different way. Stereospecific alkylation of (266) gave (267) which was transformed in four stages to (268). Baeyer-Villiger oxidation then gave (269) which was lactonized and deprotected to the alcohol (270),

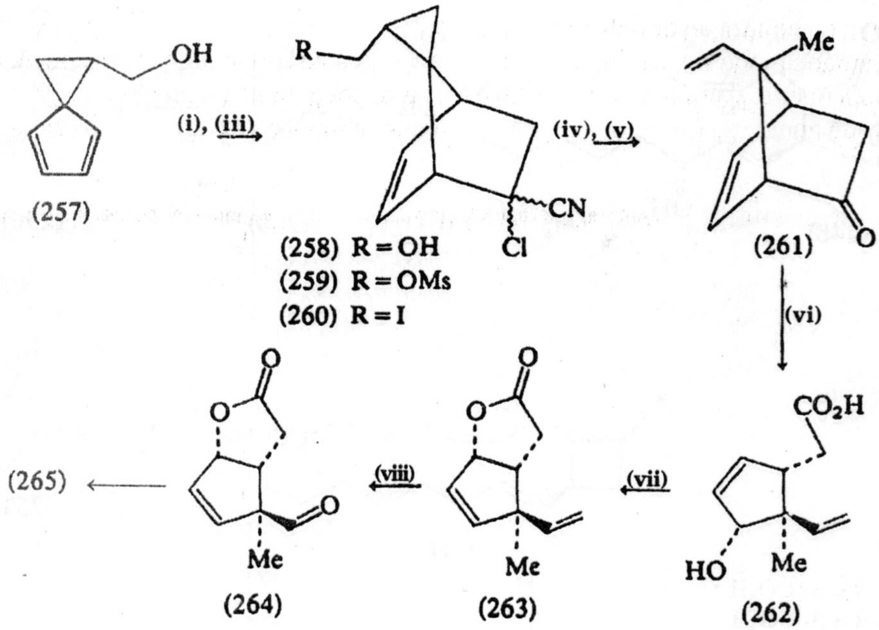

(257) (258) R=OH (259) R=OMs (260) R=I (261) (262) (263) (264) (265)

Reagents: (i) $H_2C{=}C(CN)Cl$; (ii) mesylation; (iii) NaI-Me_2CO; (iv) $Bu_3{}^nSnH$–azobisdi-isobutyronitrile; (v) hydrolysis; (vi) Baeyer–Villiger; (vii) BF_3—ether; (viii) selective vinyl cleavage (IO_4—OsO_4)

Scheme 34

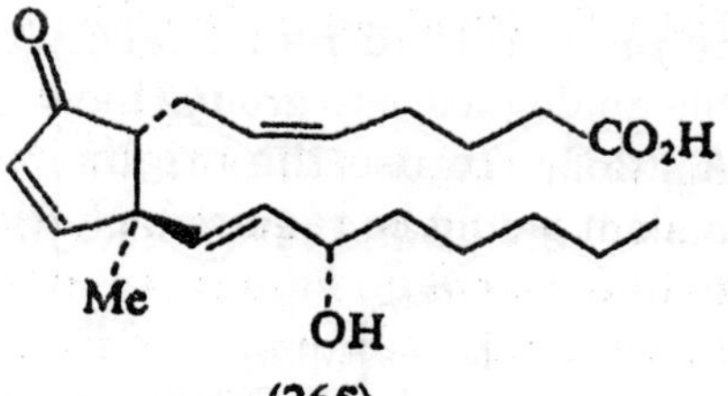

(265)

a precursor of (264) and thence (265). In a separate paper, Grieco *et. al.*, outlined the synthesis of (±)-12-methyl-PGE_2 by the Corey route from (272). Iodolactonization of (269) to (271) and then benzoylation, deprotection, and oxidation of the primary alcohol gave the key stable aldehyde (272)

10, 10-*Dimethylprostaglandins.* The synthesis of 10, 10-dimethyl-PG's in the F_1 and E_1 and series has been reported by Plantema and collaborators from the key intermediate (274), which was obtained in six steps from (273). Catalytic hydrogenation of (274) gave (275*a*) and (275*b*). (275*a*) was further reduced to (276) and (277) (presumed 8, 12-*trans*) which were transformed by standard procedures to the ethyl esters of 10, 10-dimethyl-$PGF_{1\alpha}$ (278) and -$PGF_{1\beta}$ (279). (275*b*) similarly gave the C-11 epimer (279*b*). The corresponding PGE_1's were obtained from (275*a*). Protection of the carbonyl group as the thiophenoxymethyloxime and the hydroxy-group as the THP ether furnished (280) which was transformed to (281) in several steps. Separation of the C-15 epimer and stepwise deprotection via the acetoxymethyloxime

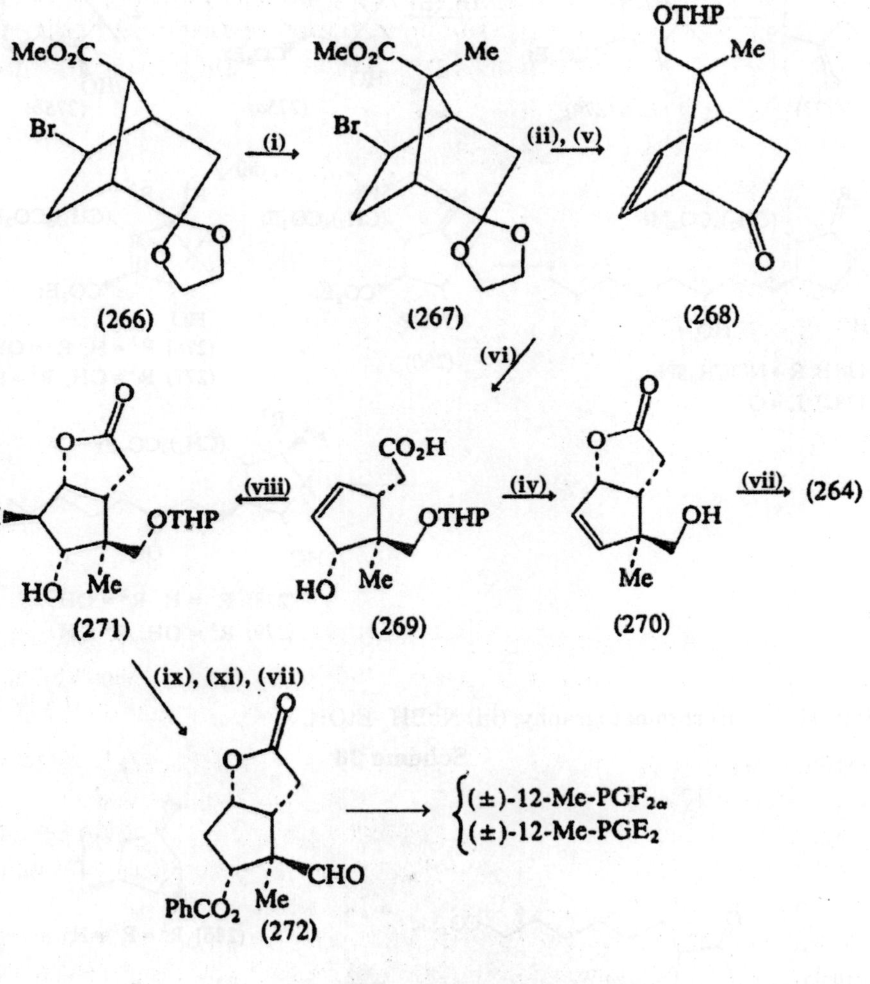

Reagents: (i) $LiNPr_2^i$ – THF – MeI, – 78°C; (ii) $LiAlH_4$; (iii) 1, 5 – diazobicyclo [5, 4, 0] undec-5 enc, – HBr; (iv) H^+; (v) DHP – H^+; (vi) Baeyer-Villiger; (vii) Collins oxidation; (viii) iodolactonization; (ix) PhCOCl – py; (x) Bu^n_3SnH(-I); (xi) –THP

Scheme 35

and the oxime gave 10,10-dimethyl-PGE_1 (282) and the C-15 epimer as the methyl esters.

10, 11-*Methyleneprostaglandins.* 11-Deoxy-10, 11-methylene-PGE_1's (285) and (286) of unambiguous stereochemistry have been prepared by Guzman and Muchowski utilizing their previously published intermediates (193*a*) and (193*b*). The mesylate (283) derived from (193*a*) on treatment with DBN yielded (284) which on hydrolysis gave (285). (286) was similarly produced from (193*b*).

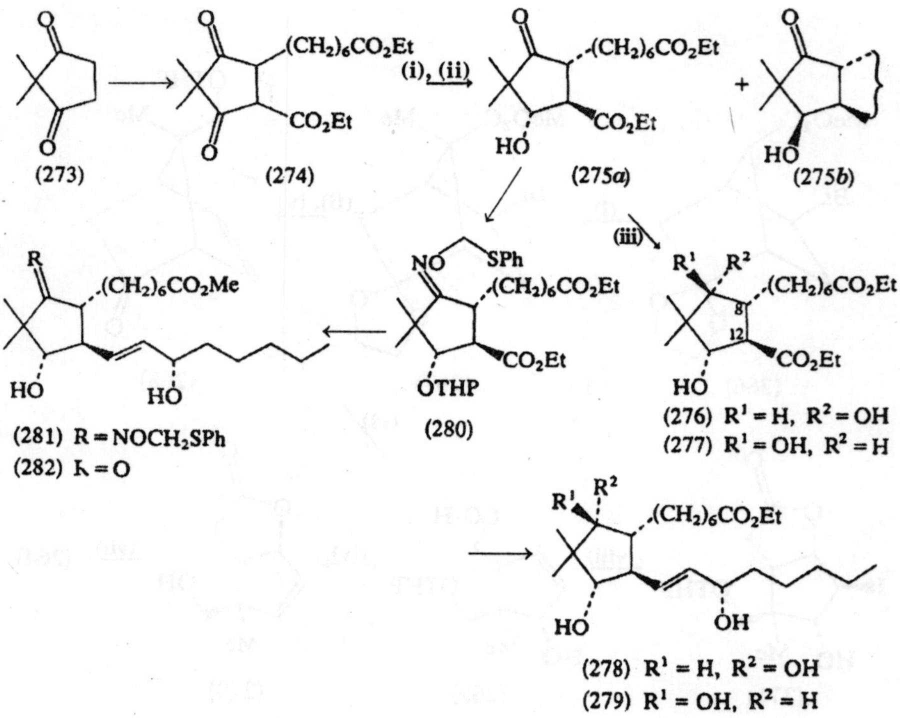

Reagent: (i) H_2-Pt; (ii) chromatography; (iii) $NaBH_4$-EtOH.

Scheme 36

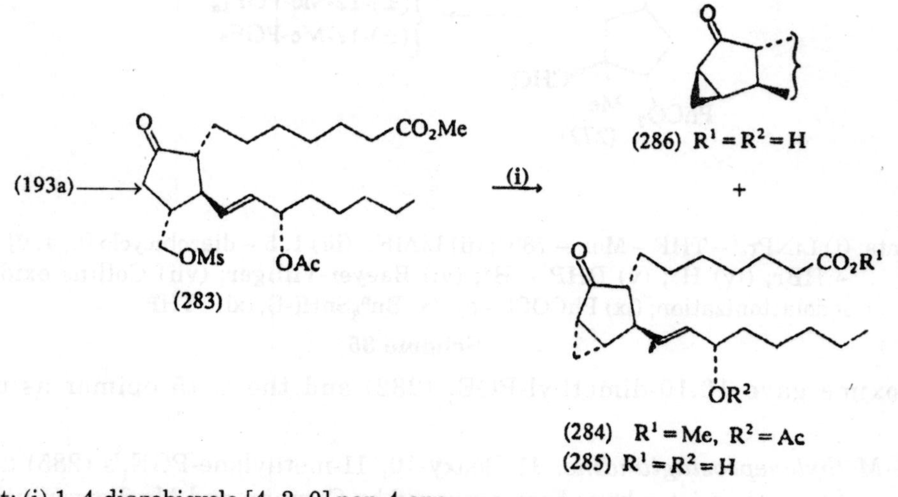

Reagent: (i) 1, 4-diazobicyclo [4, 3, 0] non-4-ene.

Scheme 37

Upper Side-chain Variants. —2-Decarboxy-2-(tetrazol-5-yl)-$PGF_{2\alpha}$, -PGE_2, -$PGF_{2\alpha}$, and -$PGE_{1\alpha}$ have been prepared by Nelson and co-workers by using the appropriate ylide (287) in the standard Wittig step in the Corey sequence. Iguchi *et al.* have similarly prepared 2-carboxy-2-sulphonic acid analogues of PG's $F_{1\alpha}$, E_2, E_1, and A_2 using (288).

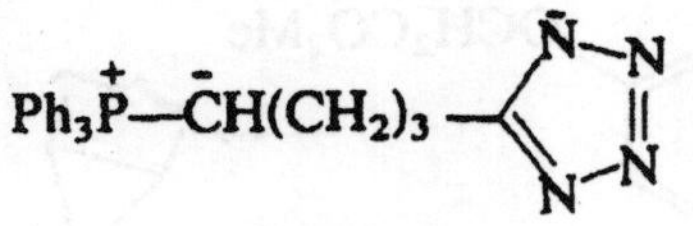

$Ph_3\overset{+}{P}—\overset{-}{C}H(CH_2)_3$—(tetrazolyl)

(287)

$Ph_3\overset{+}{P}—\overset{-}{C}H—(CH_2)_3—S\overset{-}{O}_3$

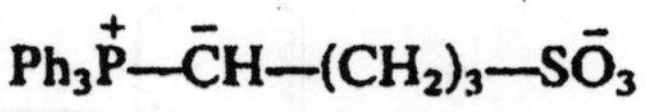

(288)

From studies on the regiospecific acylation of organocopper enolates, Tanaka *et al.* have obtained (±)-7-oxo-PGE$_1$ ethyl ester (289) by acylation of the enolate produced by the addition of (291) to (290) with (292) and deprotection. 11-Deoxy-(289) was obtained in a similar way from cyclopentanone.

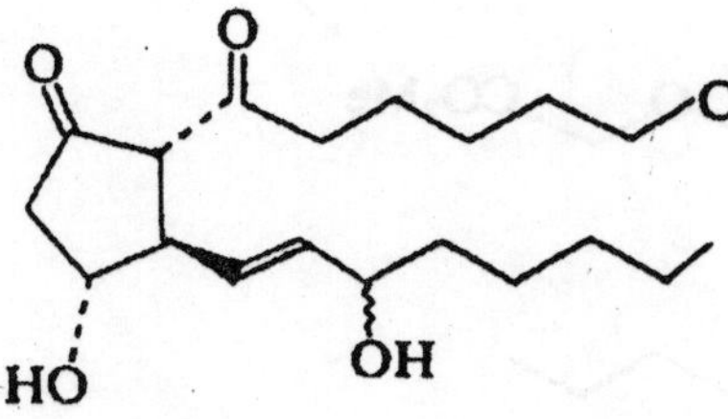

(289)

(290)

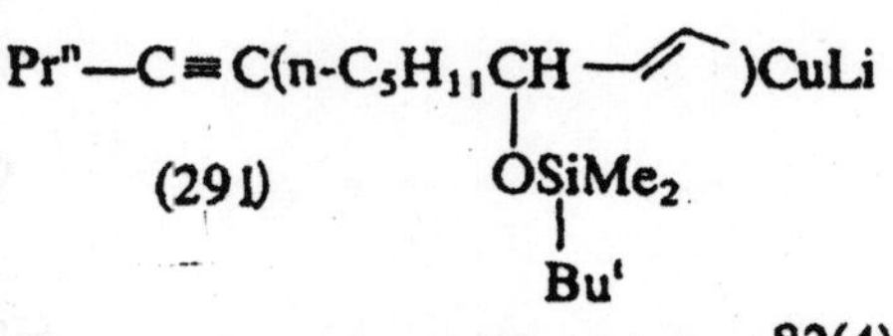

$Pr^n—C{=}C(n\text{-}C_5H_{11}CH(OSiMe_2Bu^t)—CH{=}CH)CuLi$

(291)

82(4)

$EtOCO(CH_2)_5COCl$

(292)

Van Dorp and Christhave prepared several 2-substituted PGE$_1$ analogues (293) by incubation of the appropriate racemic *cis*-8-*cis*-11-*cis*-14-eicosatrienoic acid with a particulate fraction of sheep vesicular glands. Conversions, though all lower than in the case of the natural precursors, were up to 50% for the 2-monosubstituted examples and up to 70% for the disubstituted cases. In two cases studied in an accompanying paper (293: R^1 = Ph, R^2 = H; R^1 = Me, R^2 = H) the enzyme was shown to have produced a slight excess of the *R* over the *S* configuration. Interestingly, because of the sequence rule, this means that the products were antipodal. Thus steric factors alone were not operating. The synthesis of the starting materials were outlined in two adjoining papers.

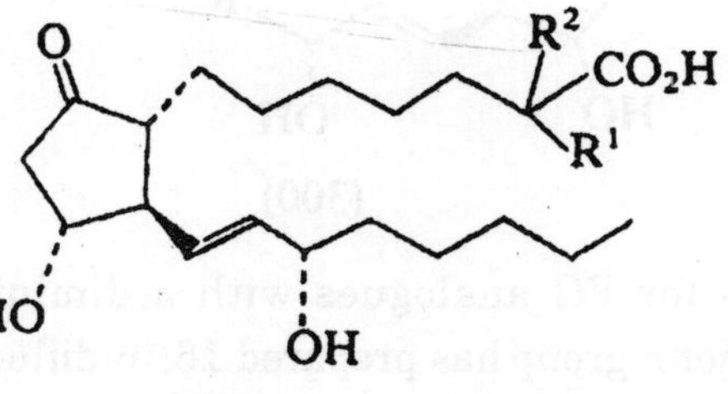

(293)

An Upjohn group has prepared the inter-*m*-phenylene-3-oxa-PGE$_1$(298) and –PGF$_{1\alpha}$ (299) by alkylation of the enamine (294) with (295) to afford (296), which was then transformed by known methods *via* solvolysis of the bis-mesylate (297) to (298) and (299).

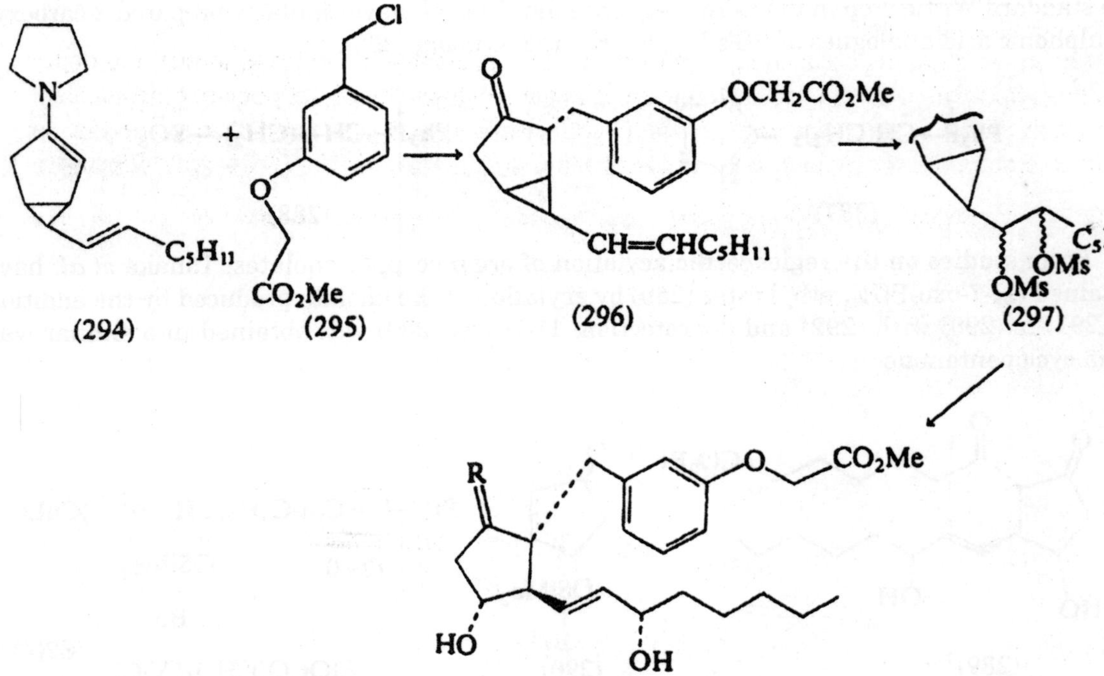

Cl

N

OCH_2CO_2Me

C_5H_{11}

$CH{=}CHC_5H_{11}$

C_5H_{11}

OMs

OMs

CO_2Me

(294) (295) (296) (297)

CO_2Me

HO OH

(298) R = O

(299) R = α-OH,β-H

Scheme 38

Lower Side-chain Variants : In a brief outline of some of the I.C.I, work, Crossley has discussed the activity of and strategy for the synthesis of various analogues, mainly in the F series, of the type (300). A more detailed account of some of this work on alkoxy-analogues (300; R =CH_2CH_2OBu, $CH_2OC_5H_{11}$, *etc.)* has appeared separately.

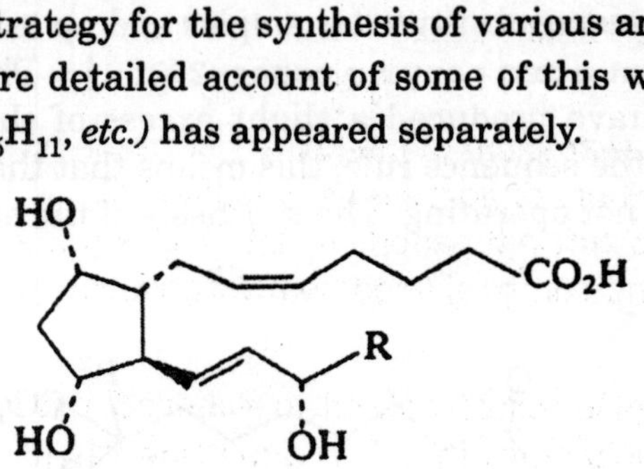

HO

CO_2H

R

HO OH

(300)

In a continuing search for PG analogues with a diminished affinity for the 15-dehydrogenase enzyme, an Upjohn group has prepared 16,16-difluoro-$PGF_{2\alpha}$ (301) and 16-fluoro-$PGF_{2\alpha}$ (302) and -PGE_2 (303) by standard PG synthesis. A further two papers from Upjohn have described the synthesis of 17-(substituted phenyl)-18, 19, 20-trinor-PG's in the 1 and 2 series (304) by both the Corey and Kelly routes.

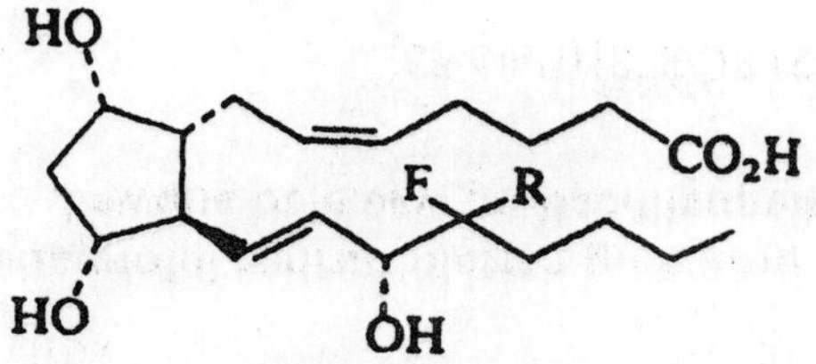

HO, CO_2H, F, R, HO, OH

(301) R = F
(302) R = H

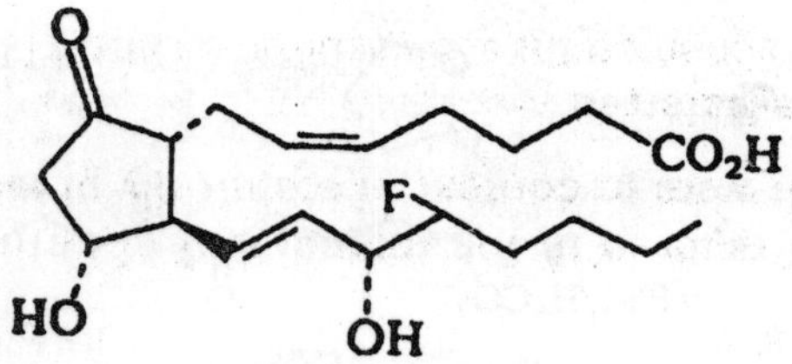

O, CO_2H, F, HO, OH

(303)

R^4, CO_2H, R^1, R^2, HO, OH, R^3

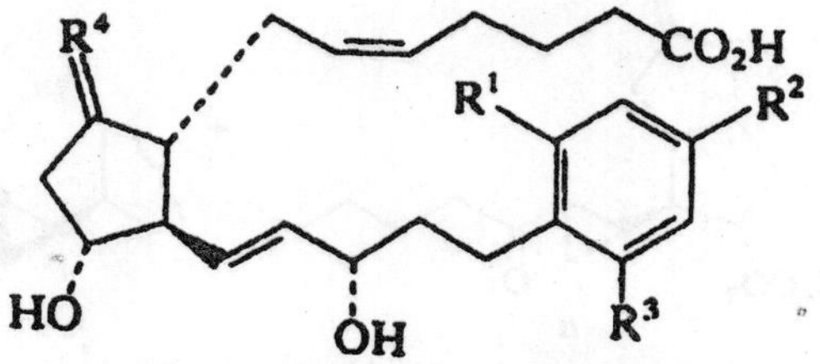

(304) R^4 = O or α-OH,β-H

The 13, 14-methylene-13, 14-dihydro-analogues of $PGF_{2\alpha}$ (308*a* and *b*) and PGE_2 (309*a* and *b*) have been reported by Radiichel and collaborators (Scheme 39). The familiar Corey route enone (305) was converted in one step into a separable mixture (2:1) of isomers (306*a* and *b*), the structures of which were assigned by c.d. on the cor-responding alcohols obtained on de-esterification. Reduction of (306*a*) gave (307*a*) as a C-15 epimer mixture in which the less polar isomer was demonstrated to have the (15S) configuration, as it was identical to one of the products obtained directly (albeit in-efficiently) from the (15S)-enol (310) (307*b*) was likewise obtained from (306*b*). (307*a* and *b*) were separately transformed to (308*a* and *b*) and (309*a* and *b*) by standard procedures.

A group from Miles Laboratories has prepared analogues in which the lower side-chain incorporates the 13,14-double bond and the 15-OH group as a cyclohexenol. Addition of the mixed cuprate (311) to (312), (313), and (314) gave the (±)-11-deoxy-and-PGE_2 analogues (315) and (316) and the PGE_1 analogue (317) respectively.

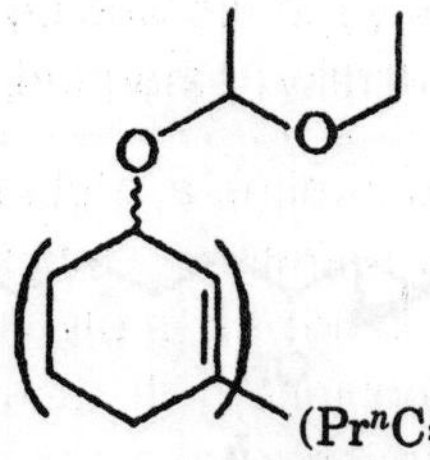

O, O, ($Pr^nC{=}C$)Cu-Li^+

(311)

O, X, CO_2Me, R

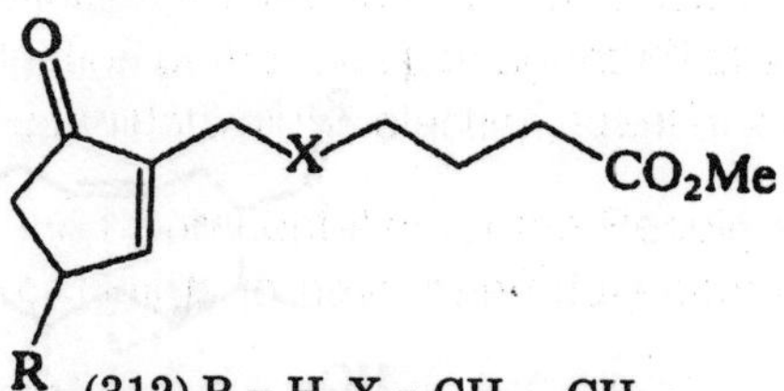

(312) R = H, X = CH_2—CH_2
(313) R = H, X = CH=CH
(314) R = OTHP, X = CH=CH

Endoperoxide Analogues: Bundy has described the elaboration of (197) to the PFH_2 analogue (319) and of PGE_2 to the isomeric compound (323). (319) was obtained on hydrolysis

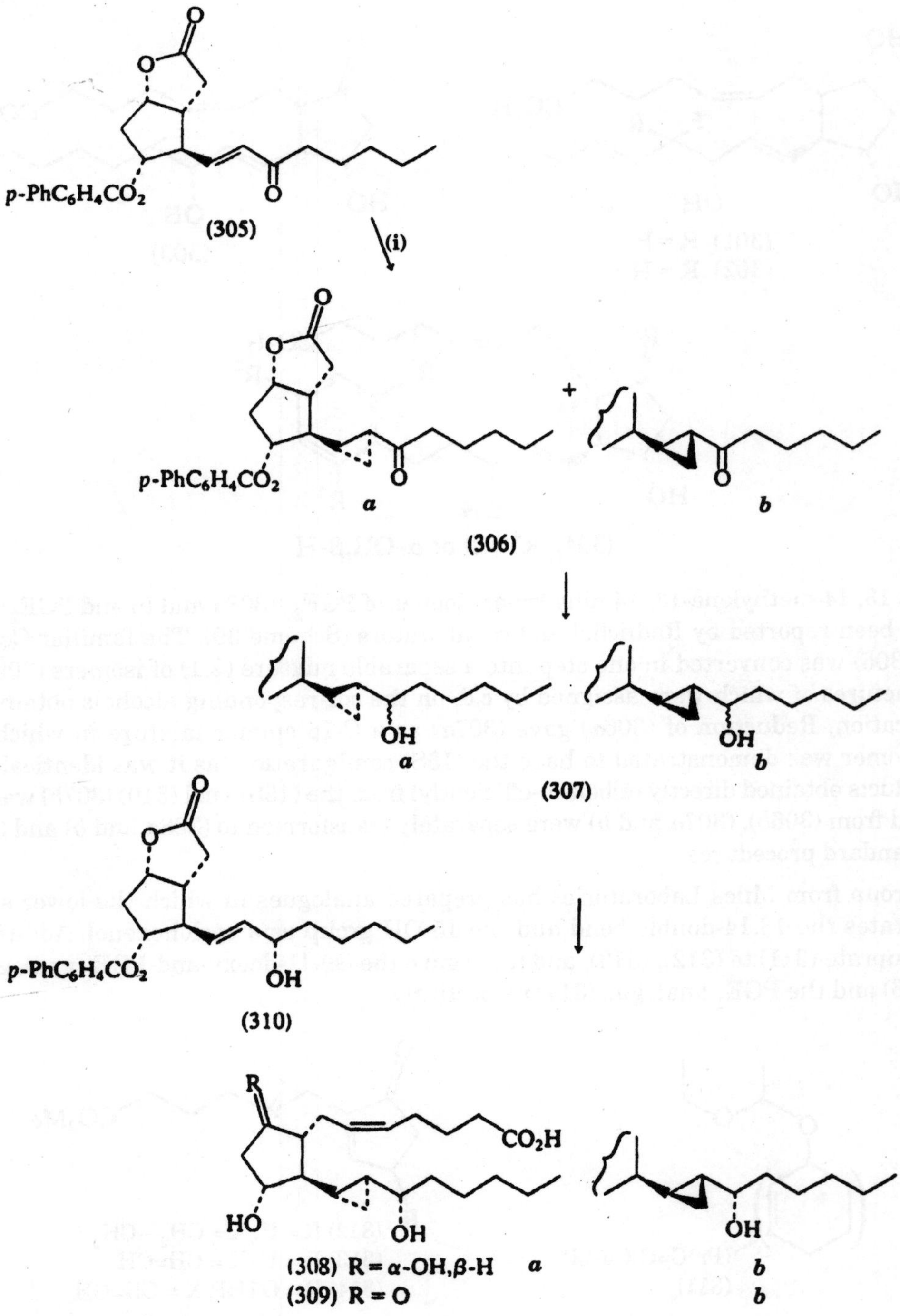

p-PhC$_6$H$_4$CO$_2$

(305)

(i)

p-PhC$_6$H$_4$CO$_2$

a *b*

(306)

OH OH

a *b*

(307)

p-PhC$_6$H$_4$CO$_2$ OH

(310)

R

CO$_2$H

HO OH OH

(308) R = α-OH,β-H *a* *b*

(309) R = O *b*

Reagent: (i) Pd(AcO)$_2$—CH$_2$N$_2$

Scheme 39

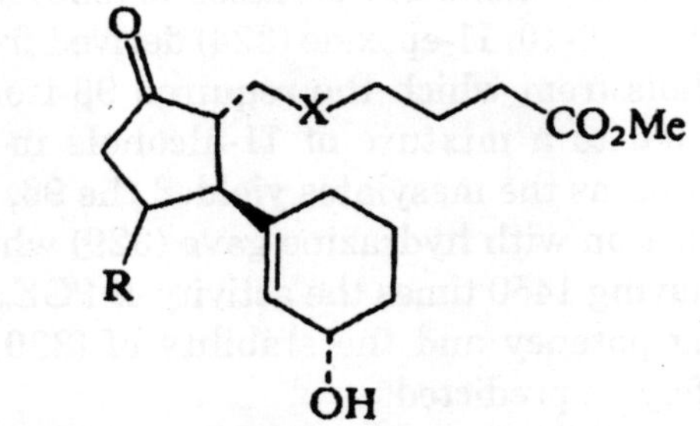

O

X CO$_2$Me

R

OH

(315) R = H, X = CH$_2$—CH$_2$
(316) R = H, X = CH=CH
(317) R = OH, X = CH$_2$—CH$_2$

of the tosylate (318) derived from (197). (323) was prepared from PGE_2 by protection to (320) and elaboration of the carbonyl groun *via* the exocyclic methylene intermediate (321) to the merylate (322) which on deprotection and hydrolysis gave (323).

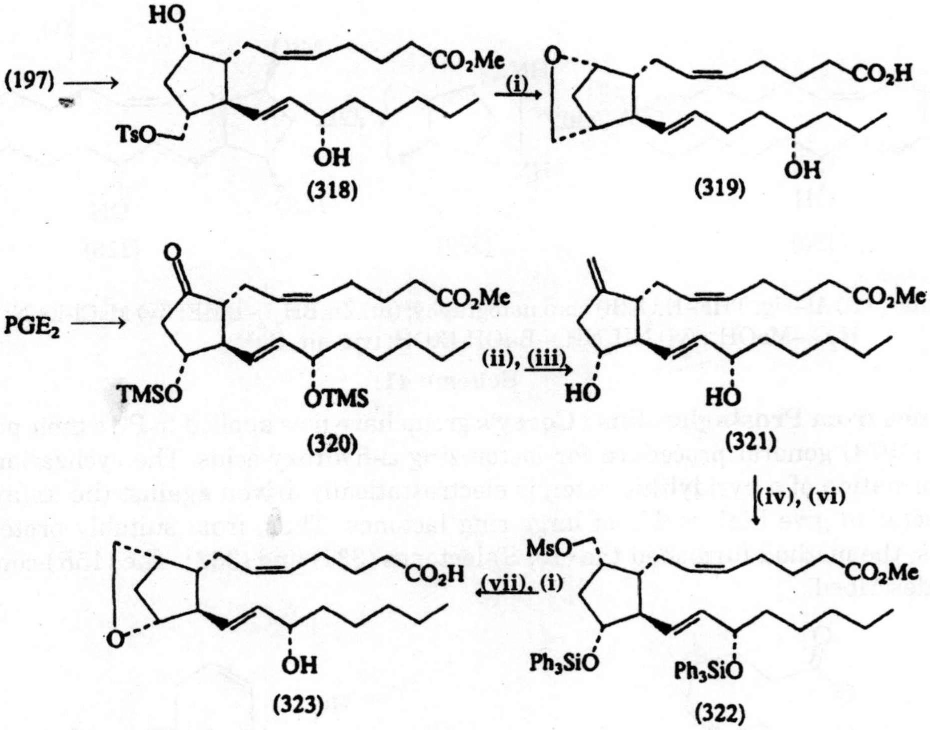

(197) ⟶ HO CO$_2$Me TsO OH (318) —(i)→ O CO$_2$H OH (319)

PGE_2 ⟶ O CO$_2$Me TMSO OTMS (320) —(ii), (iii)→ CO$_2$Me HO HO (321)

(321) —(iv), (vi)→ MsO CO$_2$Me Ph$_3$SiO Ph$_3$SiO (322) —(vii), (i)→ CO$_2$H O OH (323)

Reagents: (i) KOH-H_2O-MeOH; (ii) *N*-methyl sulphonimidoylmethyl-MgBr; (iii) reductive elimination; (iv) OH-protection; (v) selective hydroboration (9-BBN); (vi) mesylation; (vii) –Ph_3Si.

Scheme 40

An intriguing, incredibly potent, stable diaza-analogue (330) has been described by Corey and colleagues. Reduction of the α, β-10, 11-epoxide (324) derived from 15-acetoxy-PGA_2 methyl ester gave a mixture of alcohols from which the required 9β-isomer (325) was obtained by chromatography. Reduction led to a mixture of 11-alcohols in which the β-isomer (326) predominated (90%). Purification as the mesylates yielded the 9β, l lβ-dimesylate (327) which was deacetylatcd to (328). Reaction with hydrazine gave (329) which was oxidized to (330), a stable, crystalline compound having 1450 times the activity of PGE_2 in contracting rabbit aorta. As a result of this staggering potency and the stability of (330), the preparation of more analogues in this area can safely be predicted.

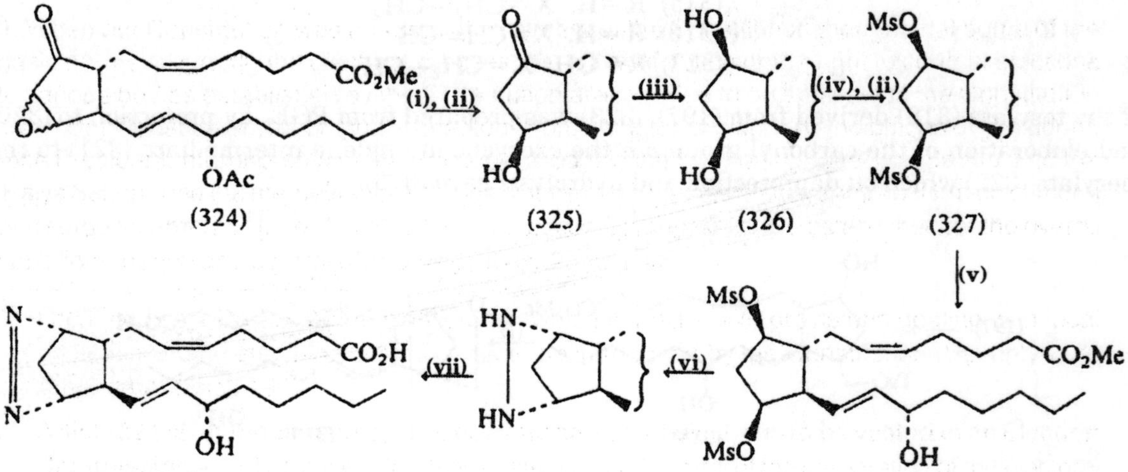

(324) (325) (326) (327)

(330) (329) (328)

Reagents: (i) Al–Hg; THF–H,O; (ii) chromatography; (iii) $Zn(BH_4)_2$–DME; (iv) MsCl-Et_3N; (v) LiOH-H_2O –$MeOH^-$; (vi) NH_2NH_2– BuOH-EtOH; (vii) air, Cu^{2+}

Scheme 41

Lactones from Prostaglandins : Corey's group have now applied to PC's their previously published (1974) general procedure for lactonizing ω-hydroxy-acids. The cyclization, which involves formation of a pyridylthio-ester, is electrostatically driven against the unfavourable entropy factor to give high yields of large-ring lactones. Thus, from suitably protected $F_{2\alpha}$ derivatives, the method furnished the (15, *S*)-lactones (331) and (332). The (15*R*)-compounds were also described.

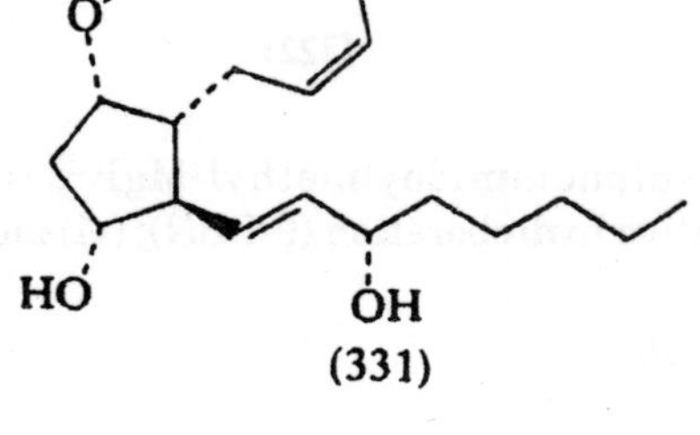

(331) 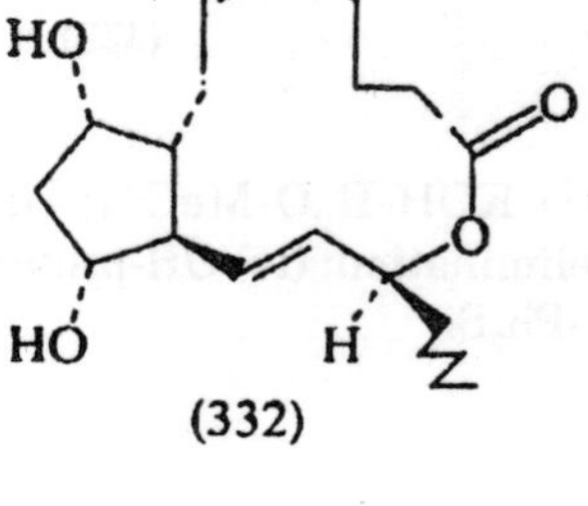 (332)

Prostaglandin Analogues Related to Steroids: Two groups have been attracted by the similarities between certain steriods and PG's in the 'hairpin' configuration. Venton *et al.* compared PG $F_{1\alpha}$ and the steroidal molecule (333) and synthesized several steroid types, two of which, (334) and (335), showed some, though non-specific antagonism towards the smooth muscle activity of PG's. A Merck group, working along similar lines, has prepared enantiomeric tetrahydro-PGA_1 analogues (336) and (337) from steroid starting materials.

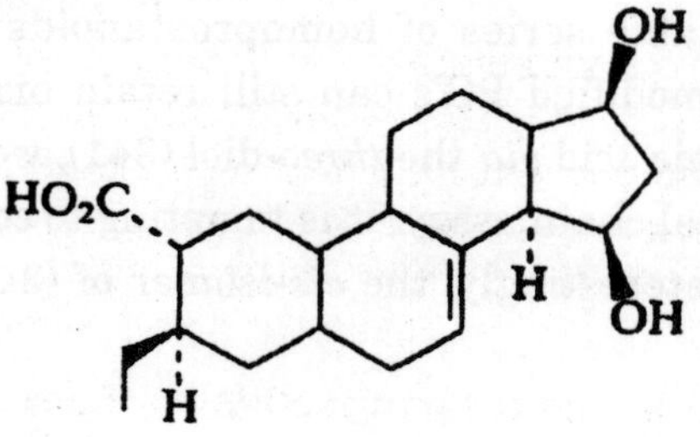

(333)

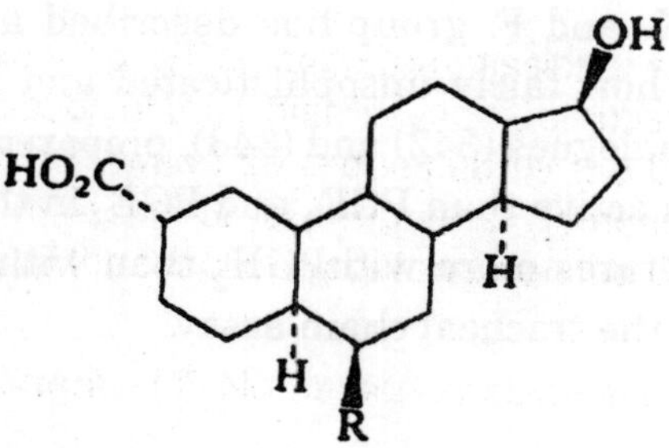

(334) R = H
(335) R = OH

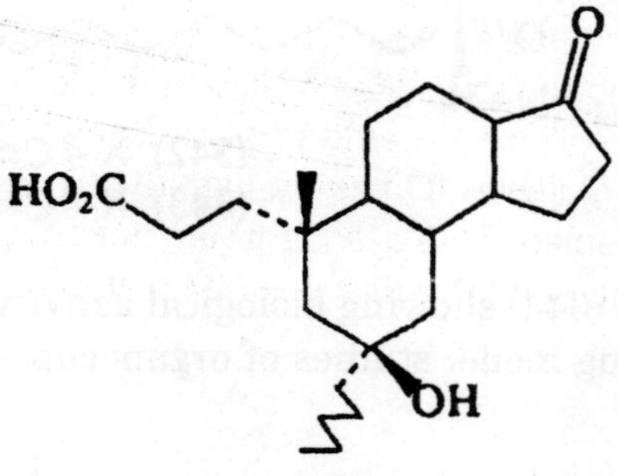

(336)

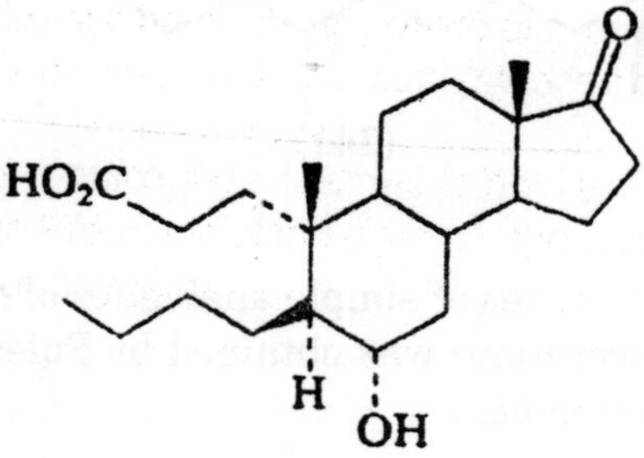

(337)

Simple Prostanoic Acid Analogues: An Italian group has outlined the synthesis of 19,20-dinor-9-hydroxyprostanoic acid and some derivatives. As an example (338), which showed some activity in blood platelet aggregation, was obtained from (339) *via* hydrogenation and borohydride reduction.

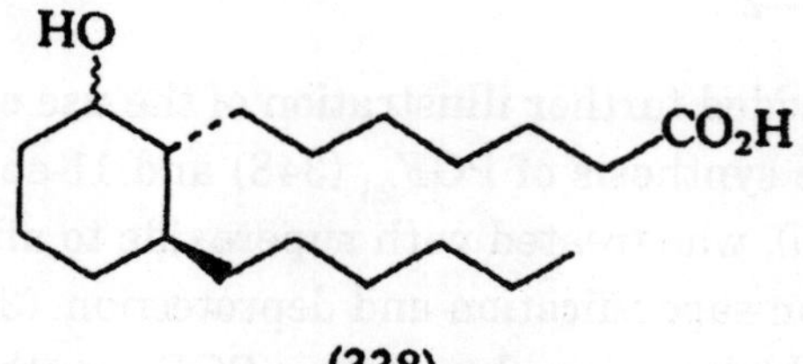

(338)

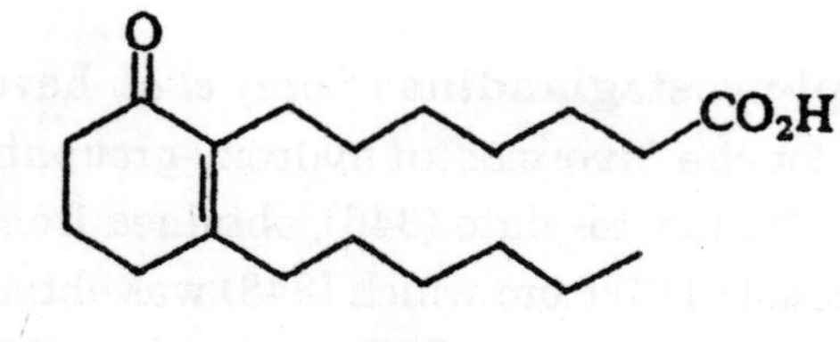

(339)

Hamon *et. al.*, have obtained both optical isomers of prostanoic acid itself from the resolved key aldehyde (340). Racemic (340) was oxidized to the acid which, after resolution as the (+)- or (–)-ephedrine salt, was reduced back to enantiomeric (340).

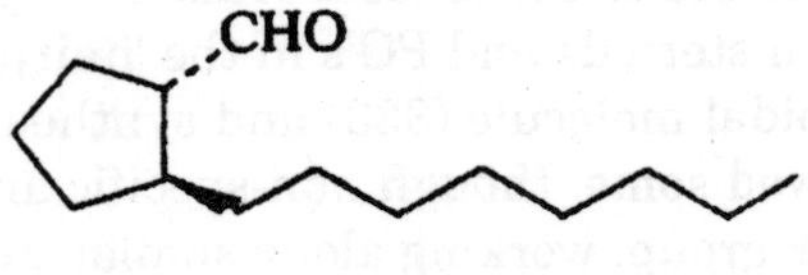

CHO

(340)

An S.K. and F. group has described a remarkable series of homoprostanoids which illustrates how fairly unsophisticated and heavily modified PG's can still retain biological potency. Analogues (342) and (343), prepared from oleic acid *via* the *threo*-diol (341), were 100 times more active than PGE_1 and PGE_2 in the tracheal chain assay. It is tempting to compare these structures more withPGH_2 than withPGE_2. Interestingly, the *cis*-isomer of (343) was inactive in the tracheal chain assay.

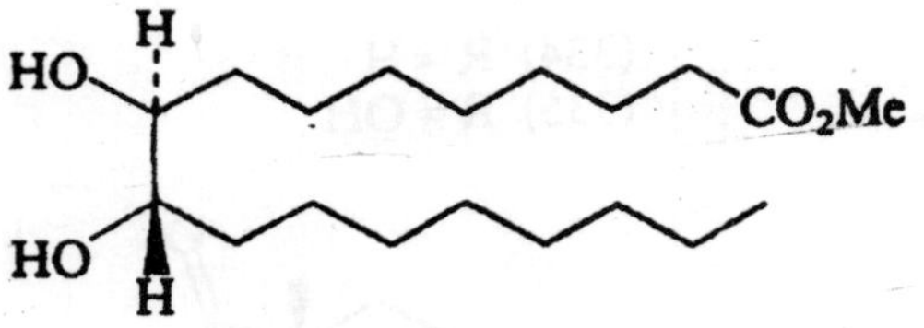

(341)

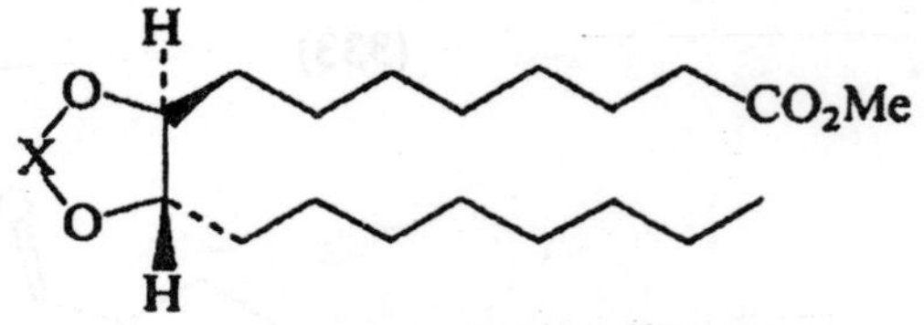

(342) X=C=O
(343) X=C=S

Another series of simple analogues of the type (344) showing biological activity (inhibition of gastric secretion) was obtained by Poletto during model studies of organocuprate addition to cyclopentenone.

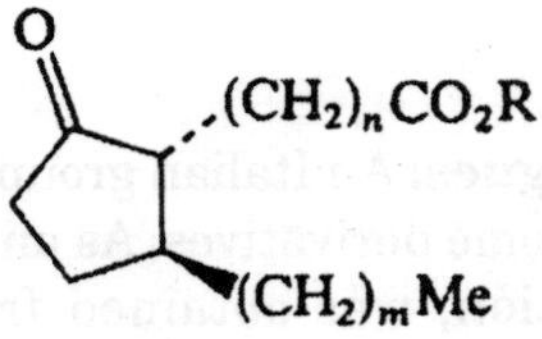

$(CH_2)_nCO_2R$
$(CH_2)_mMe$

(344) $n = 4, 6,$ or 8
$m = 0$—2

Epi-prostaglandins : Corey *et al.* have provided further illustration of the use of superoxide for the inversion of hydroxy-groups by the synthesis of $PGF_{2\beta}$ (348) and 11-epi-$PGF_{2\alpha}$ (351). The key tosylate (346), obtained from (345), was treated with superoxide to afford the 9β-alcohol (347) from which (348) was obtained on saponification and deprotection. (351) was similarly obtained *via* the monotosylate (350), which was produced from $PGF_{2\alpha}$ methyl ester *via* the 9,11-boronate (349).

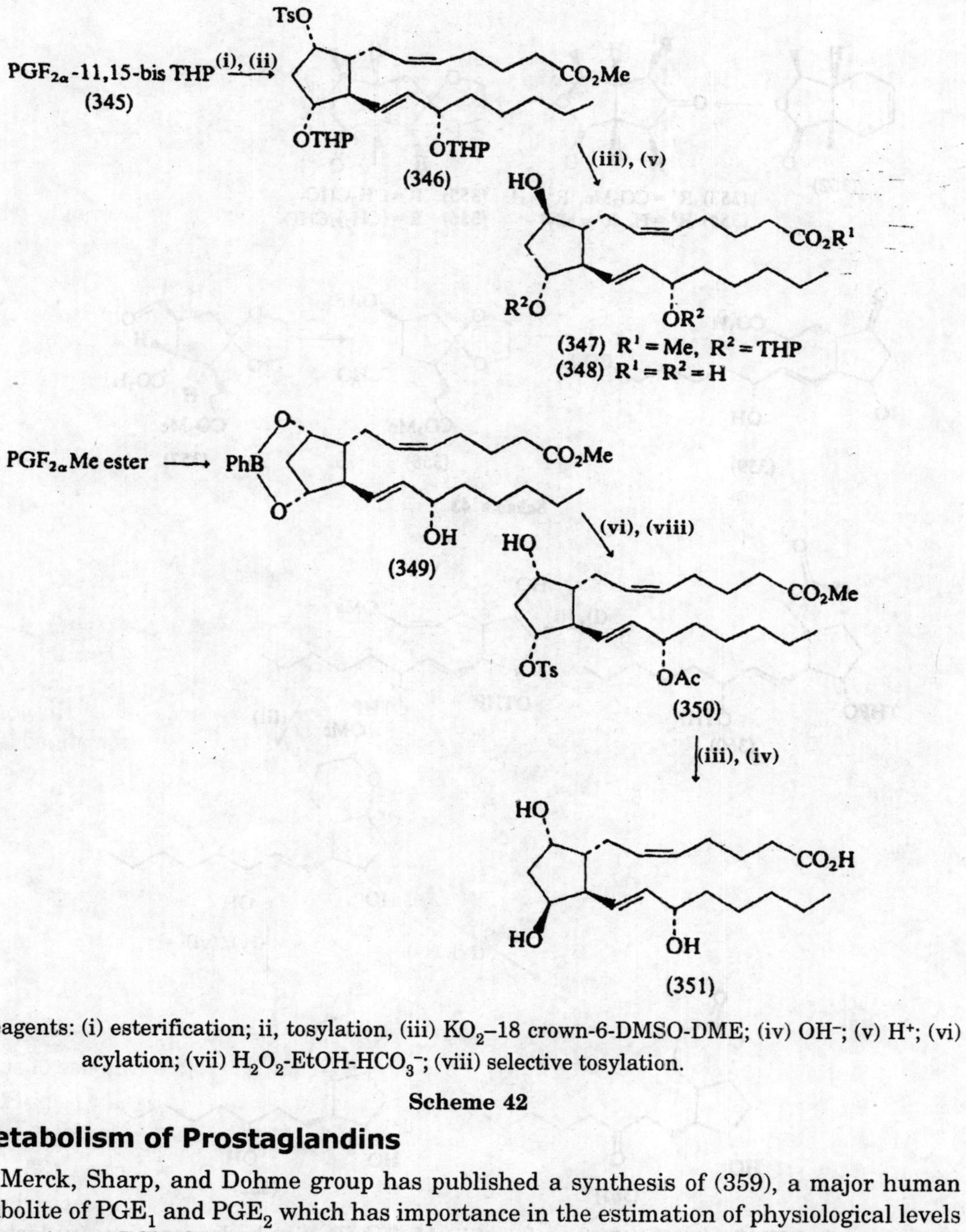

PGF$_{2\alpha}$-11,15-bis THP (345) →(i), (ii) (346) →(iii), (v) (347) $R^1 = Me$, $R^2 = THP$; (348) $R^1 = R^2 = H$

PGF$_{2\alpha}$ Me ester → (349) →(vi), (viii) (350) →(iii), (iv) (351)

Reagents: (i) esterification; ii, tosylation, (iii) KO_2–18 crown-6-DMSO-DME; (iv) OH^-; (v) H^+; (vi) acylation; (vii) H_2O_2-EtOH-HCO_3^-; (viii) selective tosylation.

Scheme 42

5. Metabolism of Prostaglandins

A Merck, Sharp, and Dohme group has published a synthesis of (359), a major human methabolite of PGE_1 and PGE_2 which has importance in the estimation of physiological levels of these primary PG's. Space permits only a bare outline of this interesting synthesis (Scheme 43). Oxidative ring cleavage of (352) and Dieckmann cyclization gave (353), which was subsequently transformed to (359) *via* the key intermediates (354) — (358). Similar metabolites

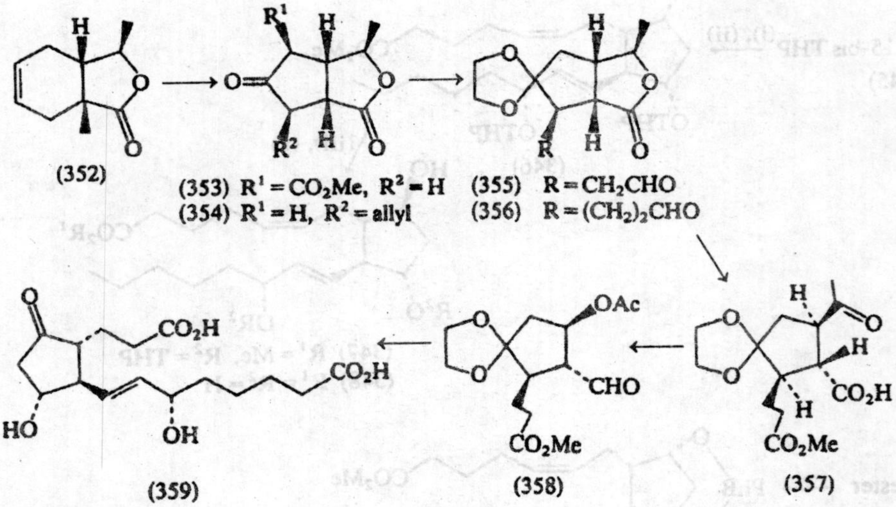

(352)

(353) $R^1 = CO_2Me$, $R^2 = H$
(354) $R^1 = H$, $R^2 = allyl$

(355) $R = CH_2CHO$
(356) $R = (CH_2)_2CHO$

(357)

(358)

(359)

Scheme 43

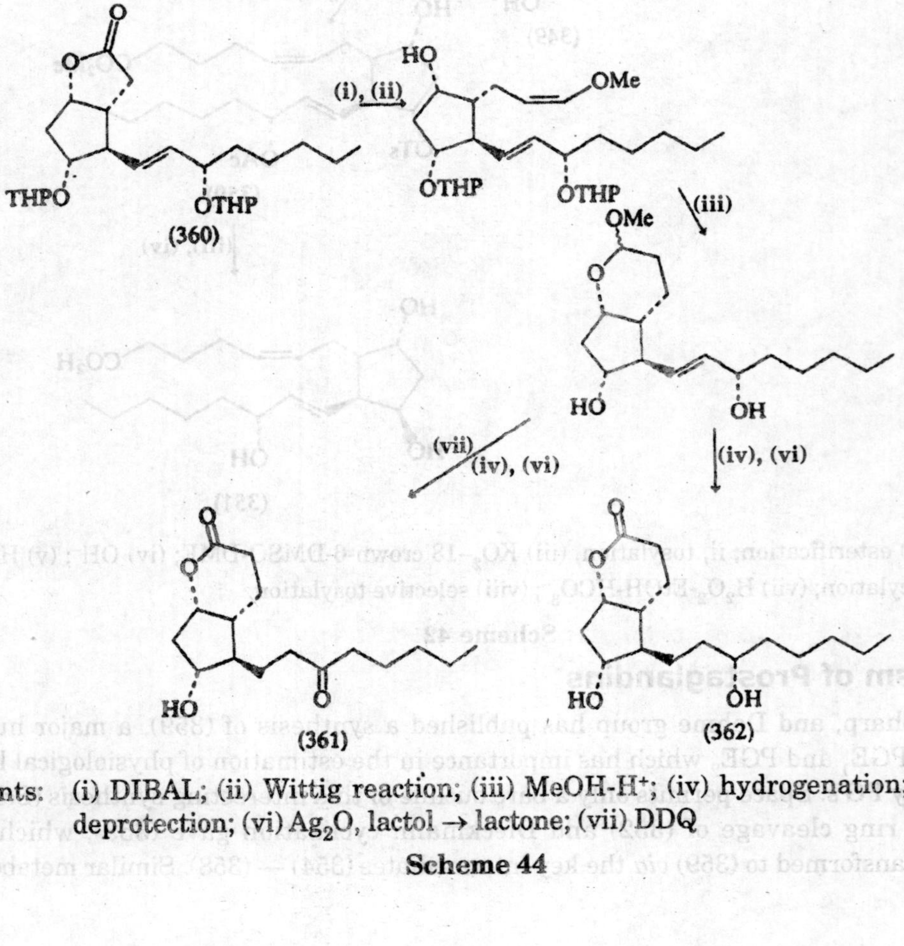

(360)

(i), (ii)

(iii)

(vii) (iv), (vi)

(iv), (vi)

(361)

(362)

Reagents: (i) DIBAL; (ii) Wittig reaction; (iii) MeOH-H$^+$; (iv) hydrogenation; (v) lactol deprotection; (vi) Ag_2O, lactol → lactone; (vii) DDQ

Scheme 44

with C-20 unoxidized, (361) and (362), have been prepared by Nidy and Johnson from the optically active Corey route lactone (360) as outlined in Scheme 44.

Biosynthesis and Biochemistry of Prostaglandins

Funk and co-workers have published some model work on peroxy-radical cyclizations of relevance to the biosynthesis of PC's and in a later communication have applied this to (363), which was generated from γ-linolenic acid with soya bean lipoxidase. Hydrogen abstraction from the hydroperoxide in (363) with di-t-butyl peroxyoxalate in O_2-saturated benzene gave the peroxy-radical, which cyclized to a mixture shown by mass spectrometry to contain the expected $PGF_{1\alpha}$, analogues. These results, obtained through a format more controlled than random autoxidation, further support the now generally accepted radical cyclization *via* peroxy-radicals in PG biosynthesis. In an accompanying paper, Pryor and Stanley provided evidence for PG-type endoper-oxides during autoxidation of methyl linolenate.

$(CH_2)_4CO_2Me$

C_5H_{11}

HOO

(363)

Some fascinating differences between the biosynthesis of PGA_2 in *P. homomalla* and mammalian systems have been disclosed by Corey and colleagues. PGE_2, PGH_2, and PGG_2, which are precursors of PGA_2 in mammalian systems, were not transformed to PGA_2 by the PGA_2-synthetase in *P. homomalla.* Another possible substrate for this enzyme, *viz.* 11-epi-PGE_1, obtained in the 10-tritio-form from PGA_2 epoxide, was also ruled out. No speculations on the possible identity of the elusive PGA_2 precursor were presented.

In possibly one of the most stimulating papers in 1975, Hamberg *et al.* have reported the structure of the very unstable Thromboxane A_2 (364) (TXA_2) which is an intermediate in the biosynthesis of 'PHD' [now called Thromboxane B_2 (365) (TXB_2)] from PGG_2. The structure was deduced largely from the series of isolable, stable TXB_2 derivatives (366) obtained by trapping TXA_2 with various nucleophiles. (The half-life of TXA_2 at 37 °C in solution is 32 s.) (For a useful summary of this whole area see Kolata note TXA_2 is incorrectly formulated in this reference.) The mechanism by which the endoperoxide system of PGG_2 rearranges to the bicyclic oxetan system in TXA_2 is obscure. No speculations on this point appeared in the original paper and none will be ventured by the present Reporter either!

TXA_2, a potent aggregator of blood platelets (a process where it appears that the primary PC'S may be of little direct importance), has also been identified as the highly potent, unstable component of rabbit aorta contracting substance (RCS). Clearly, the general area of PG endoperoxides, which has already yielded some exciting results, can confidently be expected to include TXA_2 analogues in the near future. Such compounds will present a fair challenge to the synthetic chemist, but some encouragement to meet this can be derived from at least two considerations. Firstly, there is the remarkable example of the PGH_2 analogue (330) where simultaneous increases in potency and stability over PGH_2 were achieved. Secondly, just as the PC's themselves have in many cases tolerated various substituents and structural alterations, usually introduced to improve metabolic stability, without detriment to their biological activity, so might TXA_2 be able to accommodate modifications which in this case would be designed to improve the chemical stability of the oxetan ring.

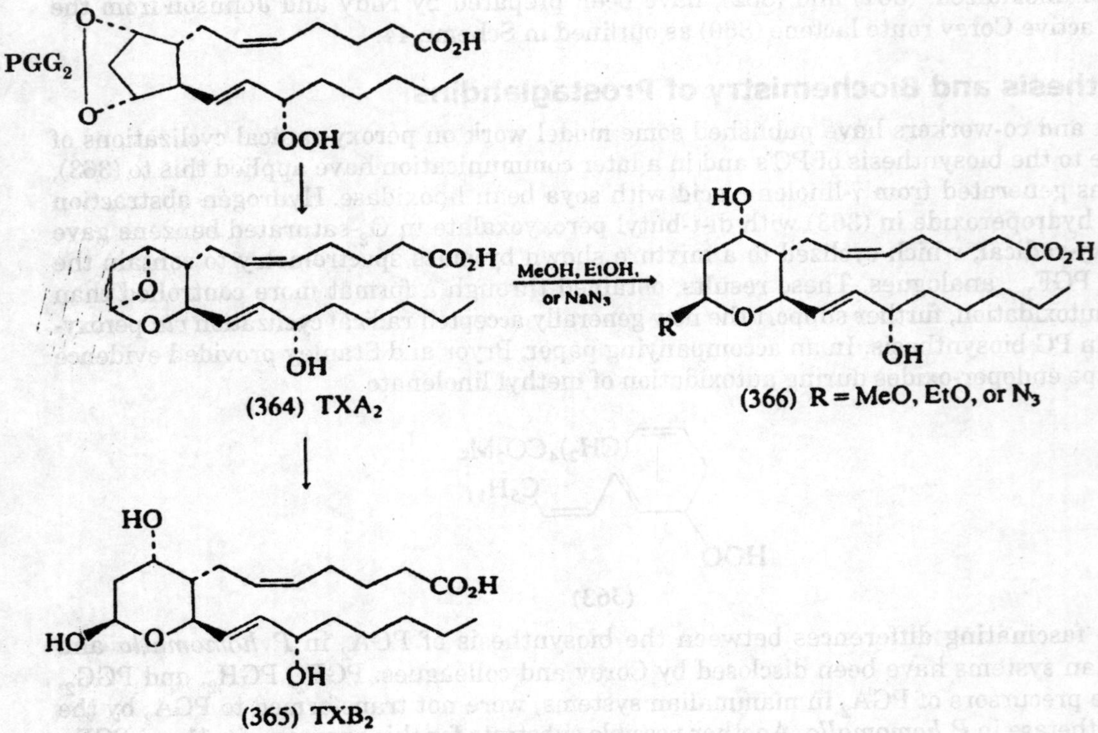

Scheme 45

The structural requirements for the binding of PG's to the lipocyte PGE receptor have been studied.

Leovey and Andersen have implied a 'hairpin' configuration (side-chains parallel), from c.d. studies, for $PGF_{2\alpha}$, $PGF_{2\beta}$, and some related compounds which showed a through-space coupling of the side-chain double bonds. Smooth muscle activity of the various PG's studied correlated with the extent of the coupling and thus with the ability of the side-chains to adopt the hairpin configuration. Further preliminary n.m.r. studies by the same authors also suggested that the side-chains are closer than would be expected from a random distribution of conformers. The hairpin configuration also appears in the orthorhombic and monoclinic forms of PGA_1.

Some reviews dealing with the biochemical role of PG's have appeared in 1975.

Analysis of Prostaglandins

A method for chromatography of PG's utilizing silicic acid-impregnated glass fibre sheets has been reported by two groups. Upjohn workers have described an excellent method for the quantitative conversion of PG's into the *p*-nitrophenacyl esters together with their resolution on h.p.l.c.

Malachite Green has been described as a new staining reagent for PG's and a quantitative separation of PG's by paper chromatography has been published. PG's A_2, B_2, E_2, and F_2 have also been determined quantitatively by t.l.c.

CHAPTER 6

Fatty Acids and Related Compounds

This Report on developments in the chemistry of fatty acids and their derivatives during 1974 and 1975 follows a similar pattern to reports in Volumes 1 and 3 of this series, and should be read in conjunction with them.

An abbreviated nomenclature now common for long-chain acids will be used. Numbers.and letters are used to indicate chain-length, the number of unsaturated centres, the position(s) of unsaturation, and its nature (α, acetylenic; *c*, *cis*-olefinic; *e*, ethylenic; and *t*, *trans*-olefinic). For example, (9*c*12*c*) is the symbol for octadeca-*cis*-9, *cis*-12-dienoic acid (linoleic). Trivial names are explained in the text or in the foot-note to this page Symbols such as ω3 or ω6 will be used only when it is necessary to emphasize the position of the first double bond with respect to the ω-methyl group.

Comments have been made on the nomenclature of acylglycerols (glycerides), alcohols and simple esters, and long-chain nitrogen compounds.

NATURAL COMPOUNDS

Fatty Acids : Most fatty acids of novel structure now reported represent an unusual conjunction of familiar structural features; only rarely is a new structural unit identified, though the unusual furan acids discovered in some fish oils probably fall into this category. In the past two decades most new acids have been C_{18} compounds of plant origin, but at the present time marine lipids (and pheromones) are producing more surprises. The following account is presented on a structural basis rather than in relation to occurrence.

Unsaturated Straight-chain Acids : The sex pheromone of the furniture carpet beetle *(Anthrenusflavipes)* has been shown to be dec-3-Z-enoic acid. The 18:1 acids in butter-fat, studied yet again, include the 7—14-*cis*-(mainly Δ^9) and the 7–*16-trans*- (mainly Δ^{11}) isomers .Two ω3 acids (20:5 and 22:5), recognized in butter oil as trace compon-ents, are the source of the caramel 'off' flavour that develops in the presence of toco-pherol and copper. New plant sources have been discovered for two uncommon 18:2 acids: the 9cl2*a* isomer (crepenynic) in *Saussurea candicans and* the 5*c*9*c* isomer in *Taxus Baccata.*

Octadecapentaenoic acid (18:5 ω3) occurs in eleven species of photosynthetic marine dinoflagellates. The evidence that this acid is not formed by desaturation of the common 18:4

acid but by chain-shortening of the 20:5 homologue means that it is not necessary to postulate the existence of a 3-desaturase:

18:5(3,6,9,12,15) ↑18:4 (6,9,12,15) -→20:4(8,11,14,17) →20:5 (5,8,11,14,17) → 18:5 (3,6,9,12,15)

Mosses appear to be a rich source of acetylenic acids which are related to better known polyenoic acids. These include two C_{18} acids (*6a9cl2c* and *6a9cl2cl5c)* and aC_{20}acid (*8a11cl4c*).

Among hitherto unknown longer-chain acids are two C_{20} dienes (5,11 and 5,13) and two C_{22} dienes (7,11 and 7,13) from the American oyster, *Crossostrea virginica,* two C_{26} polyenes (*5c9c* and *5c9cl9c)* from the marine sponge *Microciona prolifera*, and a series of ω7 monoenoic C_{20}— C_{30} acids from *Lactobacillus heterohiochii,* which grows in sake. All these acids could be formed by standard elongation and desaturation sequences.

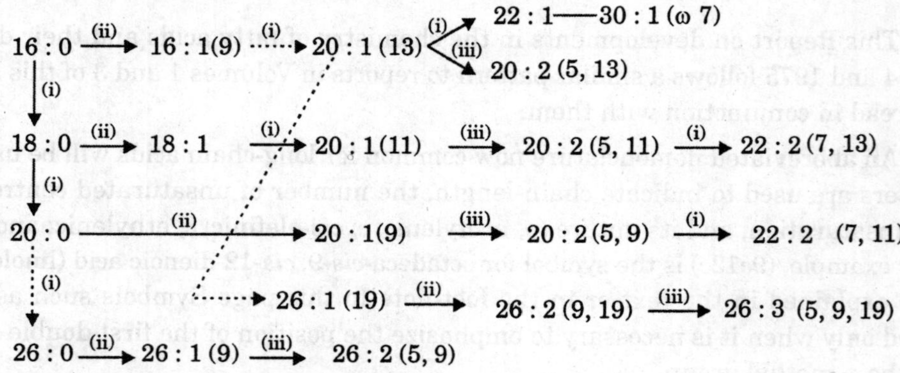

(i) chain extension; (ii) 9-desaturation; (iii) 5-desaturation

Scheme 1

Oxygenated Acids. Additional hydroxy-acids present in cutin monomers include a series of 16-hydroxy-*n*-oxohexadecanoic acids *(n* = 7—10).[11] Still more acids have been recognized in milk fat, and the oxo-acids (~ 1% of milk fatty acids) number at least 60 (C_{10}— C_{24}). Practically all of them can be included in a scheme in which 4- or 5-oxo-acids undergo chain-extension, suggesting that these may be the only positions in which the oxo-group is introduced. Careful study of soybean lipids shows them to have ~2% of oxidized glycerides which contain several epoxy- and hydroxy-C_{18} acids, including 9,10-epoxy-derivatives of 18:0, 18:1 (12c), and 18:2 (I2*c*l5*c),* 9-hydroxy-derivatives of 18:1 and 18:2, 12-hydroxy-derivatives of 18:1 and 18:2, and 13-hydroxy-18:2. Among the component acids of *Salya nilotica* are three a-hydroxy-C_{18} acids (Δ^9, Δ^9, $\Delta^{9,\ 12,\ 15}$) and three C_7 acids (Δ^8, $\Delta^{8,11}$, $A^{8,11,14}$), which are possibly derived from the former group by α-oxidation. The bark of *Ficus hispida* contains 10-oxoeicosan-l-ol as the ester of eicosanoic acid. A proposal that argemonic acid is (+)-6-hydroxy-6-methyl-9-oxo-octacosanoic acid rests on doubtful evidence. The tribasic acid (1) is present in eggs of the tick *Dermacentor andersoni* Stiles and methyl 2,3-dicarboxy-2-hydroxystearate (2) is present in several lichens.

$$Me_2CH(CH_2)_n\ CH(CO_2H)CH(CO_2H)(CH_2)_m\ CO_2H \quad (1;\ n + m = 7)$$

$$Me(CH_2)_{14}CH(CO_2H)C(OH)\begin{cases} CO_2H \\ CO_2Me \end{cases} \quad (2)$$

Branched-chain and Cyclic Acids. Improved procedures of separation and analysis, particularly g.c.– m.s., continue to provide evidence of more branched-chain com-pounds. The chemistry and biochemistry of phytanic, pristanic, and related acids has been reviewed.[19] Branched-chain acids and alcohols in secretions from the uropygial gland of birds have been studied for taxonomic purposes. Subcutaneous glycerides from barley-fed lambs contain ~11% of branched-chain acids, including C_{11}—C_{17} monomethyl acids with the methyl in all possible even positions (2,4,6, . . . ,*n* — 2), C_{11}—C_{15} 4,8-dimethyl acids, and 2,6,10-trimethyldecanoic acid.[21] Sperm-whale oil contains 7-methylhexadec-6- and -7-enoic acid and 5-methyltetradec-4-enoic acid, whilst a *Rhizobium* culture contains a group of acids [11-methyl-18:1 (11), 12-methoxy-11-methyl-18:0, 11-methoxy-12-methyl-18:0, 11-and 13-methoxy-18:0, and *cis*-11,12-methylene-18:0] all of which are thought to result from the 18:1 (11*c*) also present :

$$\underset{18:1\ (11c)}{CH{=}CH} \longrightarrow CH\overset{CH_2}{—}CH \longrightarrow \text{other products}$$

The cyclohexyl moiety of the acids (3; *n* =10 or 12) present in acidophilic thermo-philic bacteria is probably produced from glucose *via* shikimic acid. A new study of the cyclopentenyl acids in seed oils of the Flacourtiaceae has shown the existence of the acids (4; *n* —10, 12, or 14) and (5), some of which are new. There is evidence that these acids are not produced from straight-chain acids but by chain-extension of aleprolic acid (4; *n* = 0)

(3) cyclohexyl–$(CH_2)_nCO_2H$

(4) cyclopentenyl–$(CH_2)_nCO_2H$

(5) cyclopentenyl–$(CH_2)_nCH{=}CH(CH_2)_mCO_2H$

$C_{16}(\Delta^4, \Delta^6, \Delta^9)$,
$C_{18}(\Delta^6, \Delta^9)$and
$C_{20}(\Delta^8, \Delta^9)$,

Two interesting papers report the occurrence of a new series of furan-containing acids in pike and other fish oils. A C_{18} furan acid was reported from a plant source some years ago but the C_{18}— C_{24} acids (6) with methyl branching in the furan ring are completely novel, and raise many interesting problems concerning their biosynthesis, metabolism, and purpose.

Furan ring with substituents R, Me, $Me(CH_2)_n$, $(CH_2)_mCO_2H$

(6) R = H or Me, *n* = 2 or 4, *m* = 8, 10, or 12

Natural Compounds Related to Fatty Acids : Many natural compounds are derived from common fatty acids by metabolic modification, whilst others result from variants of normal fatty acid biosynthesis.

The C_{17} tetraene hydrocarbon ($\Delta^{1,8,11,14}$) in safflower seed is probably derived from linolenic acid (18:3 $\Delta^{9,12,15}$)T and the unsaturated C_{21} hydrocarbons ($\Delta^{6,9,12,15}$ and $\Delta^{3,6,9,12,15}$) in *Polytrichum commune* Hedw. spores are probably formed from the 22:4 and 22:5 ω6 acids, respectively.[28] C_{16} and C_{17} monoene, diene, and triene aldehydes present in cucumbers have double bonds in the positions that would be expected if they result from 18:1 (ω9), 18:2 (ω6), and 18:3 (ω3) acids by α-oxidation.

Shorter-chain aldehydes and alcohols (C_6—C_9), also recognized in cucumber and water melon, are derived from linolenic acid by an oxidative process involving chain-fission.

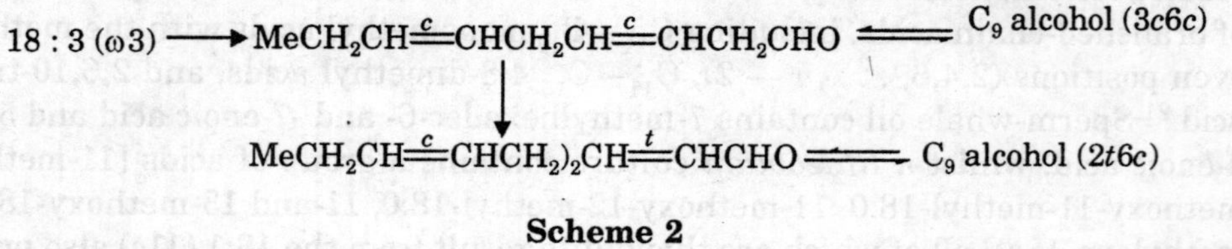

$$18:3\ (\omega 3) \longrightarrow MeCH_2CH\overset{c}{=}CHCH_2CH\overset{c}{=}CHCH_2CHO \rightleftharpoons C_9 \text{ alcohol } (3c6c)$$

$$\downarrow$$

$$MeCH_2CH\overset{c}{=}CHCH_2)_2CH\overset{t}{=}CHCHO \rightleftharpoons C_9 \text{ alcohol } (2t6c)$$

Scheme 2

Rather less obviously related to fatty acids are the C_{15} nitro-compound (7), which termites produce as a defensive compound, the marine natural product dactylyne (8), chondriol, with the revised structure (9), the fungal metabolite Aurovertine B (10), and the antibiotic pseudomonic acid (11).[5] Brefeldin A (12), despite earlier reports, is not produced directly from palmitic acid. The C_{17} and C_{21} lactones (13) present in *Lindera obtusiloba* leaves may be formed from pyruvyl-CoA and appropriate C_{14} and C_{18} acids.

$Me(CH_2)_{12}CH\overset{t}{=}CHNO_2$

(7)

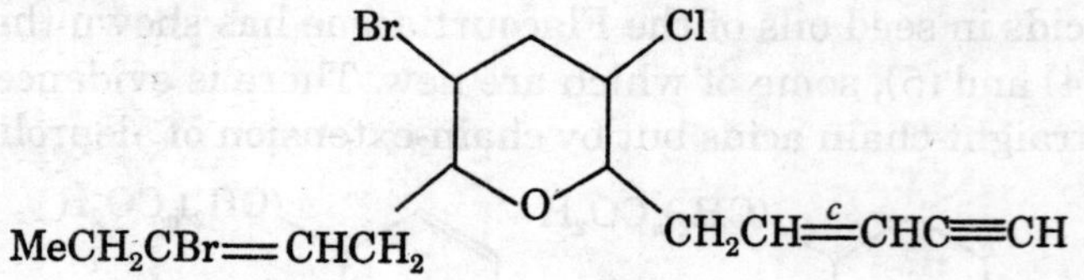

Br, Cl, $MeCH_2CBr{=}CHCH_2$, O, $CH_2CH\overset{c}{=}CHC{\equiv}CH$

(8)

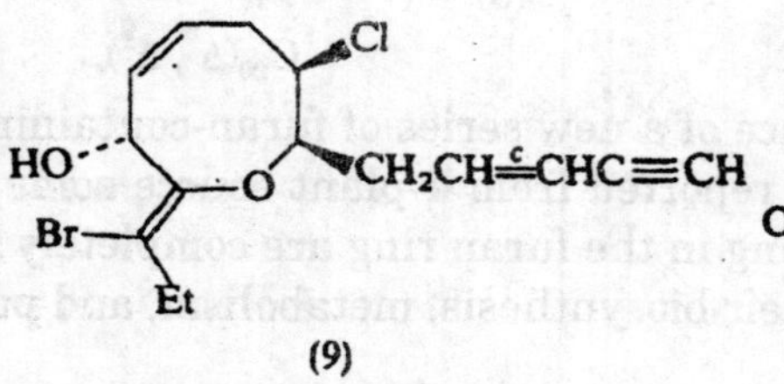

Cl, HO, O, $CH_2CH\overset{c}{=}CHC{\equiv}CH$, Br, Et

(9)

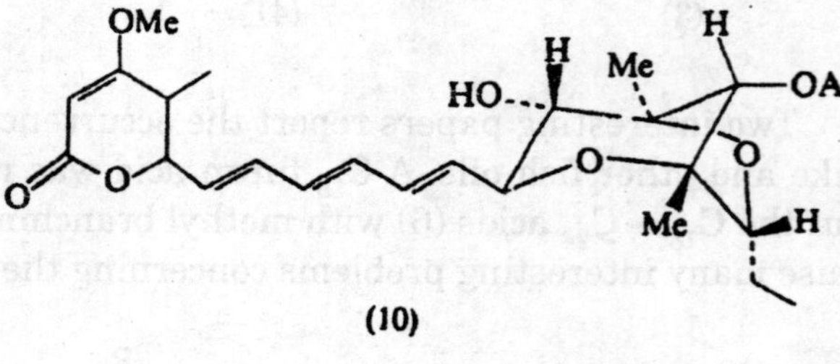

OMe, O, O, H, Me, H, HO, OAc, O, O, Me, H

(10)

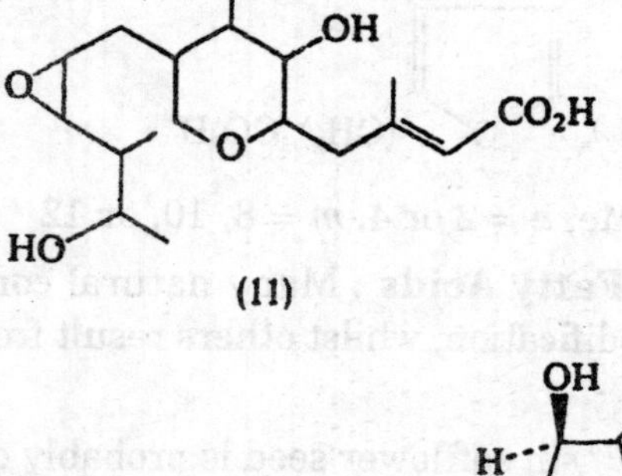

OH, OH, O, O, CO_2H, HO

(11)

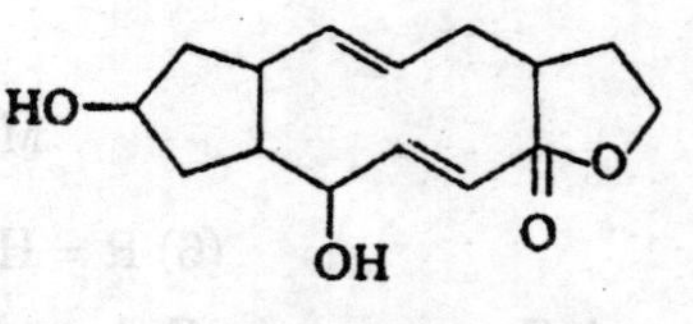

HO, OH, O, O

(12)

OH, CHR, H, CH_2, O, O

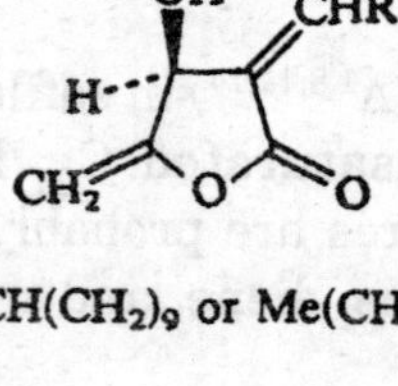

(13) $R = CH_2{=}CH(CH_2)_9$ or $Me(CH_2)_7CH{=}CH(CH_2)_5$

The lactones of 4-hydroxy-12:0 and –12:1 (6c), which have been recognized as important flavour components of animal fats, could be degradation products of oleic and linoleic acids respectively.

Modern analytical and separation procedures are the basis of new information con-cerning wax composition, and waxes from the following sources have been reported: *Tetrahymena pyriformis,* several birds, the Harderian gland of the rat, the Florida indigo snake, the rhino mutant mouse, *Ligea oceanica, dolphins,* rose flowers, rhododendrons cereals, and mosses. There is also a report on the structure of the suberin from potato tuber skin.

Δ^{15} (ω7) derivatives of spring-4-enine and 4-hydroxysphinganine represent two new long-chain bases.

Lipids—*Celastrus orbiculatus* seed oil contains acetic (but no formic), benzoic, and cinnamic acids in addition to the usual range of long-chain acids. The acetic acid is present in triglycerides but it also occurs along with the aromatic acids as esters of two sesquiterpene triols.

Further reports on the nature of melon lipids of a number of marine mammalian species lead to the general conclusion that the unusual nature of the lipids in this organ is not connected with the echolocation abilities of these animals.

Docosane-l,14-diol disulphate and several of its chloro-derivatives are present in algae, and a novel sulphono-lipid (14) has been isolated from a non-photosynthetic diatom *(Nitschia alba)*. The following new glycolipids have been reported: sitosteryl-β-D-(6-*O*-fatty acyl)glucopyranosides from maize phosphatides. Gal*f*- (I_A 2)-Gal*p*-

$$Me(CH_2)_{12}CH\overset{t}{=}CHCH(OH)CH[NHCO(CH_2)_{13}CHMe_2]CH_2OSO_2OH$$

(14)

(l→6)- GlcN- (15- methylhexadecanoyl)-(l →2)- Glc*p*-diglyceride from the extremely thermophilic *Flavobacterium thermophilum,* 3 (1)-*O*-[6^{11}-glycerol-1-phosphoryl] 6′-(1-2-diacyl-*sn*-glycerol-3-phosphoryl)-2′- *O*- (α-D-glucopyranolyl)-α-D-glucopyranosyl] - 1(3), 2-diacylglycerol from *Streptococcus faecalis,* and the glycerol glucoside (15) and its 1- (or 3-) acetate.

CH_2OHCH_2
O
$OCH(CH_2OH)_2$
OH
HO
OH

(15)

Powell and Jacobus have discussed the non-equivalence of the phosphorus atoms in cardiolipin. The novel stereoconfiguration of the lysobisphosphatidic acid of cultured BHK cells contains no *sn*-glycerol-3-phosphate but only the unusual 1-phosphate. An unusual cyclic diether lipid (16) has been recognized in very thermophilic acidophilic bacteria, and further information is available about the 2-methoxy- and 2-hydroxy-alkyl glycerol ethers which occur in several marine sources.

The age pigment isolated from human brain (lipofuscin) has a molecular weight of 6000—7000 and is of a lipidic nature.

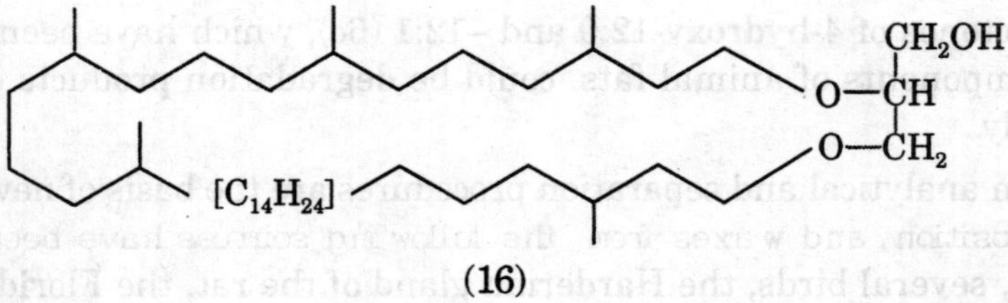

$[C_{14}H_{24}]$ CH_2OH O—CH O—CH_2

(16)

SYNTHETIC COMPOUNDS

Synthetic Procedures : New preparative procedures include a novel route to internal alkynes, convenient syntheses of alk-l-ynes ahd l-bromoalk-2-ynes, and a one-pot chain-extension of propyne at both ends.

$$R^1C \equiv CH \xrightarrow{(i)} R^2_3BC \xrightarrow{(ii)} (R^2_2B)(R^2)C{=}C(R^1)(SOMe) \longrightarrow R^2C \equiv CR^1$$

$$RMgX + CH_2 = C = CHOMe \xrightarrow{(iii)} RCH_2 = CH$$

$$HOCH_2C \equiv CH \xrightarrow{(iv)} (Me_2N)_2P(O)OCH_2C \equiv CH \xrightarrow{(v)}$$

$$(Me_2N)_2\ P(O)CH_2C = CR \xrightarrow{(vi)} BrCH_2C \equiv CR$$

$$MeC \equiv CH \xrightarrow{(vii)} LiCH_2C \equiv CLi \xrightarrow{(viii)} BuCH_2C \equiv CLi$$

$$\xrightarrow{(ix)} BuCH_2C \equiv CR(R = H, I, Bu, CH_2OH \text{ etc.})$$

Reagents: (i) BuLi, $R_3{}^2B$: (ii) MeSOCT. (iii) Cu' halidc, Et_2O; (iv) $POC1_3$, Me_2NH: (v) RI, THF; (vi) HBr or PBr; (vii) BuLi, TMhDA; (viii) LkiBr; (ix) electrophile.

Scheme 3

Chain-extension procedures applicable to fatty acids or their derivatives are sum-marized in Scheme two routes for converting *cis*-epoxides into /trans-alkenes are set out in Scheme 5.

$$RCHO \xrightarrow{(i),(ii)} RCH = CH(CH_2)_4\ CHO$$

$$RCu^-Me\overset{+}{M}gCl \xrightarrow{(iii)} R(CH_2)_n\ CO_2Et$$

$$RCH_2CH = C(R)CHO \xrightarrow{(iv)} RCH_2CH = C(R)(CH_2)_n\ CO_2H$$

$$HC \equiv CCO_2Et \xrightarrow{(v)} RCH = CHCO_2Et$$

$$RX \xrightarrow{(vi)} RCH_2CO_2H$$

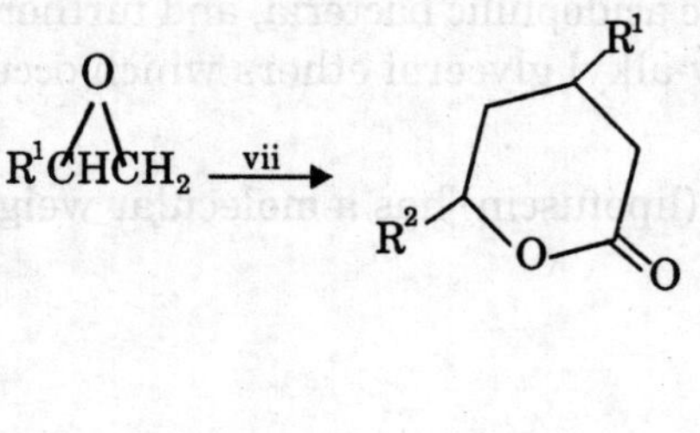

$R^1\overset{O}{CHCH_2}$ (epoxide) $\xrightarrow{vii}$ lactone (R^1, R^2, O, O)

$R^1CO_2H \xrightarrow{(viii)} RCH_2CH(OMe)CO_2Me$

Reagents : (i)  [cyclohexene], *hv*, (ii) heat (iii) $(CH_2)_n$ CO_2Et; (iv) $NaBH_4$; TsCl, either KCN and hydrolysis (n = 1) or malonation (n = 2); (v) *t*-hexylmonoalkylboranes; heat to get *trans*-isomer or allow to react with NaOEt to get *cis*-isomer; (vi) 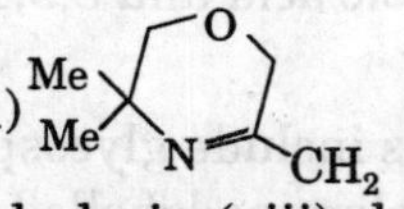 [oxazoline: Me, Me, N=, CH_2] followed by hydrolysis; (vii) 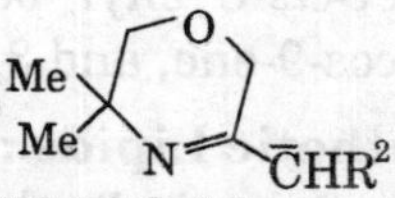 [oxazoline: Me, Me, N=, CHR^2] followed by hydrolysis; (viii) electrolysis with $HO_2CCH_2CH(OMe)CO_2Me$.

Scheme 4

Substituted propargyl iodides, important intermediates in the synthesis of polynoic and polyenoic acids, are available from the corresponding mcsylates in high yield.

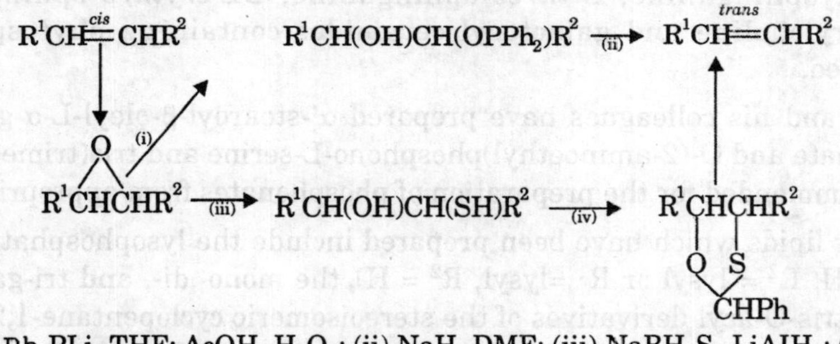

$R^1CH \overset{cis}{=} CHR^2$ → R^1CHCHR^2 (epoxide, O) $\xrightarrow{(i)}$ $R^1CH(OH)CH(OPPh_2)R^2 \xrightarrow{(ii)} R^1CH \overset{trans}{=} CHR^2$

R^1CHCHR^2 (epoxide) $\xrightarrow{(iii)} R^1CH(OH)CH(SH)R^2 \xrightarrow{(iv)} R^1CHCHR^2$ (O, S, CHPh ring) → $R^1CH \overset{trans}{=} CHR^2$

Reagents: (i) Ph_2PLi. THF; AcOH, H_2O_2: (ii) NaH, DMF: (iii) $NaBH_2S_3$ $LiAIH_4$: (iv) PhCHO. H^+; (v) R_2NLi(>2mols).

Scheme 5

New methods of making [l- ^{14} C]acids include *(a)* decarboxylation by electrolysis, followed by reaction with K^{14} CN and hydrolysis, and *(b)* the direct reaction of RCO_2Na with $^{14}CO_2$ at 300—400"C.

A new method of converting alkenes into cyclopropanes by reaction with the sulphur ylide $Ph_2S — CH_2$ is modelled on a biosynthetic router Dipyridyl disulphide is an effective reagent for preparing macrolides.

$$HO(CH_2)_nCO_2H \xrightarrow{pySSpy} HO(CH_2)_nCOSpy \xrightarrow[xylene]{heat\ with} lactone$$

n = 5 — 14

Olefin synthesis with organic phosphonate carbanions has been reviewed.

Synthetic Acids and Related Compounds : The following compounds have been synthesized using standard procedures (those which are not acids or methyl esters are usually prepared for their potential pheromone activity): nearly all the C_{10} — C_{14} acetylenic acids *(3a* to *ω-la),* all the hendecenoic (*cis*) and hendecynoic acids dodec-*trans*-2-enedioic acid (traumatic), dodeca-*trans*-7, *cis*-9-dienyl acetate, the 8*t*10*t* and 9*t*11*t* dodecadienols, 10-propyltrideca-*trans*-5, *trans*-9-dienyl acetate, methyl tetradeca-*trans*-2,4,5-trienoate,Tetradecadienyl acetates (9*t*11*c*, 9*t*11*t,* and 9*c*l2*t*), methyl tetradeca-*cis*-9, *trans*-12-dienoate, 14-methylhexadec-*cis*-8-enol and methyl 14-methylhexadec-*cis*-8-enoate, hexadeca-7, 11-dienyl acetates, The methyl *2a6a* to 13*a*17*a* octadecadiynoates containing the group — $C \equiv C(CH_2)_2C = C$ — , methyl octadeca-*cis*-12, 17-dienoate, methyl octadcca-*cis*-10-en-5-ynoate. 7, 8-epoxy-2-methyloctadecane, the C_{18}, C_{20} and C_{22} ω3-ynoic acids, eicosa-*cis*-14, 15-epoxy-*cis*-8, *cis*-11-dienoic acid, arachidonic acid (20 : 4) and its 14-methyl and 15-methyl derivatives, the heneicos-6-en-11-ones, several 2-substituted eicosa-*cis*-8,-*cis*-11, *cis*-14-trienoic acids, 2, 3-dihydroxy- and 3, 5-dihydroxy-(pentadec-*cis*-8′-enyl)-benzene, hydroxydocos-*cis*-8-enoic acid and 8,9,13-trihydroxydocosanoic acid, tricos-9-ene, and 3,1 l-dimethylnonacosan-2-one.

Synthetic Lipids : Preparations of labelled lipids include glycosphingolipids by catalytic tritiation, 2-dipalmitoyl-*sn*-glycero-3-phosphorylethanolamine labelled separately with ^{32}P, ^{14}C, and ^{3}H, and a series of phosphatidylcholines by a simplified route involving carboxylic acid rather than an acid derivative.

Interest continues in sphingolipids and 3-dehydrosphinganine, *trans*-4, 5-dihydroxysphinganine, D-*threo*-sphinganine, DL-*erythro*-sphingomyelins, ceramide-phosphorylcholine and galactosyl ceramides containing phytosphingosine have been synthesized.

Baer and his colleagues have prepared α′-stearoyl-β-oleyl-L-α-glyceryl-(2-amino-ethyl) phosphonate and O-(2-aminoethyl)phosphono-L-serine and tris(trimethylsilyl) phosphite has been recommended for the preparation of phosphonates from appropriate iodides or tosylates.

Other lipids which have been prepared include the lysophosphatidylglycerol derivatives (17; R^1 =H, R^2 = lysyl or R^1 =lysyl, R^2 = H), the mono-,di-, and tri-galactosyl-diglycerides, a series of tris-O-acyl derivatives of the stereoisomeric cyclopentane-1,2,3-triols (18), phospho- and diphospho-inositides and the complete series of l,2-dioctadecenoyl-.*sn*-glycerol-3-phosphorylcholines.The compounds (18) are interesting analogues of triglycerides where the conformation of the three-carbon system carrying the acyl groups is known.

$R^3CO_2CH_2$ CH_2OR^1
$R^3CO_2CH_2$ O $CHOR^2$
CH_2OPOCH_2
OH

(17)

Oacyl
Oacyl
Oacyl

(18)

The acylation of 1,3-diacylglycerol without migration of acyl groups has been effected with the required acid, triphenylphosphine, and diethyl azodicarboxylate.

PHYSICAL PROPERTIES

Gas-Liquid Chromatography : Useful information has been published on the g.l.c. characteristics of the *cis*-undecenoates, the undecynoates number of diunsaturated C_{18} esters with two double bonds or two acetylene groups separated by three methylene groups, several methyl epoxyoctadecen(yn)oates, arange of 18:1,18:2, and 18:3 *cis*- and *trans*-isomers, and for ω-cycloalkyl fatty acids with four- to seven-membered rings. A study has been made of the separation of monoacyl- and monoalkyl-glycerols as their diacetyl or bis(trimethylsilyl) derivatives on several polar stationary phases. Jamieson has tabulated useful data on the ECL of several C_{16}—C_{22} polyene esters on polar columns of varying polarity. Enantiomeric 2-hydroxy-alkanoates can be separated as diastereoisomeric amides with 1-phenylethylamine. Nelson claims that on SCOT columns the log plots for homologous esters are not quite parallel, and offers an equation to describe their behaviour.

High-performance Liquid Chromatography : This relatively new technique, now being used to separate fatty acid derivatives, has been applied to isomeric hydroperoxides and to *cisltrans* isomers. The elution of both phenacyl and 2-naphthacyl esters can be monitored by virtue of their u.v. absorption. Merit is also claimed for a new reverse-phase support which contains octadecylsilane chains and from which acids and esters are eluted with aqueous methanol. The liquid chromatographic separation of lipids has been reviewed.

Infrared and Raman Spectroscopy : Davies nas extended his study of Raman spectra of olefinic and acetylenic acids to a further range of C_{11} and C_{18} members, and there are also reports on the Raman spectra of dipalmitoylphosphatidylcholine and related compounds, phosphatidylethanolamine and related compounds, and a series of C_{18} and C_{36} sulphur-containing compounds. An interesting study of the i.r. spectra of solid triglycerides at temperatures down to — 180°C indicates that this technique provides a powerful tool in fat analysis. I.r. attenuated total reflectance spectrometry is a convenient way of measuring *trans* unsaturation which can be used, for example, to follow changes in *trans* content during hydrogenation. The i.r. absorption spectra of normal and substituted long-chain acids and esters in the solid state have been reviewed, and the spectra of some ω-cyclic fatty acids have been reported.

Mass Spectrometry : *t*-Butyldimethylsilyl esters (19) are recommended for giving a base peak at 25 eV at *M* — 57 (20) even when the acid has three double bonds.

$$\underset{(19)}{RC(=O)OSiMe_2Bu^t} \longrightarrow \underset{(20)}{RC(\overset{+}{\cdots}O)_2SeMe_2}$$

Anderson and Holman, and others, use acyl pyrrolidides to determine the position of unsaturated centres in mono- and poly-enoic acids. The pyrrolidides are superior to other amides, and the method has been extended to methyl-branched acids. With these acyl derivatives it is not necessary to functionalize the double bonds, as in most other methods, *e.g.* the methoxy- and methoxy-halogeno-derived products obtained by oxymercuration. Another method of locating double bonds involves the use of ion-molecule reactions occurring in the mass spectrometer itself to give products with characteristic fission fragments. For example:

$$R^1CH{=}CHR^2 + CH_2{=}CHOMe \longrightarrow \text{[cyclobutane: } R^1, R^2, OMe] \longrightarrow \text{fission products}$$

The mass spectra of methyl sterculate, methyl malvalate, and several 1,2-dialkyl-cyclopropanes, and of their reaction products with methanolic silver nitrate, have been detailed. The mass spectra of dialkylcyclopropane compounds are similar to those of isomeric alkynes and alkadienes, and fail either to demonstrate the presence of the cyclopropane unit or to locate it.

The fragmentation of ω-cyclic fatty acids and of a range of epoxy-esters has been discussed. Other significant mass spectral investigations include the study of monoacyl-, monoalkyl-, and monoalkenyl-glycerols, gangliosides before and after permethylation, bis (*O*-trimethylsilyl)-*N*-acetylsphinganinc, and the trimethylsilyl ethers of mono- and di-glycerides.

Chemical ionization mass spectrometry with hydrogen, methane, or isobutane has been applied to several methyl alkanoates and alkenoates, epoxy esters, and sphingolipids.

Nuclear Magnetic Resonance Spectroscopy : A definitive paper on the ^1H n.m.r. analyses of alkenoic and alkynoic acids and esters details the long-range deshielding effects of CO_2H, CO_2Me, Me, — CH = CH— *(cis* and *trans)*, —C = C—, — CH= $CHCH_2CH$=CH— *(cis,cis* and *trans,trans)*, CH_2 = CH—, and CH = C—, based on the 220 MHz spectra of 143 acids or esters. Among monounsaturated C_{18} acids, all the acetylenic isomers can be distinguished, and all the *cis-* and *trans-*olefinic esters except the Δ^{10}—Δ^{12} isomers. The value of these long-range deshielding effects is illustrated in the study of some monoepoxyoctadecenoates.

Several reports on the ^1H n.m.r. studies of long-chain acids or esters are concerned with the use of shift reagents, which have been employed in the study of hexanoic acid, the hexanols, hydroperoxides and alcohols, alkenes, those methyl octadecenoates not readily distinguished by high-resolution ^1H n.m.r. spectroscopy, some epoxy- and dihydroxy-esters, and glycerides containing unsaturated acyl groups and valeroyl (3-methylbutanoyl) groups. Since the effect of such reagents is different on acyl groups in the α- and β-position of triglycerides, it is possible to determine the position of acylation.

^1H n.m.r. spectroscopy continues to be used for the study of conformational problems and of the structure of micelles, bilayers, and multilayers.

There is little doubt that ^{13}C n.m.r. spectroscopy will be exploited by lipid chemists and biochemists, and such spectra may prove to be more informative than the ^1H spectra. Reports have already appeared on several isomeric 18:1 esters and on the isomeric lecithins derived from them; an attempt has been made to interpret ^{13}C chemical shifts in terms of steric effects and linear electric field effects. ^{13}C-enriched phosphatidylethanolamines have been produced from both 1- and 2-^{13}C-enriched acetate, and the ability of ^{13}C spectra to provide information in a non-destructive manner, such as an approximate fatty acid profile of an intact seed, is likely to be of value in breeding programmes.

Other Physical Properties : During the period under review, reports have appeared on the polymorphism and/or crystal structure of tritetracosanoin, single acid triglycerides containing isomeric octadecenoic acids, diacid triglycerides containing two short-chain acyl groups and either stearic or behenic acid, a number of long-chain acids, and some diglycerides. An interesting paper reports the structure of 1,2-dilauroyl-(±)-phosphatidyIethanolamine acetate. It is concluded that 'the lipid hydrocarbon chains in the molecule are essentially parallel to one another, and the dihedral angle between the planes containing the two chains is only 8°. The phosphodiester moiety has a double *gauche* conformation and the phosphorylethanolamine group lies approximately parallel to the plane of the bilayer. The phospholipid molecules pack in the form of a classical lipid bilayer, the bilayers being separated by acetic acid molecules of crystallization . . . each lipid molecule is close-packed by six others'.

The conformational analysis of phosphatidylethanolamine in multilayers has been studied by i.r. dichroism, with the conclusion that the hydrocarbon chains are inclined at ~ 75° to the film plane and the polar groups orient parallel to the plane in the built-up film.

Other physical properties reported include monolayer studies, the correlation between conformation and the ability to form urea inclusion compounds, o.r.d. and c.d. studies of optically active mono-, di-, and tri-glycerides, the mutual solubilities of some long-chain esters, the partition of fatty acids between heptane and physiological buffer at 37°C, the phase-behaviour of several binary systems, and the absorption of fatty acids at silica-benzene and silica-carbon tetrachloride interfaces.

CHEMICAL REACTIONS

Oxidation (other than Autoxidation) : A French group prefer to distinguish isomeric octadecenoic acids by ozonolysis followed by reaction with triphenylphosphine, lithium aluminium hydride, or sodium borohydride, but a North American group, interested in the isomeric monoenes formed during partial hydrogenation, prefer epoxidation followed by cleavage with periodic acid.

A series of octadecadienoic acids have been converted into their mono- and di-epoxides, and it has been shown that none of the epoxy-dienes obtained from 20:3 (8,11,14) produce prostaglandins with microsomes from sheep seminal vesicle. The structure of the transition state for peracid epoxidation of styrene is believed to differ somewhat from structures previously proposed by Bartlett or by Kwart and Hoffman. The proposed structure (21) is characterized by (i) *sp*^{2} geometry and partial positive charge at C(l), (ii) extensive C(2)—O bond formation, but with retention of some Ti-bonding between C(l) and C(2), (iii) a nearly equal amount of partial negative charge, on the peracid carboxylate, and (iv) little change in the bonding of the peracid proton.

(21)

$RCH(OCH_2)_2 \longrightarrow RCOOCH_2CH_2OH$

(22)

The ozonolysis of long-chain cyclic acetals (22) gives esters, and a safer procedure for the oxidation of oleic acid to 9,10-dioxostearic acid with potassium permanganate and cold acetic anhydride has been described.

In the oxidation of methyl stearate or methyl behenate with CrO_3— Ac_2O, all the possible oxo-esters are formed, but the 5- and 9-oxostearates predominate among the C_{18} esters and the 7- and 20-oxobehenates among the C_{22} esters. The authors claim that 'while the mechanism of these reactions remains uncertain it seems reasonable to suggest that the partially selective oxidation of fatty acid esters may be associated with preferred conformations of the compounds in the reaction medium'.

Hydroperoxide Formation and Reactions : Hydroperoxides form and decompose under both enzymic and non-enzymic conditions, and because these two areas overlap they will be considered together in this section. The enzymic and non-enzymic reactions of linoleic acid hydroperoxide have been compared in a review.

It is claimed that the cold iodometric method of measuring peroxide value gives more accurate results than either the hot iodometric procedure or the ferric thiocyanate method. An apparatus which measures oxygen usage and pentane production has been described.

The preparation of octadecyl, octadecenyl, and octadecadienyl hydroperoxide (and the corresponding *t*-butylperoxides) from the corresponding alcohol has been described, and methyl oleate has been converted into methyl 9(10)-hydroperoxystearate by a hydroboration reaction.

$$ROH \rightarrow ROMs \rightarrow RO_2H \text{ or } RO_2Bu^t$$

The formation of hydroperoxide by autoxidation is still not fully understood, and it has been suggested that chemical reaction between unsaturated acid (FA) and oxygen is preceded by formation of a complex between oxygen and two double bonds, and that $FA^+ : O_2^-$ exists as an excited state. Another group studying the oxygenation of arachidonic acid by soybean lipoxygenase conclude that the enzyme catalyses a productactivated and substrate-deactivated oxygenation accompanied by self-catalysed destruction of enzyme activity.

The compositions of the methyl hydroperoxyoctadecenoates resulting from oxidation of methyl oleate over a range of temperature have been compared. Erythrosine is the only one of nine EEC-approved colouring matters to catalyse the autoxidation of methyl linolenate in a photochemical process involving singlet oxygen.

Enzymic oxidation of linoleic acid gives a mixture of 9- and 13-hydroperoxyocta-decadienoic acids in which one isomer may predominate. Pure isomers can be obtained by chromatography on silica.

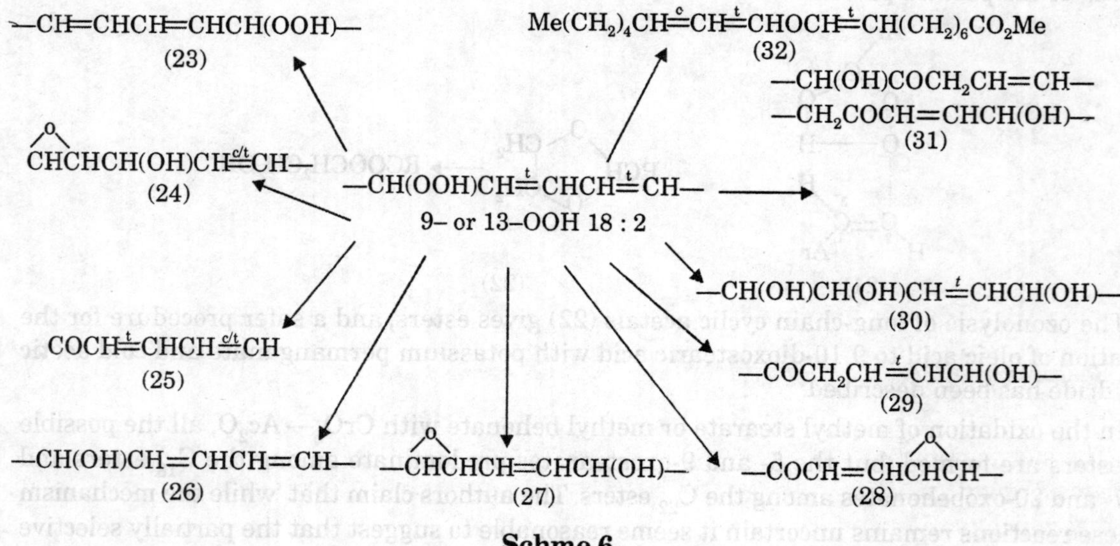

Schme 6

Three lipoxygenase isoenzymes have been isolated from peanut and partially purified. Partially purified lipoxygenase in ungerminated and in germinating barley grain converts linoleic acid into its 9-hydroperoxide. Despite earlier contrary reports, soybean lipoxygenase contains Fe^{3+} (one atom per mol), which is essential for enzyme activity. The lipoxygenase in blood platelets is most effective with arachi-donic acid (20 : 4 ω6) and furnishes the 12-hydroperoxy-acid. The enzyme also operates, though less effectively, on 20:2 (ω6) and on some C_{18} polyene acids. This specificity differs from that observed with plant lipoxygenases.

The hydroperoxy-dienoic acids produced from linoleic acid furnish a wide range of secondary products (23) — (32) under appropriate conditions (Scheme 6). The recognition that a hydroperoxyoctadecadienoic acid produced specifically by enzymes isomerizes to (23) in hexane at — 20°C by a radical process raises questions about the purity of any hydroperoxide sample that is not freshly prepared.

Heated in aqueous ethanol (100°C, 3 h), the 13-hydroperoxy-acid furnishes the epoxy-hydroxy-compound (24) along with its ethyl ether. There is evidence that one oxygen atom comes from the hydroperoxide and the other is solvent-derived. The same hydroperoxide decomposes under the influence of haemoglobin to give (24) (24%), (25) (11%), (26) (25%), and (27) (31%). Gardner *et al.* identified the products (24)—(26) and (28)—(30) and a monoethyl ether of the last. He further concluded that the α- and γ-ketols (31) are the products of an isomerase activity, and that the reaction involves a substitution at C(13) or C(9) by OH,OMe, or $OC_{18}H_{35}$ in the presence of appropriate hydroxy-compounds.

The divinyl ether acid (32) results only from the 9-hydroperoxide at pH > 7, and not from the 13-hydroperoxide. The ether oxygen is not derived from the hydroperoxide. At pH < 7 (24), (25), and (30) are found among the products.

A peroxy-radical cyclization which can serve as a model for prostaglandin biosynthesis has been described. Oxidation of ammonium γ-linolenate in the presence of soybean lipoxygenase gives the 9- and 13-hydroperoxy-acids, which can be separated by high-performance liquid chromatography after esterification. The 9-hydroperoxide, treated with di-*t*-butyl peroxyoxalate followed by sodium borohydride, gave a number of compounds, including some with mass spectral properties expected of C_{18} prostaglandin-like molecules.

18 : 3(6*c*9*c*12*c*) ⟶ 9—OOH(6*c*10*t*12*c*) —(i)⟶ peroxy radical —(ii), (iii)⟶

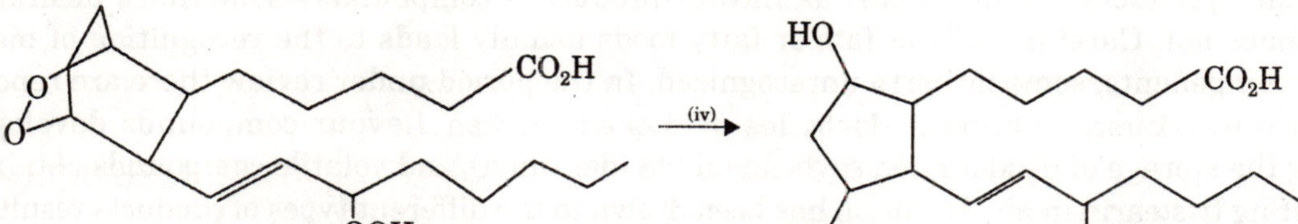

CO2H, O, O, OOH —(iv)⟶ HO, CO2H, HO, OOH

Reagents: (i) $Bu^{+}O_3CCO_3Bu^{+}$; (ii) cyclization, (iii) O_2; (iv) $NaBH_4$

Scheme 7

The suggestion that endoperoxides result from polyenes with at least three double bonds (Scheme 8) and that such compounds are the precursors of both prostaglandin and the propane-1,3-dial which is thought to be the basis of the thiobarbituric acid test is interesting.

18 : 3(9*c*12*c*15*c*)
- 9-OO• 10*t*12*c*15*c* }
- 16-OO• 9*c*12*c*14*t* } ⟶ hydroperoxides
- 12-OO• 9*c*13*t*15*c* }
- 13-OO• 9*c*11*t*15*c* } ⟶ endoperoxides (⟶ hydroperoxides)

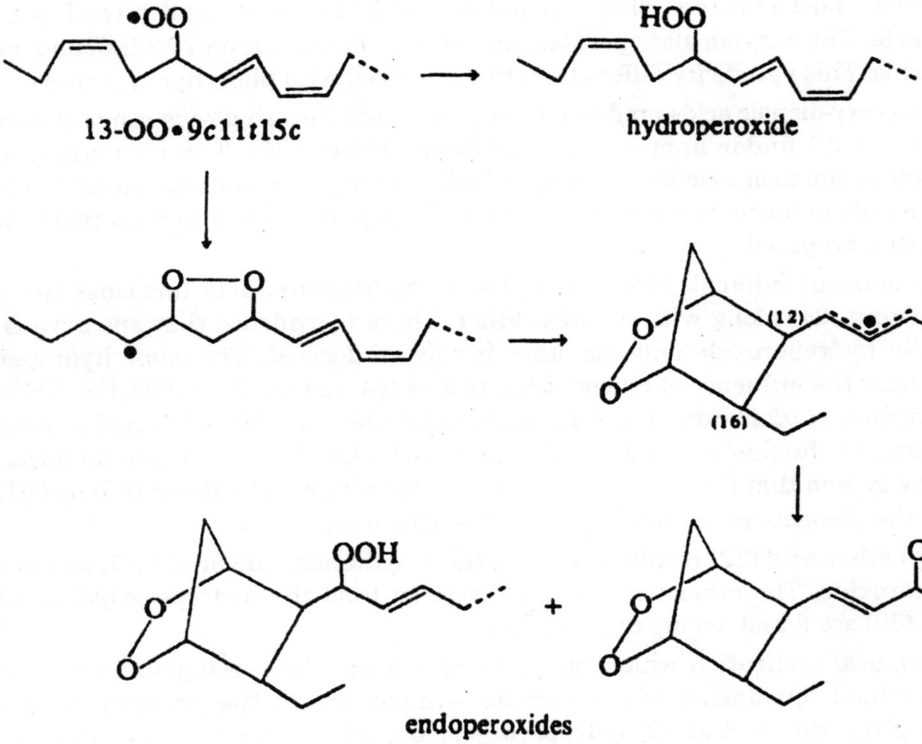

•OO HOO
13-OO•9c11t15c hydroperoxide
(12) (16)
OOH OOH
endoperoxides

Scheme 8

Autoxidized linolenate is reported to produce an unidentified product that has anti-cancer activity.

Short-chain products resulting from scission of the primary hydroperoxides or the major secondary products are important as flavour-producing compounds—sometimes desirable, sometimes not. Careful study of fats or fatty foods usually leads to the recognition of many minor components, some hitherto unrecognized. In the period under review there are reports on some previously unknown aldehydes in cooked chicken, flavour compounds developed during the storage of deodourized soybean oil (48 identified), and volatile compounds obtained by heating tristearin in air. Attention has been drawn to the different types of products resulting from homolytic and heterolytic fission.

$$R^1CH{=}CHCH(OOH)R^2 \nearrow R^1CH{=}CHCH(O)R^2 \longrightarrow R^1CH{=}CHCHO + R^2$$

$$R^1CH{=}CHCH(OOH)R^2 \searrow R^1CH{=}CHCH(\overset{+}{O})R^2 \longrightarrow R^1CH{=}CH{-}O{-}CHR^2 \longrightarrow R^1CH_2{=}CHO + R^2CHO$$

The behaviour of allylic hydroperoxides in the presence of transition-metal catalysts has been examined, and it is reported that oxidation of the antioxidant (33) furnishes the product (34).

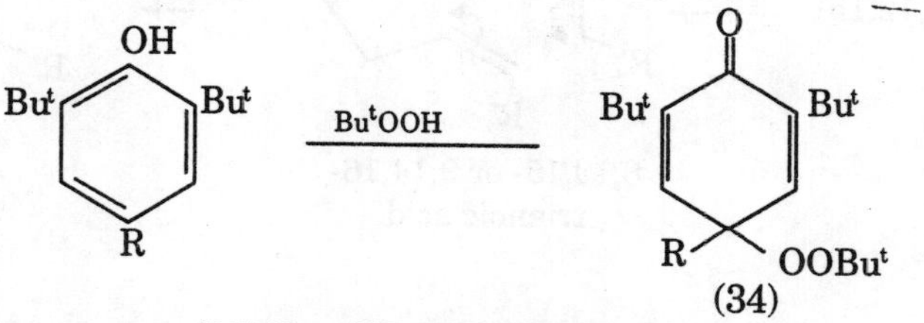

(33) $R = CH_2CH_2COOC_{18}H_{37}$

(34)

Metathesis : The metathesis reaction whereby an unsymmetrical alkene is converted into two symmetrical alkenes gives some interesting products when applied to unsaturated esters. Methyl oleate, for example, gives octadec-9-ene and octadec-9-enedioic acid. The reaction

$$2R^1CH{=}CHR^2 \xrightarrow{\text{catalyst}} R^1CH = CHR^1 + R^2CH{=}CHR^2$$

is more complex with methyl esters (or glycerides) having more than one double bond or with mixtures of two alkenes. With methyl α-linolenate a considerable amount of cyclohexa-1,4-diene is liberated, possibly through the following reaction:

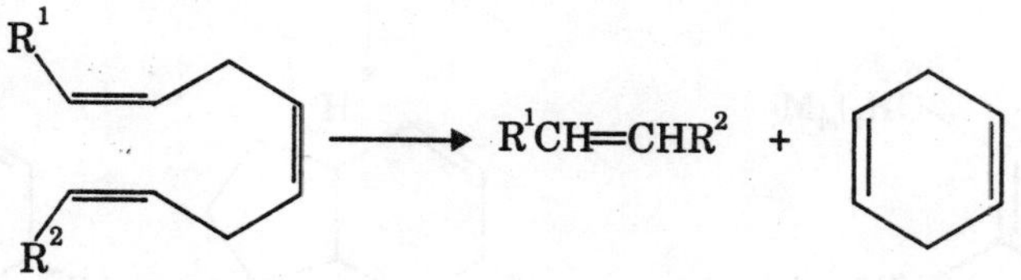

$$\longrightarrow R^1CH{=}CHR^2 +$$

18 : 3(9*c*12*c*15*c*)

Studies continue on alternative catalysts and on the mechanism of the reaction, and already there are two reviews on the subject.

Sulphur Derivatives : Thiols, sulphides, and disulphide derivatives of long-chain esters have been prepared by nucleophilic addition of hydrogen sulphide to alkene esters or by interaction of mesylates with sodium hydrogen sulphide or potassium thioacetate. Epithioesters (35) are obtained under appropriate conditions from linoleate or from unsaturated mesylates.

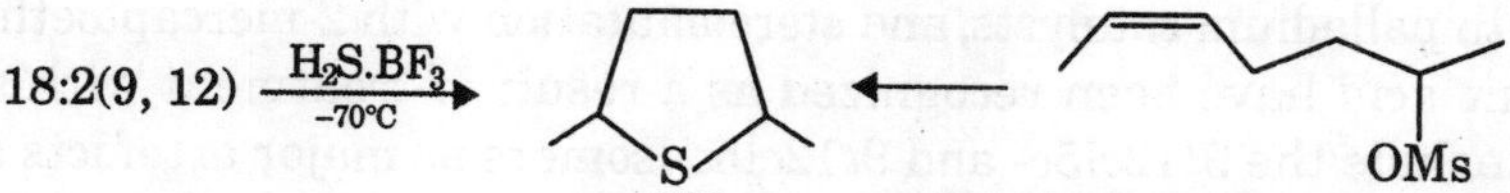

$$18{:}2(9,\ 12) \xrightarrow[-70^\circ C]{H_2S.BF_3} (35) \longleftarrow 9MsO\text{-}18{:}1(12c)$$

Cyclization, Dimerization, and Polymerization : When treated with alkali, linolenic acid undergoes migration of double bonds followed by cyclization to a bicycle system which can be dehydrogenated to give indane derivatives (36). The cyclized compound is believed to have a 1,3,8-pattern of triene unsaturation, and linolenic acid (9,12,15) rearranges, *via* 9,11,15- and 9,13,15-isomers, to the 8,10,15- and 9,14,16-trienes, which cyclize as shown. A similar study has been reported on the related hydrocarbon *trans*-deca-1,4,9-triene.

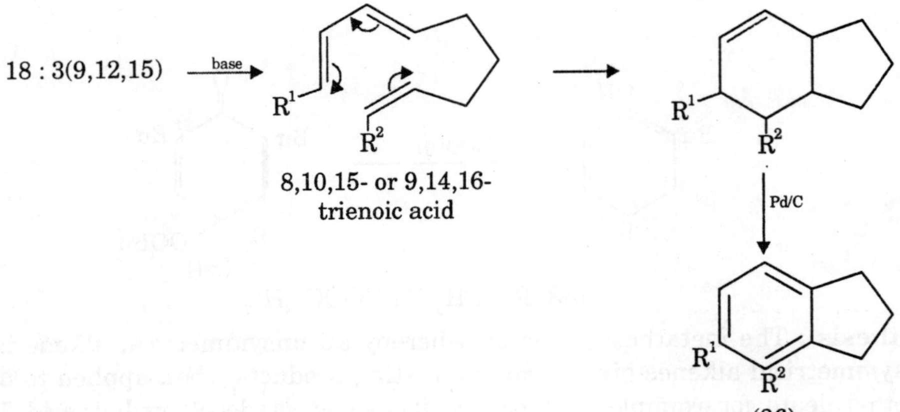

18 : 3(9,12,15) —base→ 8,10,15- or 9,14,16-trienoic acid → (36)

Tall oil contains some bicyclic acids and some polyene acids with an unusual pattern of unsaturation; the former are produced from the latter by double-bond migration (occurring under the alkaline conditions of sulphate pulping) followed by thermal cyclization.

18:3 (5*c*9*c*12*c*) —base→ 18:3 (5*c*10*t*12*c*) → 18:3 (5*c*10*t*12*t*)

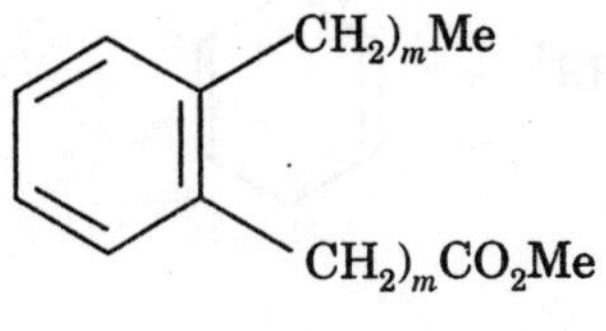

$CH_2)_mMe$ / $CH_2)_mCO_2Me$

(37, $m + n = 10$)

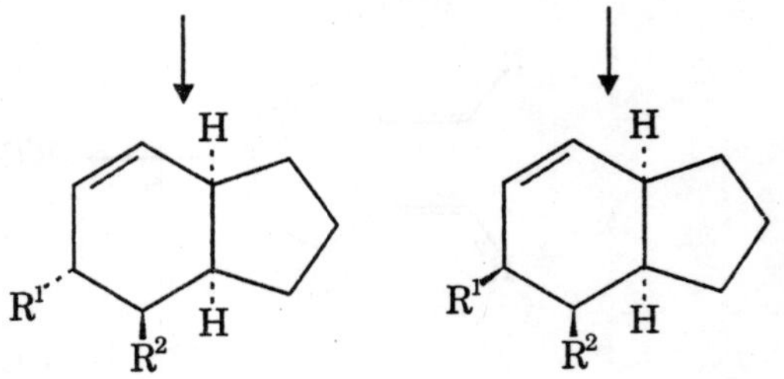

The clay-catalysed polymerization of oleic acid, claidic acid, and tall oil fatty acids produces a mixture of isomeric monoenes and polymers (mainly dimers) which may be acyclic, alicyclic (mono- or di-cyclic), or aromatic. When heated with iodine at 200 °C, methyl linoleate furnishes a range of products, including the aromatic compounds (37). The Diels-Alder adduct from a long-chain diene ester and ethylene is aromatized in the presence of 5% Pd/C and octadec-1-ene as hydrogen acceptor.

Other Reactions of Double Bonds : The following reactions of double bonds have been investigated : hydrosilylation, reaction with dibutyl phosphite, catalytic carboxylation, oxidative acetoxylation with palladium catalysts, and stereomutation with 2-mercaptoethylamine. Isomeric forms of linolenic acid have been recognized as a result of treatment with steam. They may total 25%, and include the 9*t*12*c*l5*c*- and 9*t*12*c*l5*c*-isomers as major artefacts and the 9*t*12*c*15*t*- and 9*c*12*t*15*c*-isomers as minor components.The dehydrochlorination of 9,10-dichloro- and 9,10,12,13-tetra-chloro-octadecanoic acid has been studied: the tetrachloro-ester gives mainly the 9*a*11*e*13*e* and 8*e*10*e*12*a*-acids.

Reactions of the Carboxy-group and the α-Methylene Group : Hydroxamic acids can be made directly from carboxylic acids in ethanol or aqueous ethanol, using dicyclohexyl-carbodi-imide. The reaction of fatty acid esters with α-amino-acids at 150°C to give amides may occur during food preparation. Yet another recipe for the conversion of carboxylic acids into esters by reaction of salts with alkyl halides has appeared, and branched-chain acids have been prepared by the sequence in Scheme 9.

$$RCH_2CH_2CO_2H \xrightarrow{(i)} RCH=CH_2 \xrightarrow{(ii)} RCH(Me)CO_2H$$

Reagents: (i) $PdCl_2$, 185°C or $RhCl_3$; (ii) H_2SO_4– HCO_2H or CO –H_2–$Rh(Ph_3P)$

Scheme 9

The dianions produced from carboxylic acids are being exploited increasingly, and conversions into α-methylene acids, α-hydroxy- and α-hydroperoxy-acids, and nor-aldehydes have been reported (Scheme 10). α-Chlorination occurs in 70—80% yield by reaction with chlorine and oxygen in the presence of chlorosulphonic acid and chloranil. Reaction occurs *via* $RCH = C(OH)_2$ or RCH = C = O.

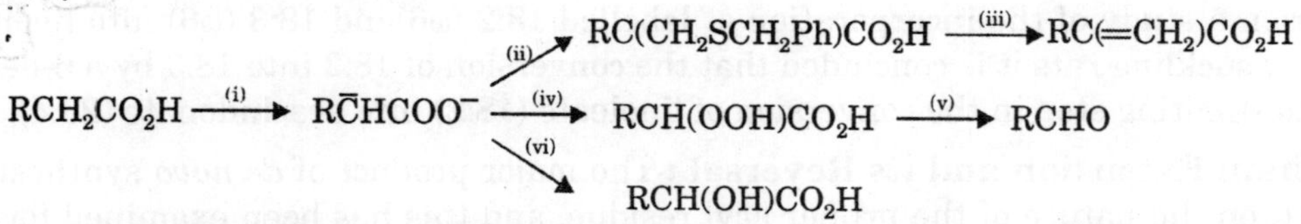

$$RCH_2CO_2H \xrightarrow{(i)} R\bar{C}HCOO^- \xrightarrow{(ii)} RC(CH_2SCH_2Ph)CO_2H \xrightarrow{(iii)} RC(=CH_2)CO_2H$$

$$R\bar{C}HCOO^- \xrightarrow{(iv)} RCH(OOH)CO_2H \xrightarrow{(v)} RCHO$$

$$R\bar{C}HCOO^- \xrightarrow{(vi)} RCH(OH)CO_2H$$

Reagents: (i) $LiNPr^i_2$ or BuLi; (ii) $PhCH_2SCH_2Br$; (iii) $NaIO_4$; (iv) O_2, -75°C; (v) TsOH or DMF acetal; (vi) O_2 r. t.

Scheme 10

Other Reactions : 3-Dihydroxy esters tautomerize under the influence of pyridine during attempts to prepare their bis(trimethylsilyl) ethers. There are further reports on nucleophilic substitution in glycerol derivatives, and a simple method for the preparation of fully deuteriated acids has been reported. The base-catalysed re-arrangement of 3-oxoceramide has been observed.

BIOLOGICAL REACTIONS

Biological oxidation has been discussed with chemical oxidation in the previous section.

Biosynthesis and Metabolism of Fatty Acids : Harwood has recognized three systems capable of synthesizing fatty acids in avocado mesocarp: a fatty acid synthetase in the cytoplasm, a palmitate elongase in the plastid stroma, and a system producing mainly 16 : 0, 18 : 0, and 18 : 1 in the cytoplasm.

In growing cultures of *P. chrysogenum,* α-linolenic acid is produced by desaturation of C_{18} acids, and also by chain-extension of shorter-chain triene acids. The former process dominates in higher and lower plants and in algae, though the chain-extension process was recognized in spinach chloroplast. The metabolism of α-linolenic acid in cultured dissociated brain cells is thought to follow a course which differs from the normal route through 18:4 :

$$18{:}3 \rightarrow 20{:}3 \rightarrow 20{:}4 \rightarrow 20{:}5 \rightarrow 22{:}5 \rightarrow 22{:}6$$

Desaturation Processes : In hen liver microsomes, 18:0-CoA is desaturated to 18 :1-CoA, which is then incorporated into phospholipid. *Candida lipolytica* contains, in addition to an 18:1-Co A desaturase, an oleyl phosphatidylcholine system by which oleic acid in the *sn*-1 or *sn-2* positions is desaturated to linoleic acid; stearic and elaidic acids, however, do not react. In leaf tissue two synthetase systems produce 18:1 and 16:0.

The latter is transferred to phosphatidylglycerol and there converted into the *3t* acid; 18:1 is transferred almost exclusively into phosphatidylcholine and desaturated to linoleic and α-linolenic derivatives before being made available to other lipids. In soybean cotyledon slices, newly formed 18:1 accumulates in the phosphatidic acid fraction and in an unidentified lipid, whilst 18:2 accumulates in the phosphatidyl-choline.

Experiments from two groups provide additional evidence for there being distinct 6-desaturase and 5-desaturase enzymes. The most usual substrate for the 5-desaturase is 20:3 (8,11,14) or 20:4 (8,11,14,17), and Sprecher *et al.* have concluded that the rate of desaturation of some methyl-branched analogues is given by the sequence 20:3 (8,ll,14) = 19Me> 18Me>17Me> 13Me > 10Me > 2Me> 5Me. The same research group conclude that there is no 8-desaturase in rat liver: 18:1 (11) is not desaturated but 19:2 (11,14), 20:2 (11,14), 20:3 (11,14,17), and 21:2 (11,14) are all converted into the Δ^5 derivatives, with no evidence of unsaturation at Δ^8.

From a study of the incorporation of labelled 18:2 (ω6)and 18:3 (ω6) into liver and brain lipids of suckling rats it is concluded that the conversion of 18:2 into 18:3 by a 6-desaturase is the rate-limiting step in the conversion of linoleate (18:2) into arachidonate (20:4).

Chain Extension and its Reversal : The major product of *de novo* synthesis depends, in part, on the nature of the primer acyl residue, and this has been examined for a range of normal, iso-, anteiso-, and cyclic acids in *Bacillus acidocaldarius*. In higher plants the chain-extension of 16:0-ACP to 18:0-ACP through the agency of malonyl-CoA requires an elongation system different from the *de novo* system and, for animals, elongation of 16:0 requires a different enzyme from that required for elongation of 18:2 and 18:3. Nervonic acid (24:1) is produced in mouse brain microsomes by elongation of 22:l-CoA, and the same enzyme will elongate 22:1 (13*t*) and poly-unsaturated C_{22} acids. Using doubly labelled substrates. Kunau and Bartnik demonstrated the presence of a 4-enoylreductase which operates in the retro-conversion of appropriate polyene acids before chain-shortening thus:

$$22{:}5(\omega 6) \rightarrow 22{:}4\ (\omega 6) \rightarrow 20{:}4\ (\omega 6)$$

Biosynthesis of Less Common Fatty Acids and Related Compounds : From a study of the metabolism of aleprolic acid (38) Spener concluded that the long-chain cyclopentene acids are produced not by cyclization of unsaturated acids but by chain-extension followed by desaturation. The unusual position of desaturation is possibly related to the configuration of the ring.

$$\underset{(38)}{RCO_2H} \longrightarrow \longrightarrow \longrightarrow \underset{\substack{\downarrow \\ \Delta^4,\ \Delta^6,\ \Delta^9 \\ C_{16}\text{ acids}}}{R(CH_2)_{10}CO_2H} \longrightarrow \underset{\substack{\\ \Delta^6,\ \Delta^9 \\ C_{18}\text{ acids}}}{R(CH_2)_{12}CO_2H} \longrightarrow \underset{\substack{\downarrow \\ \Delta^8,\ \Delta^9 \\ C_{20}\text{ acids}}}{R(CH_2)_{14}CO_2H}$$

R = (cyclopent-2-enyl)

The anacardic acids in *Ginkgo biloba* are wholly acetate-derived, though the aromatic ring and the alkyl chain may be produced in separate processes.

Separate studies on the cyclopropane acids in *E. coli* have shown that formation of the *cis*-cyclopropane system is independent of the production of the alkenoic acid pre-cursor, but that there is a fatty acid specificity such that, among 16:1 acids, 9*c* > 10*c* > 11*c* > 6*c* ≈ 7*c* with 18:1 (9*c*) and 18:1 (11*c*) reacting as effectively as 16:1 (10c). Folic acid is not a direct C_1 donor in the biosynthesis of the cyclopropane acids in *Streptococcus faecalis.* In an interesting study of the metabolism of cyclopropane acids by *Tetrahymena pyriformis* 11,12-methyleneoctadecanoic acid labelled in the methylene bridge furnishes methyl-labelled acetic acid, possibly by the following sequence:

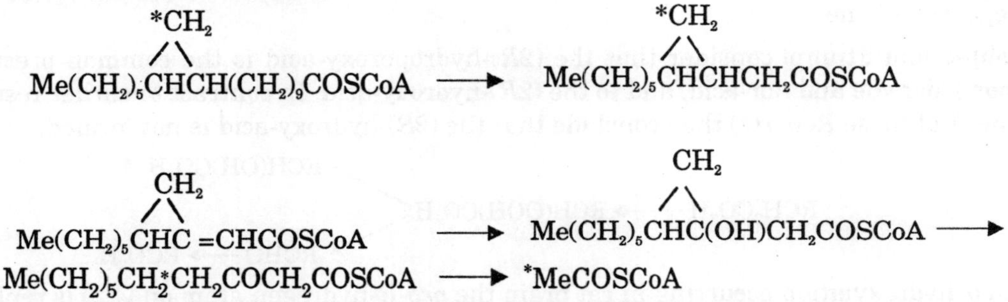

$$\mathrm{Me(CH_2)_5\overset{*CH_2}{CHCH}(CH_2)_9COSCoA} \longrightarrow \mathrm{Me(CH_2)_5\overset{*CH_2}{CHCH}CH_2COSCoA}$$

$$\mathrm{Me(CH_2)_5\overset{CH_2}{CHC}{=}CHCOSCoA} \longrightarrow \mathrm{Me(CH_2)_5\overset{CH_2}{CHC}(OH)CH_2COSCoA} \longrightarrow$$

$$\mathrm{Me(CH_2)_5CH_2^{*}CH_2COCH_2COSCoA} \longrightarrow \mathrm{{}^{*}MeCOSCoA}$$

The microbe *Mycobacterium fortuitum* uses the C_{20} hydrocarbon pristane (39) as the sole carbon source, which it converts into the compounds shown.

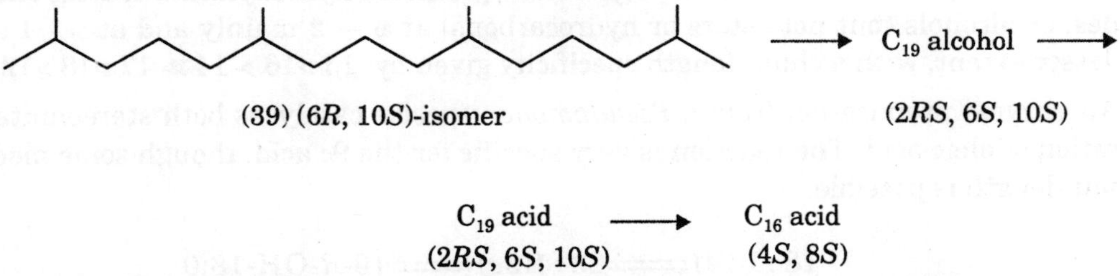

(39) (6*R*, 10*S*)-isomer ⟶ C_{19} alcohol (2*RS*, 6*S*, 10*S*) ⟶

C_{19} acid (2*RS*, 6*S*, 10*S*) ⟶ C_{16} acid (4*S*, 8*S*)

In contrast, long-chain acids are decarboxylated in leaves to furnish hydrocarbons of odd chain-lengths.

Mercer *et al.* have confirmed that chlorination is the last stage in the biosynthesis of the several polychloro- C_{22} and -C_{24} disulphates [such as (40)] which occur in *Ochromonas danica* and in other sources. Elovson considers that the oxygen of the primary alcohol comes from water and not oxygen, probably through reduction of a carboxy-group which undergoes exchange with water, and that the oxygen of the secondary alcohol is derived from oxygen, and not water, and results, not from hydration of a double bond, as previously considered (*see* Volume 3 of these reports), but by direct hydroxylation of a saturated intermediate

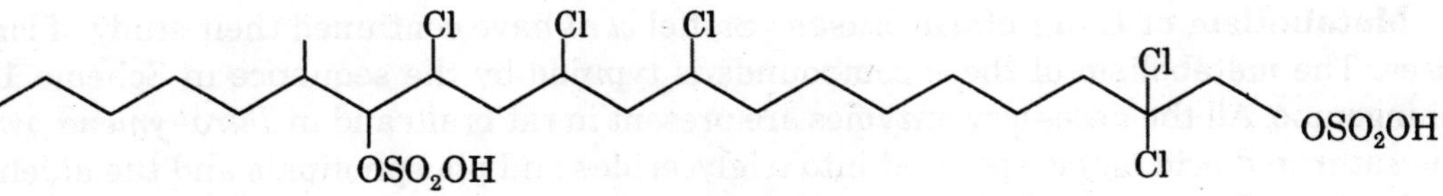

Cl Cl Cl Cl Cl OSO₂OH OSO₂OH

(40)

Croteau and Kolattukudy consider that oleic and linoleic acids are the starting materials for the more extensively oxygenated compounds found in cutins:

18:1 → ω-OH-18:l → epoxide → 9,10,18-tri-OH-18:0
18:2 → ω-OH-18:2 → 9,10-epoxide → 9,10,18-tri-OH-18:1 → 12,13-epoxide → 9,10,12,13,18-penta-OH-18:0

Fusariwn solani pisi contains a soluble epoxyhydratase which is unusual in that it converts *cis*-epoxides into *erythro*-diols.

Pseudomonas oleavorans promotes the selective epoxidation of octa-1, 7-diene to (*R*)-(+)-1, 2-epoxyoct-7-ene.

Shine and Stumpf consider that the (2*R*)-hydroperoxy-acid is the common precursor to the nor-aldehyde and nor-acid, and to the (2*R*)-hydroxy-acid. In contrast to earlier results (see Volume 1 of these Reports) they conclude that the (2*S*)-hydroxy-acid is not formed:

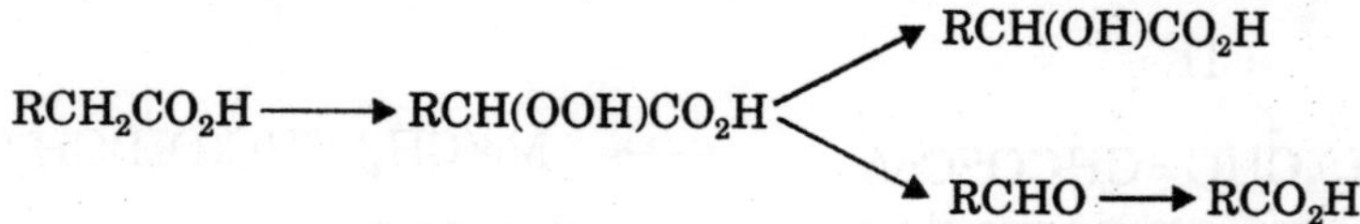

In α-hydroxylation occurring in rat brain the *pro-R*-hydrogen atom on C(2) is replaced by OH.

A soluble system from *Bacillus megaterium* promotes hydroxylation of long-chain acids, amides, or alcohols (but not esters or hydrocarbons) at ω — *2* mainly and at ω – 1 and ω – 3 to a lesser extent, with a chain-length specificity given by: 15 >16 > 14 ≫ 17 >18 >12.

An enzyme preparation from a *Pseudomonas* species catalyses both stereomutation and hydration of oleic acid. The reaction is very specific for the 9*c* acid, though some modification of chain-length is possible:

$$18:1(9t) \rightleftharpoons 18:1(9c) \rightleftharpoons 10\text{-}R\text{-OH-}18:0$$

Lipases : There are reports or reviews on the specificity of lipases derived from *Geotrichum candidum, Rhizopus arrhizus,* milk lipoprotein, and adipose tissue lipoprotein. Milk lipoprotein lipase shows a preference for reaction at the *sn*-1-position, independently of the acyl group present. Fish dietary lipids are rich in wax esters and triglycerides, and it is considered that the former are hydrolysed to acids and alcohols (which are oxidized to acids) before metabolism. Using fish intestinal fluid, it has been shown that wax esters and triglycerides are hydrolysed and that, unlike pancreatic lipase activity, both 2-monoglycerides and esters of lipase-resistant acids (20:4, and 20:5 *etc.)* are hydrolysed. Cephalins and lecithins in rat liver based on C_2, C_3, and C_4 diols are cleaved by phospholipase.

Metabolism of Long-chain Bases : Stoffel *et al* have continued their study of long-chain bases. The metabolism of these compounds is typified by the sequence in Scheme 11 for 4*t*-sphingenine. All the necessary enzymes are present in rat brain and in *Tetrahymena pyriformis.* The saturated acid is incorporated into triglycerides and phospholipids and the aldehyde and alcohol into alkenyl and alkyl ethers. The enzyme responsible for reduction of the unsaturated aldehyde has been isolated and studied. The desaturation step in the biosynthesis of sphingenine occurs on N-acylsphinganines (dihydroceramides).

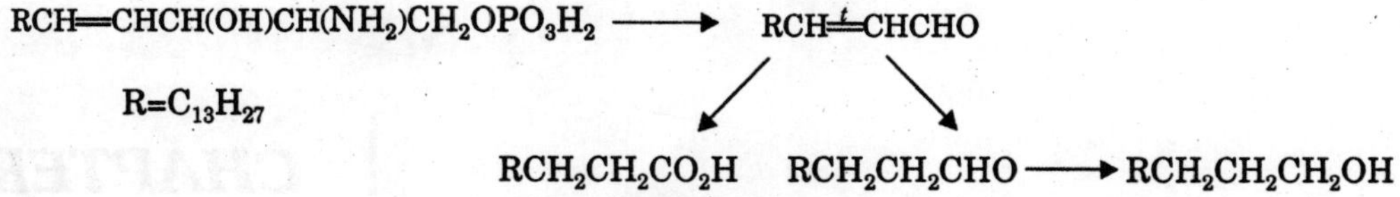

$$RCH{=}CHCH(OH)CH(NH_2)CH_2OPO_3H_2 \longrightarrow RCH{=}CHCHO$$

$R{=}C_{13}H_{27}$

$$RCH{=}CHCHO \longrightarrow RCH_2CH_2CO_2H \quad RCH_2CH_2CHO \longrightarrow RCH_2CH_2CH_2OH$$

Scheme 11

Miscellaneous : Lands *el al.* have studied the effect of various unsaturated C_{18} acids on the growth of mutants of *E. coli* and *Saccharomyces cerevisiae.* Other topics examined include: specificity in ether lipid biosynthesis in myelinating rat brain, the biosynthesis of alkane-2,3-diols, the incorporation of ^{18}O into secondary alcohols occurring as wax esters in the grasshopper, a model for the interaction of polar lipids, cholesterol, and proteins in biological membranes, and a proposal that the ratio 22:5 (ω6)/22:6 (ω3) be used as an index of linolenate (ω3) deficiency.

REVIEWS

During the period covered by this Report the following topics have been reviewed: general aspects of lipid chemistry and biochemistry, the chromatography of fatty acids and lipids, i.r. spectroscopy, the use of phosphonate carbanion and heterocyclic compounds in syntheses of long-chain compounds, olefin metathesis, the formation of hydrocarbons by micro-organisms, the biosynthesis of saturated and unsaturated acids by maturing safflower seeds. Lipolytic enzymes, lipid metabolism in the mammary gland, monolayers, membranes, fungal lipids, and recent work on diol lipids, tumour lipids, and liposome studies.

CHAPTER 7

Membrane Lipids

This chapter examines the biosynthesis of three important components of biological membranes—phospholipids, sphingolipids, and cholesterol. Triacylglycerols also are considered here because the pathway for their synthesis overlaps that of phospholipids. Cholesterol is of interest both as a membrane component and as a precursor of many signal molecules, including the steroid hormones progesterone, testosterone, estrogen, and cortisol. The biosynthesis of cholesterol exemplifies a fundamental mechanism for the assembly of extended carbon skeletons from five-carbon units.

The transport of cholesterol in blood by the low-density lipoprotein and its uptake by a specific receptor on the cell surface vividly illustrate a recurring mechanism for the entry of metabolites and signal molecules into cells. The absence of this receptor in people with *familial hypercholesterolemia,* a genetic disease, leads to markedly elevated cholesterol levels in the blood, cholesterol deposits on blood vessels, and childhood heart attacks. Indeed, cholesterol is implicated in the development of atherosclerosis in individuals without genetic defects. Thus, the regulation of cholesterol synthesis and transport can be a source of especially clear insight into the role that our understanding of biochemistry plays in medicine.

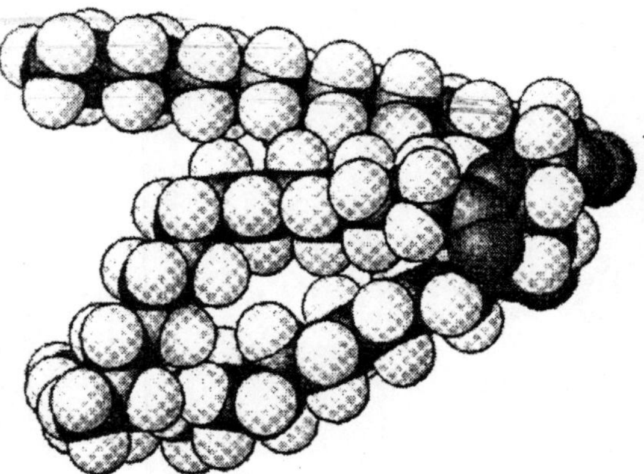

Fats such as the triacylglcerol molecule are widely used to store excess energy for later use and to fulfill other purposes, illustrated by the insulating blubber of washles. The natural tendency of fats to exist in nearly water-free forms makes these molecules well-suited for these roles.

PHOSPHATIDATE

The first step in the synthesis of both phospholipids for membranes and triacylglycerols for energy storage is the synthesis of *phosphatidate* (diacylglycerol 3-phosphate). In mammalian cells, phosphatidate is synthesized in the endoplasmic reticulum and the outer mitochondrial membrane. It is formed by the addition of two fatty acids to *glycerol 3-phosphate,* which in turn is formed primarily by the reduction of dihydroxyacetone phosphate, a glycolytic intermediate, and to a lesser extent by the phosphorylation of glycerol. Glycerol 3-phosphate is acylated by acyl CoA to form *lysophosphatidate,* which is again acylated by acyl CoA to yield phosphatidate.

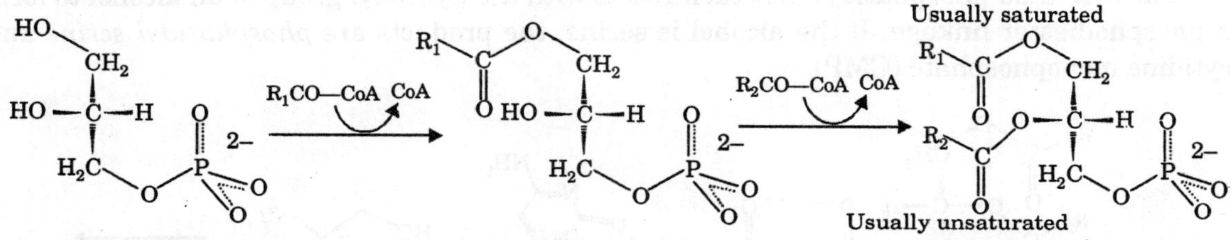

R_1CO—CoA CoA → ; R_2CO—CoA CoA →

Usually saturated; Usually unsaturated

Glycerol 3=phosphate **Lysophosphatidate** **Phosphatidate**

These acylations are catalyzed by *glycerol phosphate acyltransferase.* In most phosphatidates, the fatty acyl chain attached to the C-l atom is saturated, whereas the one attached to the C-2 atom is unsaturated.

The pathways diverge at phosphatidate. In the synthesis of triacylglycerols, phosphatidate is hydrolyzed by a specific phosphatase to give a *diacylglycerol (DAG).* This intermediate is acylated to a *triacylglycerol* in a reaction that is catalyzed by *diglyceride acyltransferase.* Both enzymes are associated in a *triacylglycerol synthetase complex* that is bound to the endoplasmic reticulum membrane.

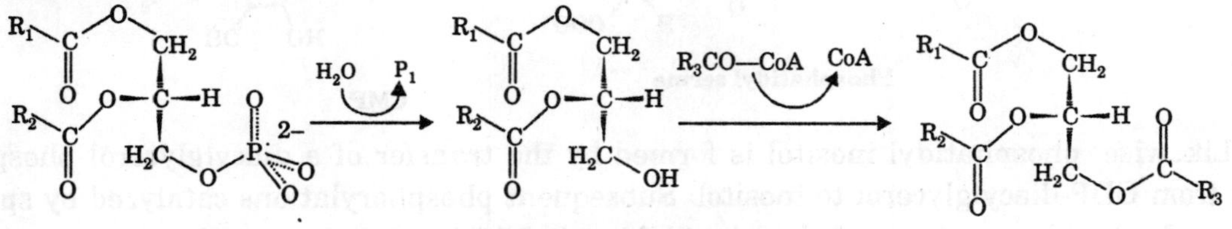

H_2O P_i → ; R_3CO—CoA CoA →

Phosphatidate **Diacylglycerol (DAG)** **Triacylglycerol**

The liver is the primary site of triacylglycerol synthesis. From the liver, the triacylglycerols are transported to the muscles for energy conversion or to the adipocytes for storage.

Activated Intermediate

Phospholipid synthesis requires the combination of a diacylglyceride with an alcohol. As in most anabolic reactions, one of the components must be activated. In this case, either of the two components may be activated, depending on the source of the reactants.

Activated Diacylglycerol

The de novo pathway starts with the reaction of phosphatidate with cytidine triphosphate (CTP) to form *cytidine diphosphodiacylglycerol (CDP-diacylglycerol).* This reaction, like those of many biosyntheses, is driven forward by the hydrolysis of pyrophosphate.

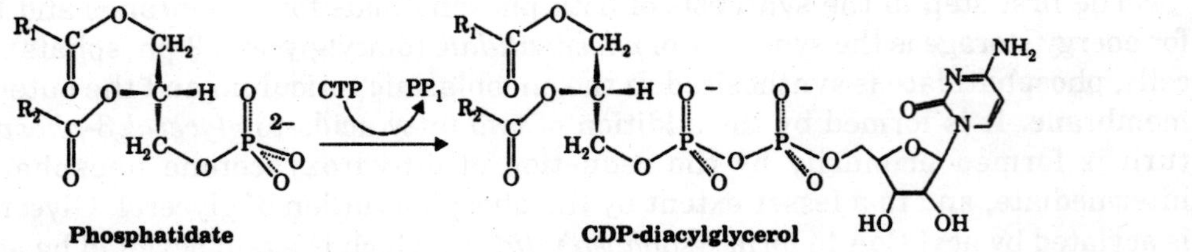

Phosphatidate CTP PP_i CDP-diacylglycerol

The activated phosphatidyl unit then reacts with the hydroxyl group of an alcohol to form a phosphodiester linkage. If the alcohol is serine, the products are *phosphatidyl serine* and cytidine monophosphate (CMP).

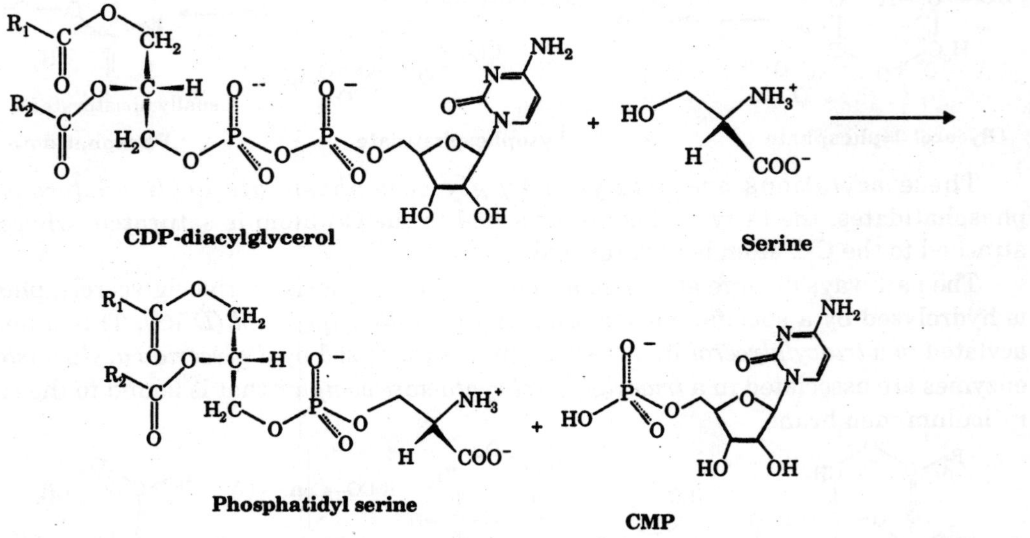

CDP-diacylglycerol + Serine →

Phosphatidyl serine + CMP

Likewise, phosphatidyl inositol is formed by the transfer of a diacylglycerol phosphate unit from CDP-diacylglycerol to inositol. Subsequent phosphorylations catalyzed by specific kinases lead to the synthesis *of phosphatidyl inositol 4,5-bisphosphate,* an important molecule in signal transduction. Recall that hormonal and sensory stimuli activate *phospholipase C,* an enzyme that hydrolyzes this phospholipid to form two intracellular messengers — diacylglycerol and inositol 1,4,5-trisphosphate.

The fatty acid components of a phospholipid may vary, and thus phosphatidyl serine, as well as most other phospholipids, represents a class of molecules rather than a single species. As a result, a single mammalian cell may contain thousands of distinct phospholipids. Phosphatidyl inositol is unusual in that it has a nearly fixed fatty acid composition. Stearic acid usually occupies the C-1 position and arachidonic acid the C-2 position.

In bacteria, the decarboxylation of phosphatidyl serine by a pyridoxal phosphate-dependent enzyme *yields phosphatidyl ethanolamine,* another common phospholipid. The amino group of this phosphoglyceride is then methylated three times to form *phosphatidyl choline. S-Adenosylmethionine* is the methyl donor.

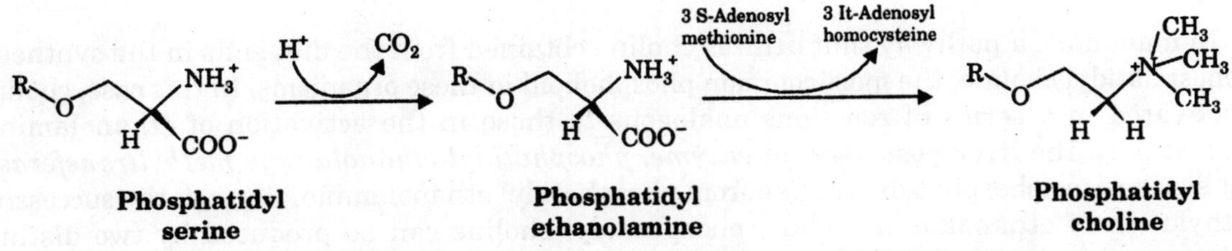

3 S-Adenosyl methionine
3 It-Adenosyl homocysteine

Phosphatidyl serine → (H+, CO_2) → Phosphatidyl ethanolamine → Phosphatidyl choline

In mammals, phosphatidyl ethanolamine can be formed from phosphatidyl serine by the enzyme-catalyzed exchange of ethanolamine for the serine moiety of the phospholipid.

Synthesis from an Activated Alcohol

In mammals, phosphatidyl ethanolamine can also be synthesized from ethanolamine through the formation of CDP-ethanolamine. In this case, the alcohol ethanolamine is phosphorylated by ATP to form the precursor, *phosphorylethanolamine.* This precursor then reacts with CTP to form the activated alcohol, *CDP-ethanolamine.* The phosphorylethanolamine unit of CDP-ethanolamine is then transferred to a diacylglycerol to form *phosphatidyl ethanolamine.*

Ethanolamine

ATP → ADP

Phosphorylethanolamine

CTP → PP_1

Cytidine

CDP-ethanolamine

Diacylglycerol → CMP

Phosphatidyl ethanolamine

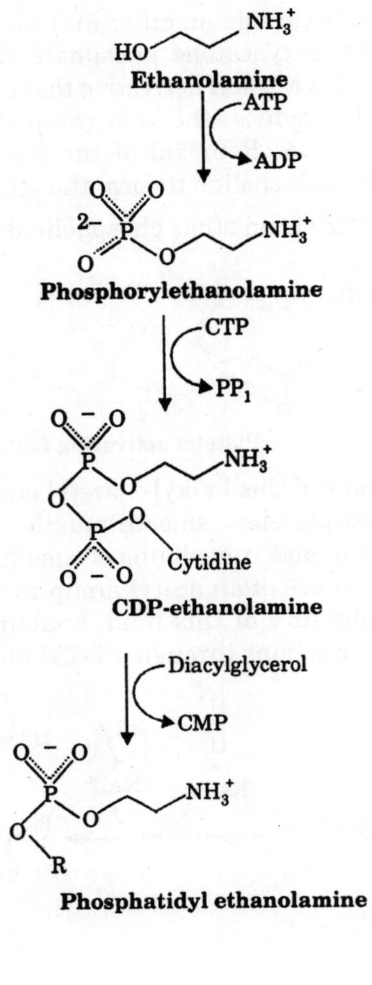

In mammals, a pathway that utilizes choline obtained from the diet ends in the synthesis of phosphatidyl choline, the most common phospholipid in these organisms. In this case, choline is activated in a series of reactions analogous to those in the activation of ethanolamine. Interestingly, the liver possesses an *enzyme, phosphatidyl ethanolamine methyltransferase,* that synthesizes phosphatidyl choline from phosphatidyl ethanolamine, through the successive methylation of ethanolamine. Thus, phosphatidyl choline can be produced by two distinct pathways, ensuring that this phospholipid can be synthesized even if the components for one pathway are in limited supply.

Note that a cytidine nucleotide plays the same role in the synthesis of these phosphoglycerides as a uridine nucleotide does in the formation of glycogen. In all of these biosyntheses, an activated intermediate (UDP-glucose, CDP-diacylglycerol, or CDP-alcohol) is formed from a phosphorylated substrate (glucose 1-phosphate, phosphatidate, or a phosphorylalcohol) and a nucleoside triphosphate (UTP or CTP). The activated intermediate then reacts with a hydroxyl group (the terminus of glycogen, the side chain of serine, or a diacylglycerol).

Plasmalogens and Other Ether Phospholipids

Glyceryl ether phospholispids contain an ether unit instead of an acyl unit at C-l and are synthesized starting with dihydroxyacetone phosphate rather than glycerol 3-phosphate. Acylation by a fatty acyl CoA yields a 1-acyl derivative that exchanges with a long-chain alcohol to form an ether at C- l. NADPH reduces the keto group at C-2, and the resulting alcohol is acylated by a long-chain acyl CoA. Removal of the 3-phosphate group yields l-alkyl-2-acylglycerol, which reacts with CDP-choline to form the ether analog of phosphatidyl choline.

Platelet-activating factor (PAF) is an ether phospholipid implicated in a number of allergic and inflammatory responses.

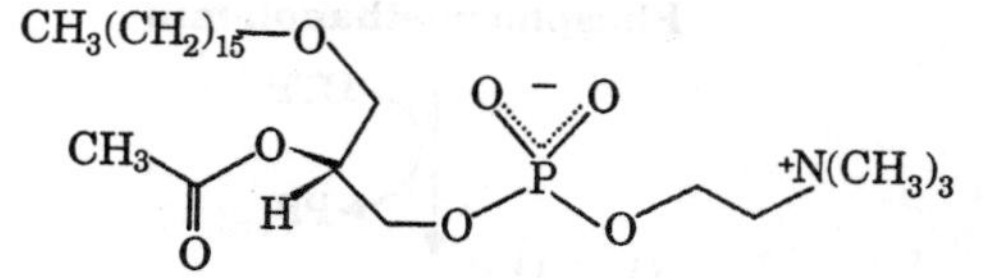

Platelet activating factor (RF)

Subnanomolar concentrations of this l-alkyl-2-acetyl ether analog of phosphatidyl choline induce the aggregation of blood platelets, smooth muscle contraction, and the activation of cells of the immune system. It is also a mediator of anaphylactic shock, a severe and often fatal allergic response. The presence of an acetyl group rather than a long-chain acyl group at C-2 increases the water solubility of this lipid, enabling it to function in the aqueous environment of the blood. PAF functions through a 7-TM receptor.

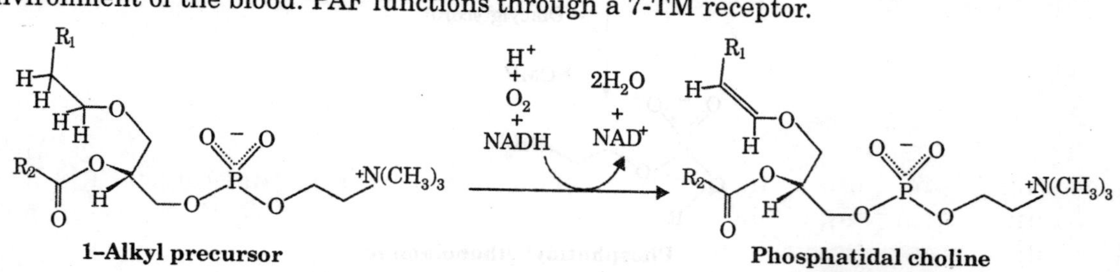

1–Alkyl precursor **Phosphatidal choline** (a plasmalogen)

Plasmalogens are phospholipids containing an χ, δ-unsaturated ether at C-1. Phosphatidal choline, the plasmalogen corresponding to phosphatidyl choline, is formed by desaturation of a 1-alkyl precursor.

The desaturase catalyzing this final step in the synthesis of a plasmalogen is an endoplasmic reticulum enzyme akin to the one that introduces double bonds into long-chain fatty acyl CoA molecules. In both cases, O_2 and NADH are reactants, and cytochrome b_5 participates in catalysis

Sphingolipids

We turn now from glycerol-based phospholipids to another class of membrane lipid — the *sphingolipids.* These lipids are found in the plasma membranes of all eukaryotic cells, although the concentration is highest in the cells of the central nervous system. The backbone of a sphingolipid is *sphingosine,* rather than glycerol. Palmitoyl CoA and serine condense to form

Palmitoyl CoA + Serine —(H^+; CO_2 + CoA)→ Dehydro-Sphingosine —(H^+ + NADPH; $NADP^+$)→ Dihydro-sphingosine —(FAD; $FADH_2$)→ Sphingosine

dehydrosphingosine, which is then converted into sphingosine. The enzyme catalyzing this reaction requires pyridoxal phosphate, revealing again the dominant role of this cofactor in transformations that include amino acids.

In all sphingolipids, the amino group of sphingosine is acylated: a long-chain acyl CoA reacts with sphingosine to form *ceramide (N-acyl sphingosine).* The terminal hydroxyl group also is substituted. In *sphingomyelin,* a component of the myelin sheath covering many nerve fibers, the substituent is phosphorylcholine, which comes from phosphatidyl choline. In a *cerebroside,* the substituent is glucose or galactose. UDP-glucose or UDP-galactose is the sugar donor. In a *ganglioside,* an oligosaccharide is linked to the terminal hydroxyl group of ceramide by a glucose residue

Carbohydrate-Rich Sphingolipids

In gangliosides, the most complex sphingolipids, an *oligosaccharide chain* attached to the ceramide contains at least one acidic sugar. The acidic sugar is *N-acetylneuraminate* or *N-glycolylneuraminate.* These acidic sugars are called *sialic acids.* Their nine-carbon backbones are synthesized from phosphoenolpyruvate (a three-carbon unit) and *N*-acetylmannosamine 6-phosphate (a six-carbon unit).

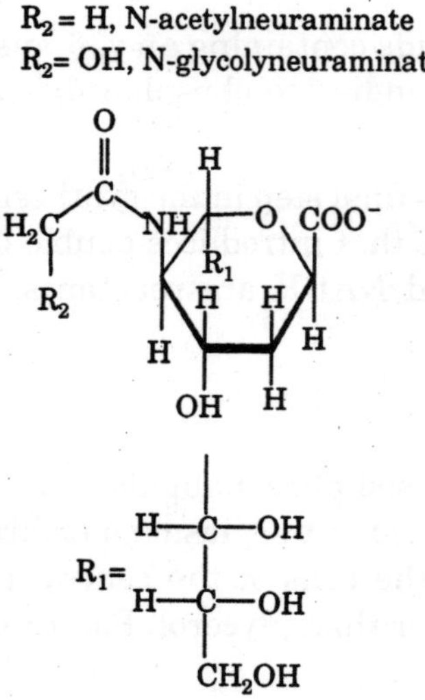

R_2 = H, N-acetylneuraminate
R_2= OH, N-glycolyneuraminate

R_1=

Gangliosides are synthesized by the ordered, step-by-step addition of sugar residues to ceramide. The synthesis of these complex lipids requires the activated sugars UDP-glucose, UDP-galactose, and UDP-*N*-acetylgalactosamine, as well as the CMP derivative of *N*-acetylneuraminate. CMP-*N*-acetylneuraminate is synthesized from CTP and *N*-acetylneuraminate. The structure of the resulting ganglioside is determined by the specificity of the glycosyltransferases in the cell. More than 60 different gangliosides have been characterized.

Diversity on Lipid Structure and Function

The structures of sphingolipids and the more abundant glycerophospholipids are very similar. Given the structural similarity of these two types of lipids, why are sphingolipids required at all? Indeed, the prefix "sphingo" was applied to capture the "sphinxlike" properties of this enigmatic class of lipids. Although the precise role of sphingolipids is not firmly established, progress toward solving the riddle of their function is being made. The most notable function attributed to sphingolipids is their role as a source of second messengers. For instance, ceramide derived from a sphingolipid may initiate programmed cell death in some cell types.

Respiratory Distress Syndrome and Tay-Sachs Disease

Respiratory distress syndrome is a pathological condition resulting from a failure in the biosynthetic pathway for dipalmitoyl phosphatidyl choline. This phospholipid, in conjunction with specific proteins and other phospholipids, is found in the extracellular fluid that surrounds the alveoli of the lung, where it decreases the surface tension of the fluid to prevent lung collapse at the end of the expiration phase of breathing. Premature infants may suffer from respiratory distress syndrome because their immature lungs do not synthesize enough dipalmitoyl phosphatidyl choline.

Tay-Sachs disease is caused by a failure of lipid degradation: an inability to degrade gangliosides. Gangliosides are found in highest concentration in the nervous system, particularly in gray matter, where they constitute 6% of the lipids. Gangliosides are normally degraded inside lysosomes by the sequential removal of their terminal sugars but, in Tay-Sachs disease, this degradation does not occur. As a consequence, neurons become enormously swollen with lipid-filled lysosomes An affected infant displays weakness and retarded psychomotor skills before 1 year of age. The child is demented and blind by age 2 and usually dead before age 3.

The ganglioside content of the brain of an infant with Tay-Sachs disease is greatly elevated. *The concentration of ganglioside G_{M^2} is many times as high as normal because its terminal N-acetylgalactosamine residue is removed slowly or not at all.* The missing or deficient enzyme is a specific 8 *-N-acetylhexosaminidase.*

GalNAc — Gal — Glc — Ceramide $\xrightarrow{H_2O}$ GalNAc + Gal — Glc — Ceramide

NAN | Gangliloside G_{M2} | NAN | Gangliloside G_{M3}

Tay-Sachs disease can be diagnosed in the course of fetal development. Amniotic fluid is obtained by amniocentesis and assayed for S-TV-acetylhexosaminidase activity.

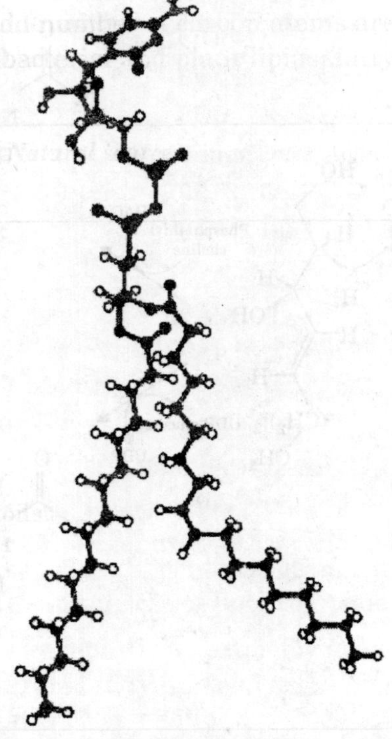

Fig. 7.1. Structure of CDP-Diacylglycerol. A key intermediate in the synthesis of phospholipids consists of phosphatidate and CMP joined by a pyrophosphate linkage

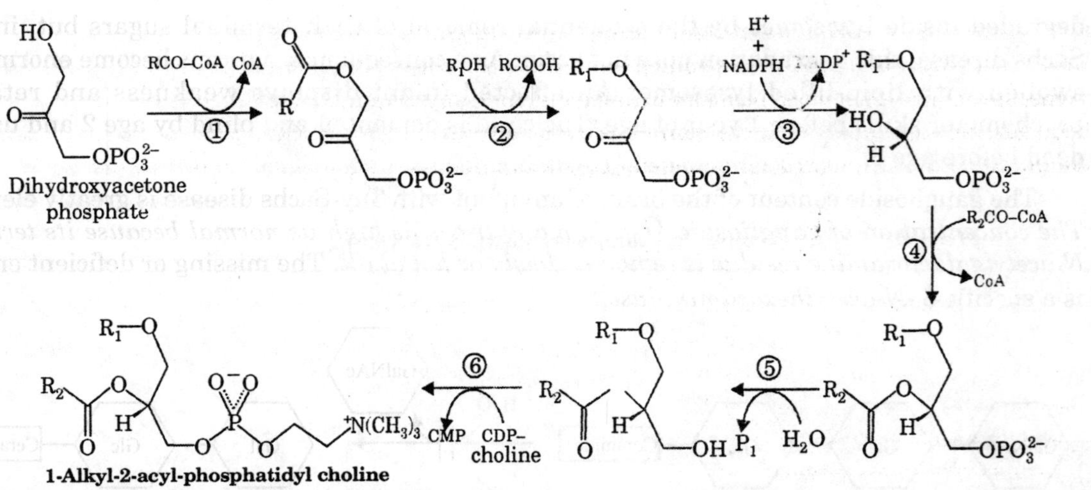

Dihydroxyacetone phosphate

1-Alkyl-2-acyl-phosphatidyl choline

Fig. 7.2. Synthesis of an Ether Phospholipid. Steps in the synthesis include (1) acylation of dihydroxyacetone phosphate by acyl CoA, (2) exchange of an alcohol for the carboxylic acid, (3) reduction by NADPH, (4) acylation by a second acyl CoA, (5) hydrolysis of the phosphate ester, and (6) transfer of a phosphocholine moiety.

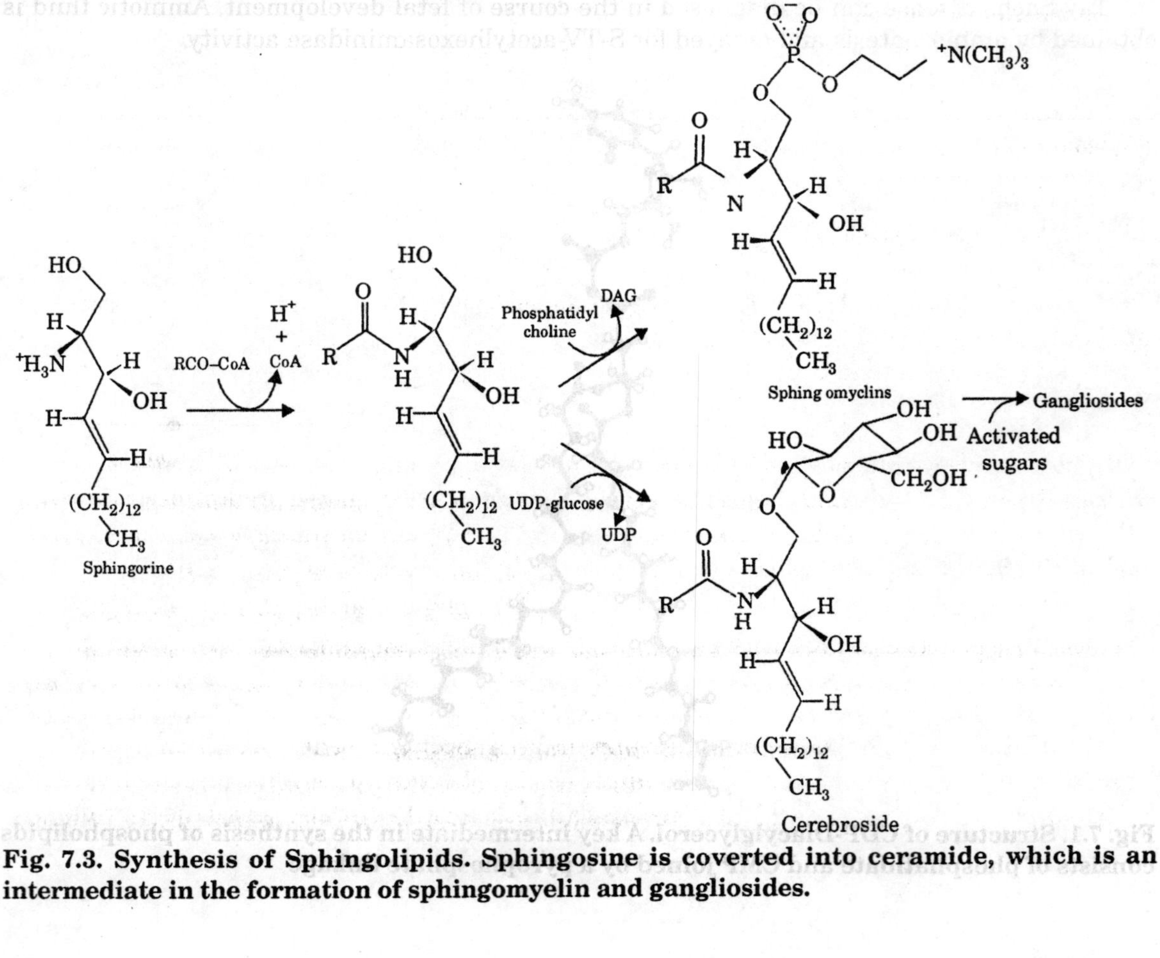

Sphingorine

Sphing omyclins

Gangliosides

Activated sugars

Cerebroside

Fig. 7.3. Synthesis of Sphingolipids. Sphingosine is coverted into ceramide, which is an intermediate in the formation of sphingomyelin and gangliosides.

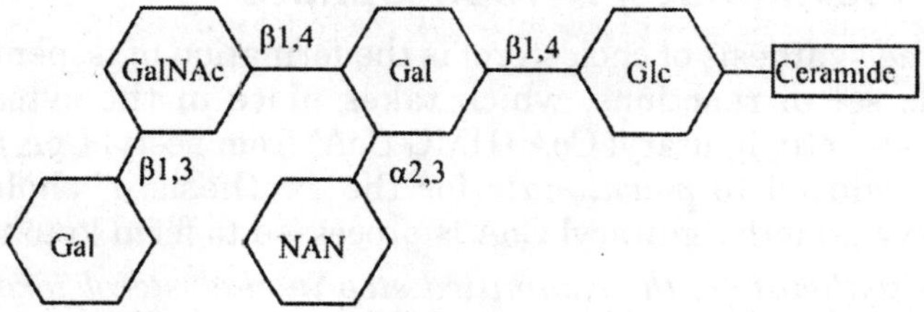

Fig. 7.4. Ganglioside G_{M1}. This ganglioside consists of five monosaccharides linked to ceramide: one glucose (Glc) molecule, two galactose (Gal) molecules, one *N*-acetylgalactosamine (GalNAc) Molecule and one *N*-acetylneuraminate (NAN) molecules. The structures of the linkages are indicated.

CHOLESTEROL FROM ACETYL COENZYME A

We now turn our attention to the synthesis of the fundamental lipid *cholesterol.* This steroid modulates the fluidity of animal cell membranes and is the precursor of steroid hormones such as progesterone, testosterone, estradiol, and cortisol. *All 27 carbon atoms of cholesterol are derived from acetyl CoA* in a three-stage synthetic process

"Cholesterol is the most highly decorated small molecule in biology. Thirteen Nobel Prizes have been awarded to scientists who devoted major parts of their careers to cholesterol. Ever since it was isolated from gallstones in 1784, cholesterol has exerted an almost hypnotic fascination for scientists from the most diverse areas of science and medicine.... Cholesterol is a Janus-faced molecule. The very property that makes it useful in cell membranes, namely its absolute insolubility in water, also makes it lethal."

1. Stage one is the synthesis of isopentenyl pyrophosphate, an activated isoprene unit that is the key building block of cholesterol.
2. Stage two is the condensation of six molecules of isopentenyl pyrophosphate to form squalene.
3. In stage three, squalene cyclizes. in an astounding reaction and the tetracyclic product is subsequently converted into cholesterol.

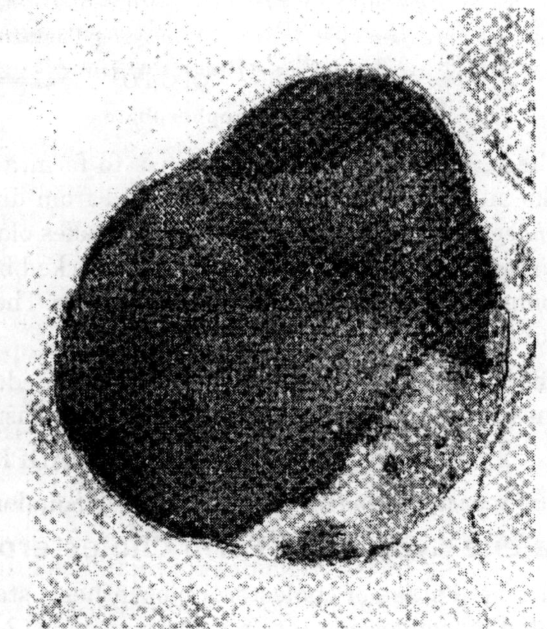

Fig. 7.5. Lysosome with Lipids. An electron micrograph of a lysosome containing an abnormal amount of lipid.

The Synthesis of Mevalonate, Pyrophosphate

The first stage in the synthesis of cholesterol is the formation of isopentenyl pyrophosphate from acetyl CoA. This set of reactions, which takes place in the cytosol, starts with the formation of 3-hydroxy-3-methylglutaryl CoA (HMG CoA) from acetyl CoA and acetoacetyl CoA. This intermediate is reduced to *mevalonate* for the synthesis of cholesterol. Recall that mitochondrial 3-hydroxy-3-methylglutaryl CoA is processed to form ketone bodies.

The synthesis of *mevalonate is the committed step in cholesterol formation.* The enzyme catalyzing this irreversible step, *3-hydroxy-3-methylglutaryl CoA reductase (HMG-CoA reductase),* is an important control site in cholesterol biosynthesis, as will be discussed shortly.

3-Hydroxy-3-methylglutaryl CoA + 2 NADPH + 2 H^+ → mevalonate + 2 $NADP^+$ + CoA

HMG-CoA reductase is an integral membrane protein in the endoplasmic·reticulum.

Mevalonate is converted into *3 -isopentenyl pyrophosphate* in three consecutive reactions requiring ATP. Decarboxylation yields isopentenyl pyrophosphate, an activated isoprene unit that is a key building block for many important biomolecules throughout the kingdoms of life. We will return to a discussion of this molecule later in the chapter.

Synthesis of Squalene (C_{30})

Squalene is synthesized from isopentenyl pyrophosphate by the reaction sequence

$$C_5 \rightarrow C_{10} \rightarrow C_{15} \rightarrow C_{30}$$

This stage in the synthesis of cholesterol starts with the isomerization *of isopentenyl pyrophosphate* to *dimethylallyl pyrophosphate.*

CH_3 | H_2C = ... $OPO_3PO_3^{3-}$ ⇌ CH_3 | H_3C ... $OPO_3PO_3^{3-}$

Isopentenyl pyrophosphate **Dimethylallyl pyrophosphate**

These isomeric C_5 units condense to form a C_{10} compound: isopentenyl pyrophosphate attacks an allylic carbonium ion formed from dimethylallyl pyrophosphate to yield *geranyl pyrophosphate.* The same kind of reaction takes place again: geranyl pyrophosphate is converted into an allylic carbonium ion, which is attacked by isopentenyl pyrophosphate. The resulting C_{15} compound is *called farnesyl pyrophosphate.* The same enzyme, *geranyl transferase,* catalyzes each of these condensations.

The last step in the synthesis of *squalene* is a reductive tail-to-tail condensation of two molecules of farnesyl pyrophosphate catalyzed by the endoplasmic reticulum enzyme *squalene synthase.*

2 Farnesyl pyrophosphate (C_{15}) + NADPH → squalene (C_{30}) + 2 PP_i +NADP + H^+

The reactions leading from C_5 units to squalene, a C_{30} isoprenoid, are summarized in

Squalene Cyclizes to Form Cholesterol

The final stage of cholesterol biosynthesis starts with the cyclization of Squalene is first activated by conversion into squalene epoxide (2,3-oxidosqualene) in a reaction that uses O_2 and NADPH. Squalene epoxide is then cyclized to *lanosterol* by *oxidosqualene cyclase.* This remarkable transformation proceeds in a concerted fashion. The enzyme holds squalene epoxide in an appropriate conformation and initiates the reaction by protonating the epoxide oxygen. The carbocation formed spontaneously rearranges to produce lanosterol. Lanosterol is converted into cholesterol in a multistep process by the removal of three methyl groups, the reduction of one double bond by NADPH, and the migration of the other double bond

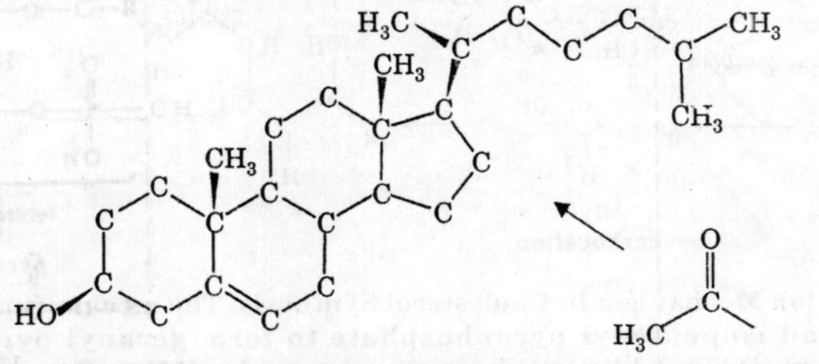

Fig. 7.6. Labeling of Cholesterol. The results of isotope-labeling experiments reveal the source of carbon atoms in cholesterol synthesized from acetate labeled in its methyl group (blue) or carboxylate atom (red).

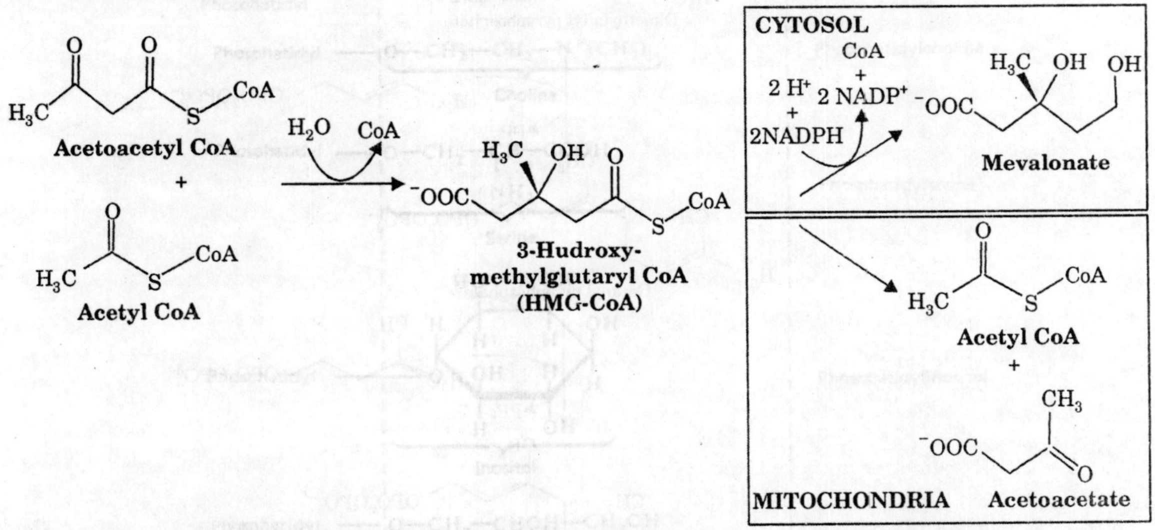

Fig. 7.7. Fates of 3-Hydroxy-3-Methylglutaryl CoA. In the cytosol, HMG-CoA is converted into mevalonate. In mitochondria, it is converted into acetyl CoA and acetoacetate.

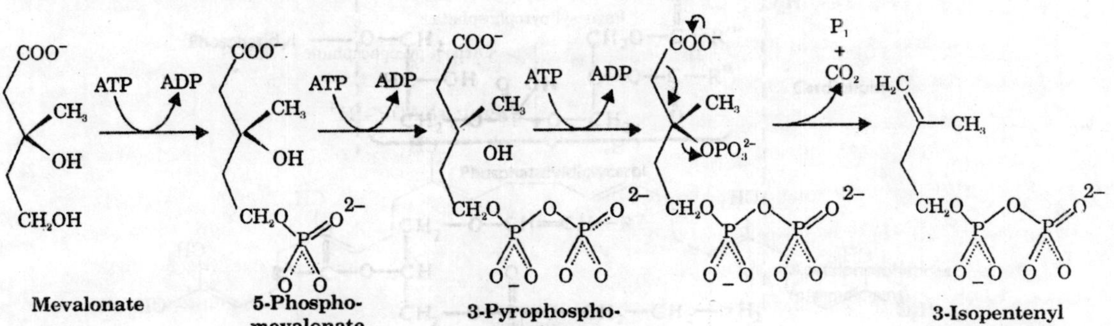

Fig. 7.8. Synthesis of Isopentenyl Pyrophosphate. This activated intermediate is formed from mevalonate in three steps, the last of which includes a decarboxylation.

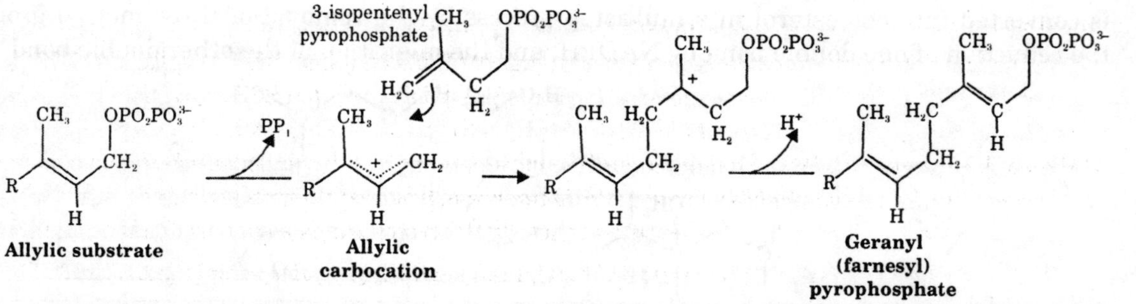

3-isopentenyl pyrophosphate

Allylic substrate

Allylic carbocation

Geranyl (farnesyl) pyrophosphate

Fig. 7.9. Condensation Mechanism in Cholesterol Synthesis. The mechanism for joining dimethylally pyrophosphate and isopentenyl pyrophosphate to form geranyl pyrophosphate. The same mechanism is used to add an additional isopentenyl pyrophosphate to form farnesyl pyrophosphate.

Dimethylallyt pyrophosphate

Geranyl pyrophosphate

Farnesyl pyrophosphate

Farneql pyrophosphate + NADPH

$2PP_3 + NADP^+ + H^+$

Squalene

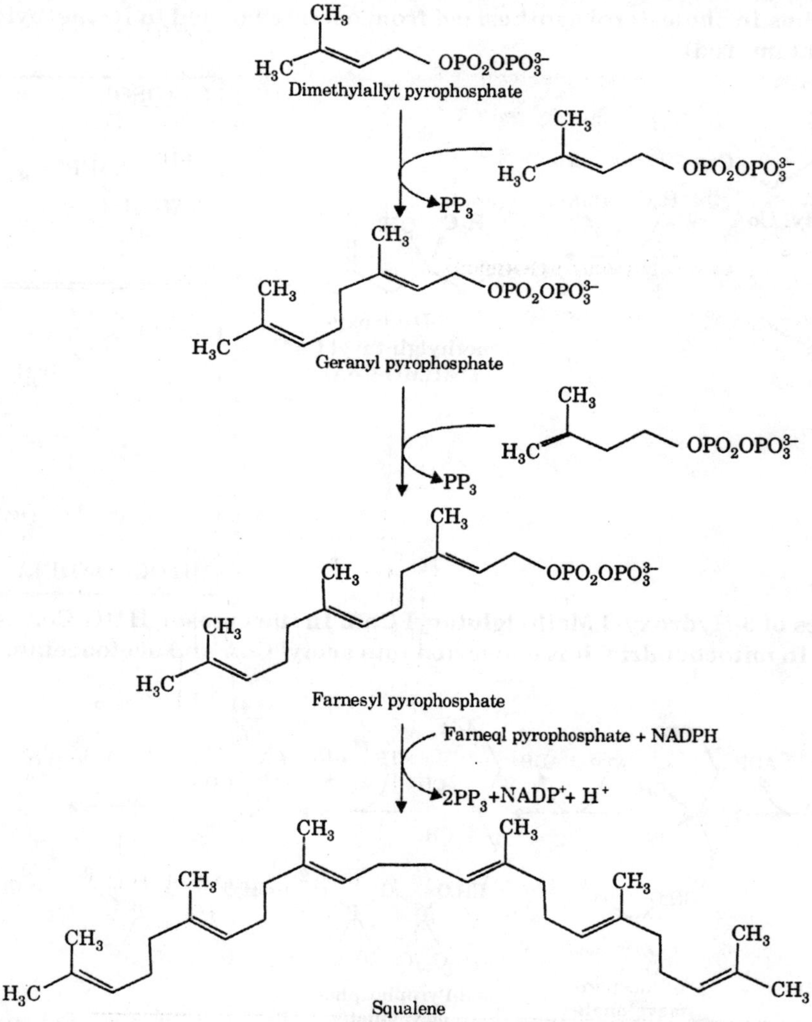

Fig. 7.10. Squalene Synthesis. One molecule of dimethyallyl pyrophosphate and two molecules of isopentenyl pyrophosphate condense to form farnesyl pyrophosphate. The tail-to-tail coupling of two molecules of farnesyl pyrophosphate yields squalene.

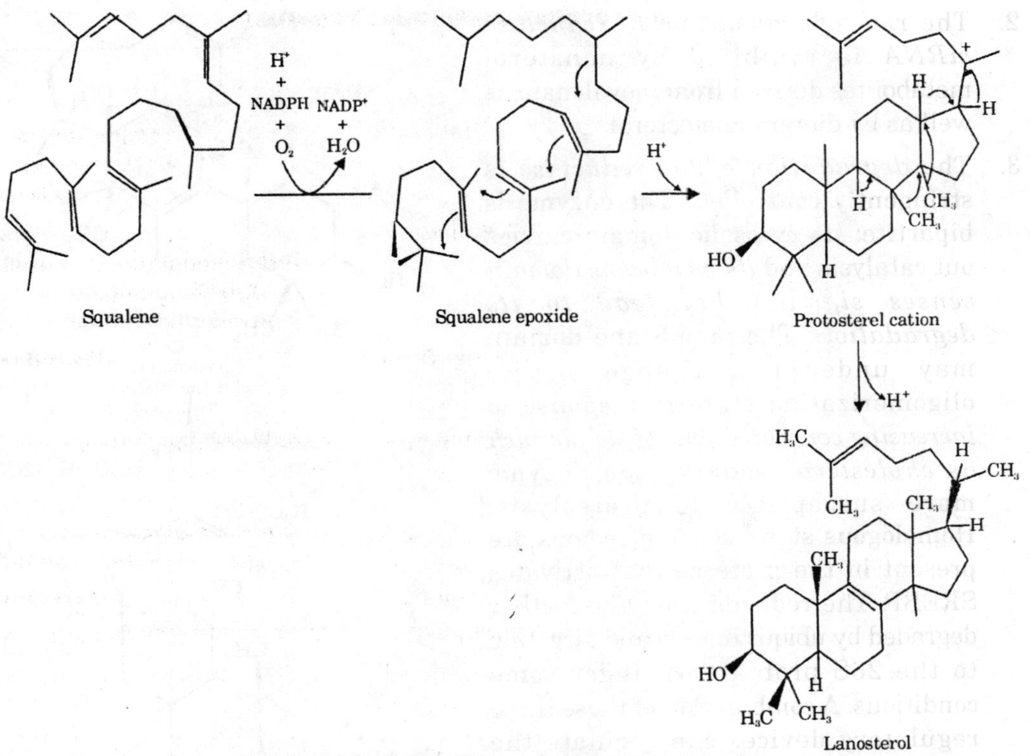

Squalene Squalene epoxide Protosterel cation

Lanosterol

Fig. 7.11. Sequalene Cyclization. The formation of the steroid nucleus from squalene begins with the formation of squalene epoxide. The intermediate is protonated to form a carbocation that cyclizes to form a tetracyclic structure, which rearranges to form lanosterol.

THE COMPLEX REGULATION

Cholesterol can be obtained from the diet or it can be synthesized de novo. An adult on a low-cholesterol diet typically synthesizes about 800 mg of cholesterol per day. The liver is the major site of cholesterol synthesis in mammals, although the intestine also forms significant amounts. The rate of cholesterol formation by these organs is highly responsive to the cellular level of cholesterol. *This feedback regulation is mediated primarily by changes in the amount and activity of 3-hydroxy-3-methylglutaryl CoA reductase.* As discussed in this enzyme catalyzes the formation of mevalonate, the committed step in cholesterol biosynthesis. HMG CoA reductase is controlled in multiple ways:

1. The rate *of synthesis of reductase mRNA* is controlled by the *sterol regulatory element binding protein (SREBP).* This transcription factor binds to a short DNA sequence called the *sterol regulatory element (SRE)* on the 5' side of the reductase gene. In its inactive state, the SREBP is anchored to the endoplasmic reticulum or nuclear membrane. When cholesterol levels fall, the amino-terminal domain is released from its association with the membrane by two specific proteolytic cleavages. The released protein migrates to the nucleus and binds the SRE of the HMG-CoA reductase gene, as well as several other genes in the cholesterol biosynthetic pathway, to enhance transcription. When cholesterol levels rise, the proteolytic release of the SREBP is blocked, and the SREBP in the nucleus is rapidly degraded. These two events halt the transcription of the genes of the cholesterol biosynthetic pathways.

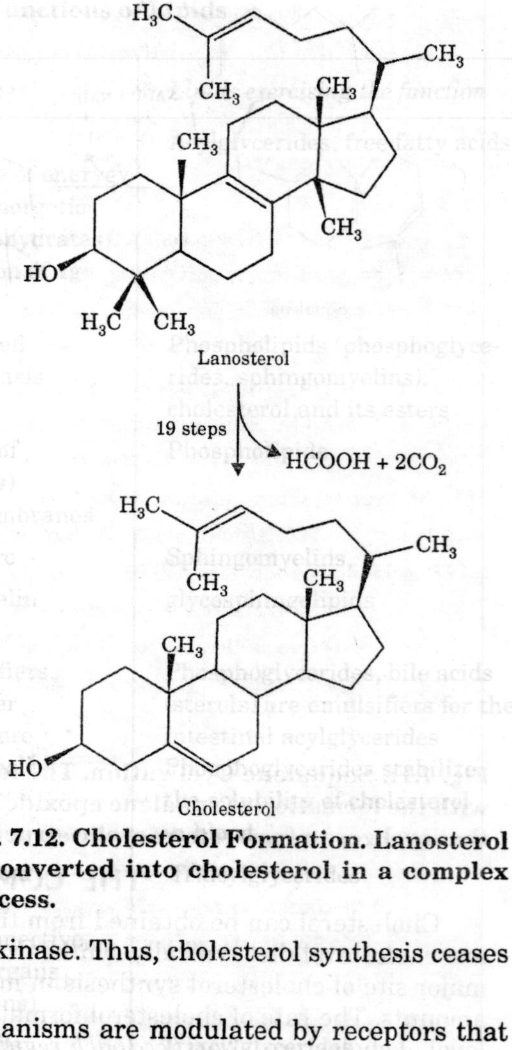

2. The rate *of translation of reductase mRNA* is inhibited by nonsterol metabolites derived from mevalonate as well as by dietary cholesterol.
3. The *degradation of the reductase* is stringently controlled. The enzyme is bipartite: its cytosolic domain carries out catalysis and *its membrane domain senses signals that lead to its degradation.* The membrane domain may undergo a change in its oligomerization state *in response to increasing concentrations of sterols such as cholesterol,* making the enzyme more susceptible to proteolysis. Homologous sterol-sensing regions are present in the protease that activates SREBP. The reductase may be further degraded by ubiquitination and targeting to the 26S proteasome under some conditions. A combination of these three regulatory devices can regulate the amount of enzyme over a 200-fold range.
4. *Phosphorylation decreases the activity of the reductase.* This enzyme, like acetyl CoA carboxylase (which catalyzes the committed step in fatty acid synthesis, is switched off by an AMP-activated protein kinase. Thus, cholesterol synthesis ceases when the ATP level is low.

Lanosterol

19 steps → $HCOOH + 2CO_2$

Cholesterol

Fig. 7.12. Cholesterol Formation. Lanosterol is converted into cholesterol in a complex process.

As we will see shortly, all four regulatory mechanisms are modulated by receptors that sense the presence of cholesterol in the blood

Transport of Cholesterol and Triacylglycerols

Cholesterol and triacylglycerols are transported in body fluids in the form of *lipoprotein particles.* Each particle consists of a core of hydrophobic lipids surrounded by a shell of more polar lipids and apoproteins. The protein components of these macromolecular aggregates have two roles: *they solubilize hydrophobic lipids and contain cell-targeting signals.* Lipoprotein particles are classified according to increasing density : *chylomicrons, chylomicron remnants, very low density lipoproteins (VLDL), intermediate-density lipoproteins (IDL), low-density lipoproteins (LDL),* and *high-density lipoproteins (HDL).* Ten principal apoproteins have been isolated and characterized. They are synthesized and secreted by the liver and the intestine.

Triacylglycerols, cholesterol, and other lipids obtained from the diet are carried away from the intestine in the form of large *chylomicrons* (180- 500 nm in diameter. These particles have a very low density (d <0.94 g cm^{-3}) because triacylglycerols constitute ~99% of their content. Apolipoprotein B-48 (apo B-48), a large protein (240 kd), forms an amphipathic spherical shell around the fat globule; the external face of this shell is hydrophilic. The triacylglycerols in chylomicrons are released through hydrolysis by *lipoprotein lipases.* These enzymes are located on the lining of blood vessels in muscle and other tissues that use fatty acids as fuels and in the synthesis of fat. The liver then takes up the cholesterol-rich residues, known as *chylomicron remnants.*

The liver is a major site of triacylglycerol and cholesterol synthesis . Triacylglycerols and cholesterol in excess of the liver's own needs are exported into the blood in the form of very low density lipoproteins *(d<1.006* g cm^{-3}). These particles are stabilized by two lipoproteins — apo B-100 and apo E (34 kd). Apo B-100, one of the largest proteins known (513 kd), is a longer version of apo B-48. Both apo B proteins are encoded by the same gene and produced from the same initial RNA transcript. In the intestine, RNA editing modifies the transcript to generate the mRNA for apo B-48, the truncated form. Triacylglycerols in very low density lipoproteins, as in chylomicrons, are hydrolyzed by lipases on capillary surfaces. The resulting remnants, which are rich in cholesteryl esters, are called *intermediate-density lipoproteins* ($1.006 < d < 1.019$ g cm^{-3}). These particles have two fates. Half of them are taken up by the liver for processing, and half are converted into low-density lipoprotein ($1.019 < d < 1.063$ g cm^{-3}) by the removal of more triacylglycerol.

Low-density lipoprotein is the major carrier of cholesterol in blood. This lipoprotein particle has a diameter of 22 nm and a mass of about 3 million daltons. It contains a core of some 1500 esterified cholesterol molecules; the most common fatty acyl chain in these esters is linoleate, a polyunsaturated fatty acid. A shell of phospholipids and unesterified cholesterols surrounds this highly hydrophobic core. The shell also contains a single copy of apo B-100, which is recognized by target cells. *The role of LDL is to transport cholesterol to peripheral tissues and regulate de novo cholesterol synthesis at these sites,* as described in. A different purpose is served by *high-density lipoprotein* ($1.063 < d < 1.21$ g cm^{-3}), which picks up cholesterol released into the plasma from dying cells and from membranes undergoing turnover. An acyltransferase in HDL esterifies these cholesterols, which are then either rapidly

Levels of Lipoproteins

High serum levels of cholesterol cause disease and death by contributing to the formation of atherosclerotic plaques in arteries throughout the body. This excess cholesterol is present in the form of the low density lipoprotein particle, so-called "bad cholesterol."The ratio of cholesterol in the form of high density lipoprotein, sometimes referred to as "good cholesterol," to that in the form of LDL can be used to evaluate susceptibility to the development of heart disease. For a healthy person, the LDL/HDL ratio is 3.5.

High-density lipoprotein functions as a shuttle that moves cholesterol throughout the body. HDL binds and esterifies cholesterol released from the peripheral tissues and then transfers cholesteryl esters to the liver or to tissues that use cholesterol to synthesize steroid hormones. A specific receptor mediates the docking of the HDL to these tissues. The exact nature of the protective effect of HDL levels is not known; however, a possible mechanism is discussed in.

Cholesterol Metabolism

Cholesterol metabolism must be precisely regulated to prevent atherosclerosis. The mode of control in the liver, the primary site of cholesterol synthesis, has already been discussed: dietary cholesterol reduces the activity and amount of 3-hydroxy-3-methylglutaryl CoA reductase, the enzyme catalyzing the committed step. The results of studies by Michael Brown and Joseph Goldstein are sources of insight into the control of cholesterol metabolism in nonhepatic cells. In general, cells outside the liver and intestine obtain cholesterol from the plasma rather than synthesizing it de novo. Specifically, *their primary source of cholesterol is the low-density lipoprotein.* The process of LDL uptake, called *receptor-mediated endocytosis,* serves as a paradigm for the uptake of many molecules.

The steps in the receptor-mediated endocytosis of LDL are as follows :

1. Apolipoprotein B-100 on the surface of an LDL particle binds to a specific receptor protein on the plasma membrane of nonhepatic cells. The receptors for LDL are localized in specialized regions called *coated pits,* which contain a specialized protein called *clathrin.*
2. The receptor-LDL complex is internalized by *endocytosis,* that is, the plasma membrane in the vicinity of the complex invaginates and then fuses to form an endocytic vesicle.
3. These vesicles, containing LDL, subsequently fuse with *lysosomes,* acidic vesicles that carry a wide array of degradative enzymes. The protein component of the LDL is hydrolyzed to free amino acids. The cholesteryl esters in the LDL are hydrolyzed by a lysosomal acid lipase. The LDL receptor itself usually returns unscathed to the plasma membrane. The round-trip time for a receptor is about 10 minutes; in its lifetime of about a day, it may bring many LDL particles into the cell.
4. *The released unesterified cholesterol can then be used for membrane biosynthesis.* Alternatively, it can be *reesterified for storage inside the cell.* In fact, free cholesterol activates *acyl CoA:cholesterol acyltransferase (ACAT),* the enzyme catalyzing this reaction. Reesterified cholesterol contains mainly oleate and palmitoleate, which are monounsaturated fatty acids, in contrast with the cholesterol esters in LDL, which are rich in linoleate, a polyunsaturated fatty acid. It is imperative that the cholesterol be reesterified. High concentrations of unesterified cholesterol disrupt the integrity of cell membranes.

The synthesis of LDL receptor is itself subject to feedback regulation. The results of studies of cultured fibroblasts show that, *when cholesterol is abundant inside the cell, new LDL receptors are not synthesized, and so the uptake of additional cholesterol from plasma LDL is blocked.* The gene for the LDL receptor, like that for the reductase, is regulated by SREBP, which binds to a sterol regulatory element that controls the rate of mRNA synthesis.

The LDL Receptor

The amino acid sequence of the human LDL receptor reveals the mosaic structure of this 115-kd protein, which is composed of six different types of domain. The amino-terminal region of the mature receptor consists of a cysteine-rich sequence of about 40 residues that is repeated, with some variation, seven times to form the LDL-binding domain. A set of conserved acidic side chains in this domain bind calcium ion; this metal ion lies at the center of each domain and, along with disulfide bonds formed from the conserved cysteine residues, stabilizes the three-dimensional structure. Protonation of these glutamate and aspartate side chains of the receptor in lysosomes leads to the release of calcium and hence to structural disruption and the release of LDL from its receptor. A second region of the LDL receptor includes two types of recognizable domains, three domains homologous to epidermal growth factor and six repeats that are similar to the blades of the transducin δ subunit. The six repeats form a propeller-like structure that packs against one of the EGF-like domains. An aspartate residue forms hydrogen bonds that hold each blade to the rest of the structure. These interactions, too, would most likely be disrupted at the low pH in the lysosome.

The third region contains a single domain that is very rich in serine and threonine residues and contains O-linked sugars. These oligosaccharides may function as struts to keep the receptor extended from the membrane so that the LDL-binding domain is accessible to LDL. The fourth region contains the fifth type of domain, which consists of 22 hydrophobic residues that span the plasma membrane. The final region contains the sixth type of domain; it consists of 50 residues and emerges on the cytosolic side of the membrane, where it controls the interaction of the receptor with coated pits and participates in endocytosis. The gene for the LDL receptor consists of 18 exons, which correspond closely to the structural units of the protein. *The LDL receptor is a striking example of a mosaic protein encoded by a gene that was assembled by exon shuffling.*

Hypercholesteremia and Atherosclerosis

The results of Brown and Goldstein's pioneering studies *of familial hypercholesterolemia* revealed the physiologic importance of the LDL receptor. The total concentration of cholesterol and LDL in the plasma is markedly elevated in this genetic disorder, which results from a mutation at a single autosomal locus. The cholesterol level in the plasma of homozygotes is typically 680 mg dl^{-1}, compared with 300 mg dl^{-1} in heterozygotes (clinical assay results are often expressed in milligrams per deciliter, which is equal to milligrams per 100 milliliters). A value of < 200 mg dl^{-1} is regarded as desirable, but many people have higher levels. *In familial hypercholesterolemia, cholesterol is deposited in various tissues because of the high concentration of LDL cholesterol in the plasma.* Nodules of cholesterol called *xanthomas* are prominent in skin and tendons. Of particular concern is the oxidation of the excess blood LDL to form oxidized LDL (oxLDL). The oxLDL is taken up by immune-system cells called macrophages, which become engorged to form foam cells. These foam cells become trapped in the walls of the blood vessels and contribute to the formation of atherosclerotic plaques that cause arterial narrowing and lead to heart attacks. In fact, *most homozygotes die of coronary artery disease in childhood.* The disease in heterozygotes (1 in 500 people) has a milder and more variable clinical course. A serum esterase that degrades oxidized lipids is found in association with HDL. Possibly, the HDL-associated protein destroys the oxLDL, accounting for HDL's ability to protect against coronary disease.

The molecular defect in most cases of familial hypercholesterolemia is an absence or deficiency of functional receptors for LDL. Receptor mutations that disrupt each of the stages in the endocytotic pathway have been identified. Homozygotes have almost no functional receptors for LDL, whereas heterozygotes have about half the normal number. Consequently, the entry of LDL into liver and other cells is impaired, leading to an increased plasma level of LDL. Furthermore, less IDL enters liver cells because IDL entry, too, is mediated by the LDL receptor. Consequently, IDL stays in the blood longer in familial hypercholesterolemia, and more of it is converted into LDL than in normal people. All deleterious consequences of an absence or deficiency of the LDL receptor can be attributed to the ensuing elevated level of LDL cholesterol in the blood.

The Clinical Management of Cholesterol-Levels

Homozygous familial hypercholesterolemia can be treated only by a liver transplant. A more generally applicable " therapy is available for heterozygotes and others with high levels of cholesterol. *The goal is to reduce the amount of cholesterol in the blood by stimulating the single normal gene to produce more than the customary number of LDL receptors.* We have already observed that the production of LDL receptors is controlled by the cell's need for cholesterol. Therefore, in essence, the strategy is to deprive the cell of ready sources of cholesterol. When cholesterol is required, the amount of mRNA for the LDL receptor rises and more receptor is found on the cell surface. This state can be induced by a two-pronged approach. First, the intestinal reabsorption of bile salts is inhibited. Bile salts are cholesterol derivatives that promote the absorption of dietary cholesterol and dietary fats. Second, de novo synthesis of cholesterol is blocked.

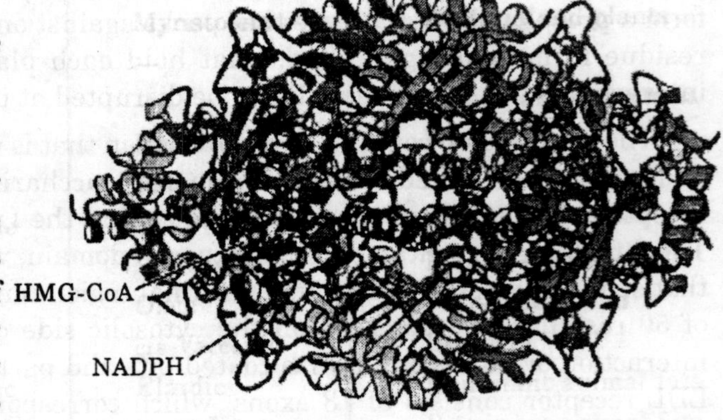

Fig. 7.13. HMG-CoA Reductase. The structure of a portion of the tetrameric enzyme is shown.

Table 7.1. Properties of Plasma Lipoproteins

Lipoproteins	*Major core lipids*	*Apoproteins*	*Mechanism of lipid delivery*
Chylomicron	Dietary triacylglycerols	B-48, C, E	Hydrolysis by lipoprotein lipase
Chylomicron remnant	Dietary cholesterol esters	B-48, E	Receptor-mediated endocytosis by liver
Very low density lipoprotein (VLDL)	Endogenous triacylglycerols	B-100, C, E	Hydrolysis by lipoprotein lipase
Intermediate-density lipoprotein (IDL)	Endogenous cholesterol esters	B-100, E	Receptor-mediated endocytosis by liver and conversion into LDL
Low-density lipoprotein (LDL)	Endogenous cholesterol esters	B-100	Receptor-mediated endocytosis by liver and other tissues
High-density lipoprotein	Endogenous cholesterol esters	A	Transfer of cholesterol esters to IDL and LDL

The reabsorption of bile is impeded by oral administration of positively charged polymers, such as cholestyramine, that bind negatively charged bile salts and are not themselves absorbed. Cholesterol synthesis can be effectively blocked by a class of compounds called *stating* (*e.g.,* lovastatin, which is also called mevacor;. These compounds are potent competitive inhibitors (K_i < 1 nM) of HMG-CoA reductase, the essential control point in the biosynthetic pathway. Plasma cholesterol levels decrease by 50% in many patients given both lovastatin and inhibitors of bile-salt reabsorption. Lovastatin and other inhibitors of HMG-CoA reductase are widely used to lower the plasma cholesterol level in people who have atherosclerosis, which is the leading cause of death in industrialized societies.

IMPORTANT DERIVATIVES OF CHOLESTEROL

Cholesterol is a precursor for other important steroid molecules: the bile salts, steroid hormones, and vitamin D.

Bile Salts

As polar derivatives of cholesterol, *bile salts* are highly effective *detergents* because they contain both polar and nonpolar regions. Bile salts are synthesized in the liver, stored and

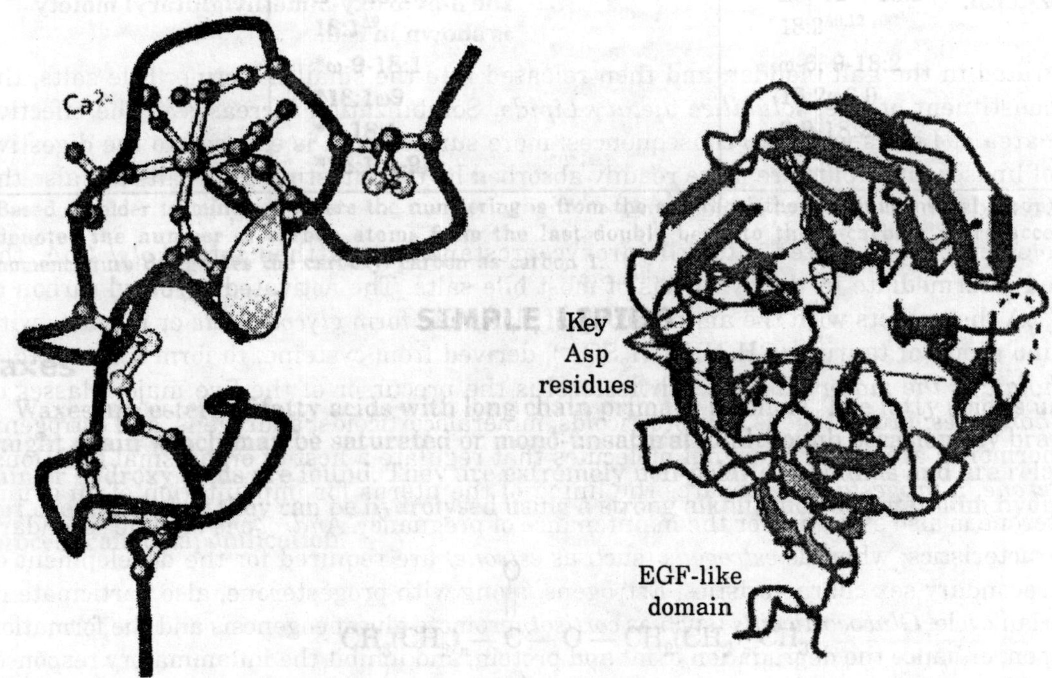

Fig. 7.14. Structure of Cysteine-Rich Domain. This calcium-bonding cysteine-rich domain is repeated seven times at the amino terminus of the LDL receptor.

Fig. 7.15. Structure of Propeller Domain. The six-bladed propeller domain and an an adjacement EGF-like domain of the LDL receptor.

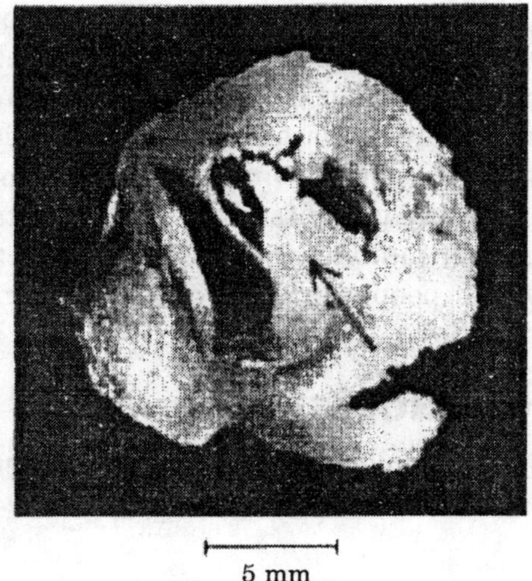

Fig. 7.16. An Atherosclerotic Plaque. A plaque (marked by an arrow) blocks most of the lumen of this blood vessel. The plaque is rich in cholesterol.

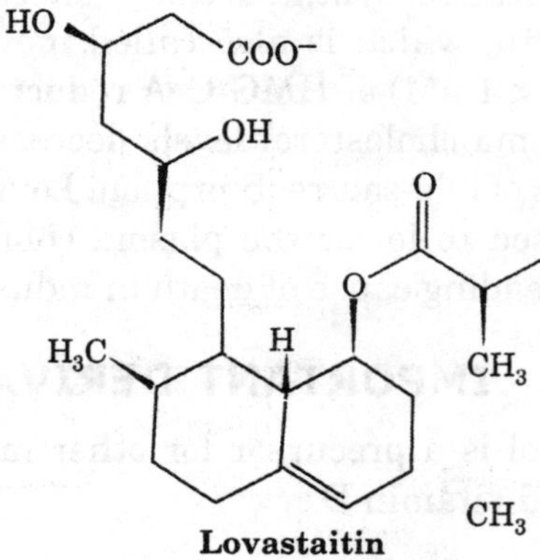

HO, COO^-, OH, O, O, CH_3, H_3C, H, CH_3, CH_3

Lovastaitin

Fig. 7.17. Lovastatin, a Competitive Inhibitor of HMG-CoA Recluctase. The part of the structure that resembles the 3-hydroxy-3-methylglutaryl moiety is shown in red.

concentrated in the gall bladder, and then released into the small intestine. Bile salts, the major constituent of bile, *solubilize dietary Lipids*. Solubilization increases in the effective surface area of Lipids with two consequences: more surface area is exposed to the digestive action of lipases and lipids are more readily absorbed by the intestine. Bile salts are also the major breakdown products of cholesterol.

Cholesterol is converted into trihydroxycoprostanoate and then into *cholyl CoA,* the activated intermediate in the synthesis of most bile salts. The activated carboxyl carbon of cholyl CoA then reacts with the amino group of glycine to form *glycocholate* or it reacts with the amino group of taurine ($CH_2NH_2CH_2SO_3^-$), derived from cysteine, to form *taurocholate. Glycocholate is the major bile salt.* Cholesterol is the precursor of the five major classes *of steroid hormones:* progestagens, glucocorticoids, mineralocorticoids, androgens, and estrogens. These hormones are powerful signal molecules that regulate a host of organismal functions. *Progesterone,* a *progestagen,* prepares the lining of the uterus for implantation of an ovum. Progesterone is also essential for the maintenance of pregnancy. *Androgens* of male secondary sex characteristics, whereas *estrogens* (such as *estrone)* are required for the development of female secondary sex characteristics. Estrogens, along with progesterone, also participate in the ovarian cycle. *Glucocorticoids* (such as *cortisot)* promote gluconeogenesis and the formation of glycogen, enhance the degradation of fat and protein, and inhibit the inflammatory response. They enable animals to respond to stress—indeed, the absence of glucocorticoids can be fatal. *Mineralocorticoids* (primarily *aldosterone)* act on the distal tubules of the kidney to increase the reabsorption of Na^+ and the excretion of K^+ and H^+, which leads to an increase in blood volume and blood pressure. The major sites of synthesis of these classes of hormones are the corpus luteum, for progestagens; the ovaries, for estrogens; the testes, for androgens; and the adrenal cortex, for glucocorticoids and mineralocorticoids.

Steroid hormones bind to and activate receptor molecules that serve as transcription factors to regulate gene expression. These small, relatively similar molecules are able to have greatly differing effects because the slight structural differences among them allow interactions with specific receptor molecules.

The Nomenclature of Steroid Hormones

Carbon atoms in steroids are numbered as shown for cholesterol in. the rings in steroids are denoted by the letters A, B, C, and D. Cholesterol contains two angular methyl groups: the C-19 methyl group is attached to C-10, and the C-18 methyl group is attached to C-13. The C-18 and C-19 methyl groups of cholesterol lie *above* the plane containing the four rings. A substituent that is above the plane is termed δ *oriented,* whereas a substituent that is below the plane is χ *oriented.*

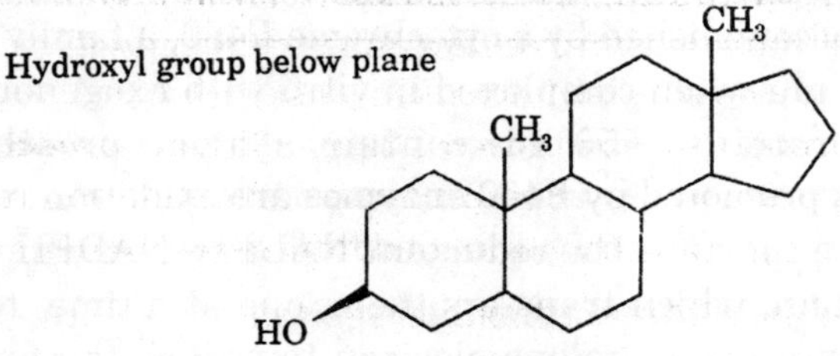

Hydroxyl group below plane

CH_3 CH_3 HO

3β-Hydroxy

Hydroxyl group below plane

CH_3 CH_3 HO

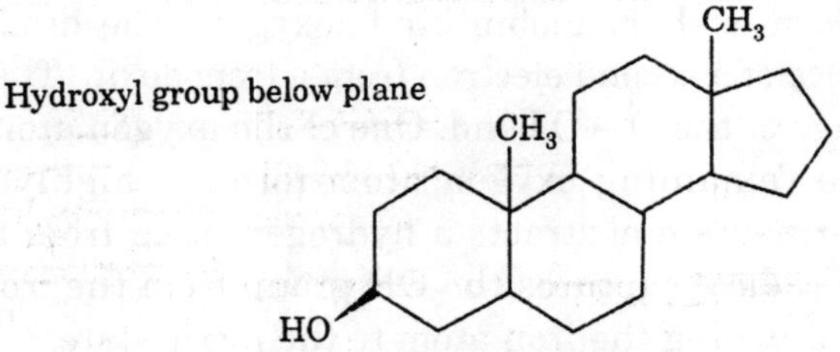

3α-Hydroxy

If a hydrogen atom is attached to C-5, it can be either χ or δ oriented. The A and B steroid rings are fused in a *trans* conformation if the C-5 hydrogen is < *oriented,* and *cis* if it is < *oriented.* The absence of a Greek letter for the C-5 hydrogen atom on the steroid nucleus implies a trans fusion. The C-5 hydrogen atom is χ oriented in all steroid hormones that contain a hydrogen atom in that position. In contrast, bile salts have a δ-oriented hydrogen atom at C-5. Thus, *a cis fusion is characteristic of the bile salts, whereas a transfusion is characteristic of all steroid hormones that possess a hydrogen atom at C-5.* A trans fusion yields a nearly planar structure, whereas a cis fusion gives a buckled structure.

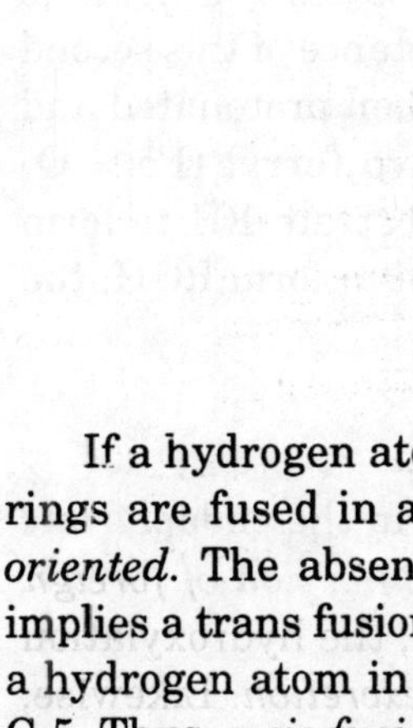

CH_3 CH_3 H

5β-Hydrogen
(cis fusion)

CH_3 CH_3 H

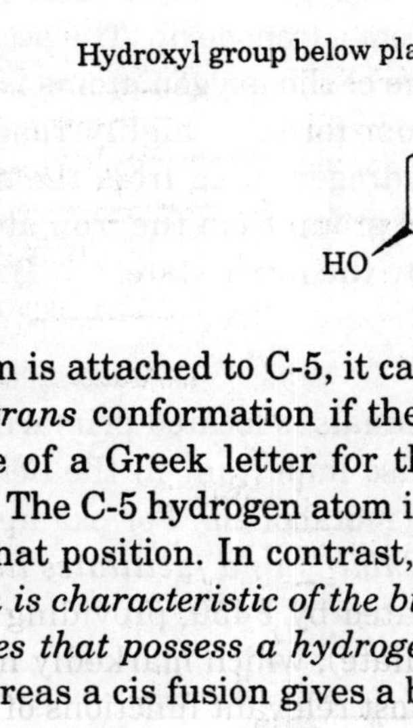

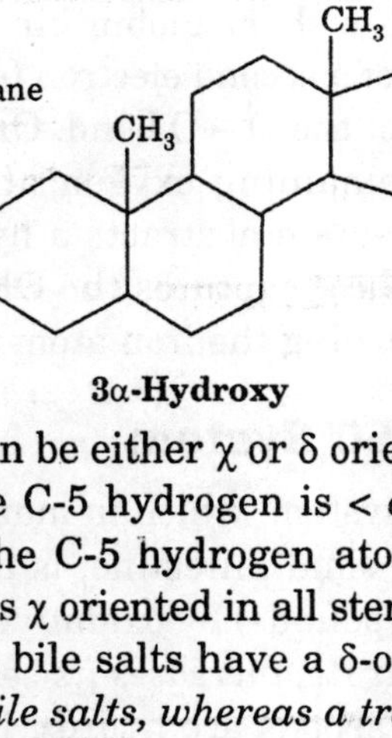

5β-Hydrogen
(cis fusion)

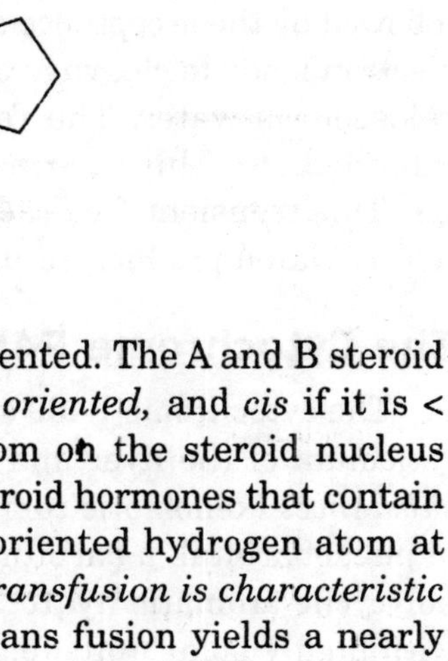

Steroids

Hydroxylation reactions play a very important role in the synthesis of cholesterol from squalene and in the conversion of cholesterol into steroid hormones and bile salts. All these hydroxylations require *NADPH and O_2*. The oxygen atom of the incorporated hydroxyl group comes from O_2 rather than from H_2O. While one oxygen atom of the O_2 molecule goes into the substrate, the other is reduced to water. The enzymes catalyzing these reactions are called *monooxygenases* (or *mixed-function oxygenases)*. Recall that a monooxygenase also participates in the hydroxylation of aromatic amino acids.

$$RH + O_2 + NADPH + H^+ \longrightarrow ROH + H_2O + NADP^+$$

Hydroxylation requires the activation of oxygen. In the synthesis of steroid hormones and bile salts, activation is accomplished by a cytochrome P450, a family of cytochromes that absorb light maximally at 450 nm when complexed in vitro with exogenous carbon monoxide. These membraneanchored proteins (~50 kd) contain a heme prosthetic group. Because the hydroxylation reactions promoted by P450 enzymes are oxidation reactions, it is at first glance surprising that they also consume the reductant NADPH. NADPH transfers its high-potential electrons to a flavoprotein, which transfers them, one at a time, to *adrenodoxin,* a nonheme iron protein. Adrenodoxin transfers one electron to reduce the ferric (Fe^{2+}) form of P450 to the ferrous (Fe^{2+}) Without the addition of this electron, P450 will not bind oxygen. Recall that only the ferrous form of hemoglobin binds oxygen. The binding of O_2 to the heme is followed by the acceptance of a second electron from adrenodoxin. The acceptance of this second electron leads to cleavage of the O – O bond. One of the oxygen atoms is then protonated and released as water. The remaining oxygen atom forms a highly reactive ferryl (Fe = O) intermediate. This intermediate abstracts a hydrogen atom from the substrate RH to form R•. This transient free radical captures the OH group from the iron atom to form ROH, the hydroxylated product, returning the iron atom to the ferric state.

The Cytochrome P450 System

The cytochrome P450 system, which in mammals is located primarily in the endoplasmic reticulum of the liver and small intestine, is also important in the *detoxification of foreign substances* (xenobiotic compounds) by oxidative metabolism. For example, the hydroxylation of phenobarbital, a barbiturate, *increases its solubility and facilitates its excretion.* Likewise, polycyclic aromatic hydrocarbons are hydroxylated by P450, providing sites for conjugation with highly polar units (e.g., glucuronate or sulfate), which markedly increase the solubility of the modified aromatic molecule. One of the most relevant functions of the cytochrome P450 system to human beings is its role in drug metabolism. Drugs such as caffeine and ibuprofen are oxidatively metabolized by these monooxygenases. Indeed, the duration of action of many medications depends on their rate of inactivation by the P450 system. Despite its general protective role in the removal of foreign chemicals, the action of the P450 system is not always beneficial. *Some of the most powerful carcinogens are generated from harmless compounds by the P450 system in vivo* in the process of *metabolic activation.* In plants, the cytochrome P450 system plays a role in the synthesis of toxic compounds as well as the pigments of flowers.

The cytochrome P450 system is a ubiquitous superfamily of monooxygenases that is present in plants, animals, and prokaryotes. The human genome encodes more than 50 members of the family, whereas the genome of the plant *Arabidopsis* encodes more than 250 members. All members of this large family arose by gene duplication followed by subsequent divergence that generated a range of substrate specificity. Indeed, the specificity of these enzymes is encoded in delimited regions of the primary structure, and the substrate specificity of closely related members is often defined by a few critical residues or even a single amino acid.

Pregnenolone

Steroid hormones contain 21 or fewer carbon atoms, whereas cholesterol contains 27. Thus, the first stage in the synthesis of steroid hormones is the removal of a six-carbon unit from the side chain of cholesterol to form *pregnenolone.* The side chain of cholesterol is hydroxylated at C-20 and then at C-22, and the bond between these carbon atoms is subsequently cleaved by *desmolase.* Three molecules of NADPH and three molecules of O_2 are consumed in this remarkable six-electron oxidation.

Cholesterol → **20α, 22β-Dihydroxycholesterol** → **Pregnenolone**

Adrenocorticotropic hormone (ACTH, or corticotropin), a polypeptide synthesized by the anterior pituitary gland, stimulates the conversion of cholesterol into pregnenolone, the precursor of all steroid hormones.

The Synthesis of Progesterone and Corticosteroids from Pregnenolone

Progesterone is synthesized from pregnenolone in two steps. The 3-hydroxyl group of pregnenolone is oxidized to a 3-keto group, and the Φ^5 double bond is isomerized to a Φ^4 double bond *Cortisol,* the major glucocorticoid, is synthesized from progesterone by hydroxylations at C-17, C-21, and C-11; C-17 must be hydroxylated before C-21 is, whereas C-11 can be hydroxylated at any stage. The enzymes catalyzing these hydroxylations are highly specific, as shown by some inherited disorders. The initial step in the synthesis of *aldosterone,* the major mineralocorticoid, is the hydroxylation of progesterone at C-21. The resulting deoxycorticosterone is hydroxylated at C-11. The oxidation of the C-18 angular methyl group to an aldehyde then yields aldosterone.

The Synthesis of Androgens and Estrogens from Pregnenolone

Androgens and estrogens also are synthesized from pregnenolone through the intermediate progesterone. Androgens contain 19 carbon atoms. The synthesis of androgens starts with the hydroxylation of progesterone at C-17. The side chain consisting of C-20 and C-21 is then cleaved to yield *androstenedione,* an androgen. *Testosterone,* another androgen, is formed by the reduction of the 17-keto group of androstenedione. Testosterone, through its actions in the brain, is paramount in the development of male sexual behavior. It is also important for maintenance of the testes and development of muscle mass. Owing to the latter activity, testosterone is referred to as an *anabolic steroid.* Testosterone is reduced by *5a-reductase* to yield *dihydrotestosterone (DHT),* a powerful embryonic androgen that instigates the development and differentiation of the male phenotype. Estrogens are synthesized from androgens by the loss of the C-19 angular methyl group and the formation of an aromatic A ring. *Estrone,* an estrogen, is derived from androstenedione, whereas *estradiol*, another estrogen, is formed from testosterone.

Vitamin D

Cholesterol is also the precursor of vitamin D, which plays an essential role in the control of calcium and phosphorus metabolism. *7-Dehydrocholesterol (provitamin D_3) is photolyzed by the ultraviolet light of sunlight to previtamin D_3, which spontaneously isomerizes to vitamin* D_3. Vitamin D_3 (cholecalciferol) is converted into *calcitriol* (1,25-dihydroxycholecalciferol), the active hormone, by hydroxylation reactions in the liver and kidneys. Although not a steroid, vitamin D acts in an analogous fashion. It binds to a receptor, structurally similar to the steroid receptors, to form a complex that functions as a transcription factor, regulating gene expression.

Vitamin D deficiency in childhood produces *rickets,* a disease characterized by inadequate calcification of cartilage and bone. Rickets was so common in seventeenth-century England that it was called the "children's disease of the English." The 7-dehydrocholesterol in the skin of these children was not photolyzed to previtamin D_3, because there was little sunlight for many months of the year. Furthermore, their diets provided little vitamin D, because most naturally occurring foods have a low content of this vitamin. Fish-liver oils are a notable exception. Cod-liver oil, abhorred by generations of children because of its unpleasant taste, was used in the past as a rich source of vitamin D. Today, the most reliable dietary sources of vitamin D are fortified foods. Milk, for example, is fortified to a level of 400 international units per quart (10 og per quart). The recommended daily intake of vitamin D is 400 international units, irrespective of age. In adults, vitamin D deficiency leads to softening and weakening of bones, a condition called *osteomalacia.* The occurrence of osteomalacia in Bedouin Arab women who are clothed so that only their eyes are exposed to sunlight is a striking reminder that vitamin D is needed by adults as well as by children.

Isopentenyl Pyrophosphate

Before this chapter ends, we will revisit isopentenyl pyrophosphate, the activated precursor of cholesterol. The combination of isopentenyl pyrophosphate (C_5) units to form squalene (C_{30}) exemplifies a fundamental mechanism for the assembly of carbon skeletons of biomolecules. *A remarkable array of compounds is formed from isopentenyl pyrophosphate, the basic five-carbon building block.* The fragrances of many plants arise from volatile C_{10} and C_{15} compounds, which are called *terpenes.* For example, myrcene ($C_{10}H_{16}$) from bay leaves consists of two isoprene units, as does limonene ($C_{10}H_{15}$) from lemon oil Zingiberene ($C_{15}H_{24}$), from the oil of ginger, is made up of three isoprene units. Some terpenes, such as gera-niol from geraniums and menthol from peppermint oil, are alcohols; others, such as citronellal, are aldehydes. We shall see later how specialized sets of 7-TM receptors are responsible for the diverse and delightful odor and taste sensations that these molecules induce.

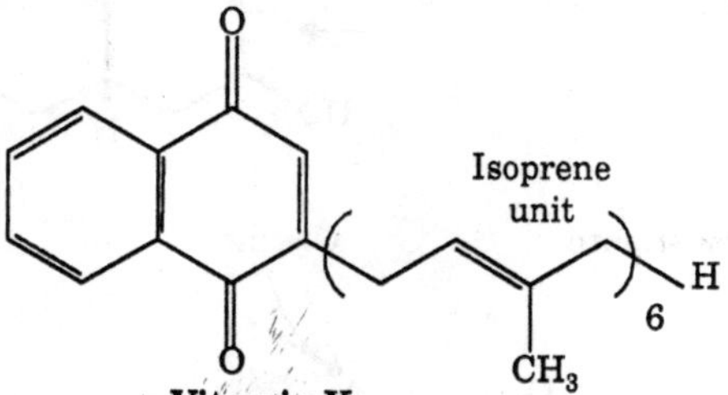

Vitamin K_2

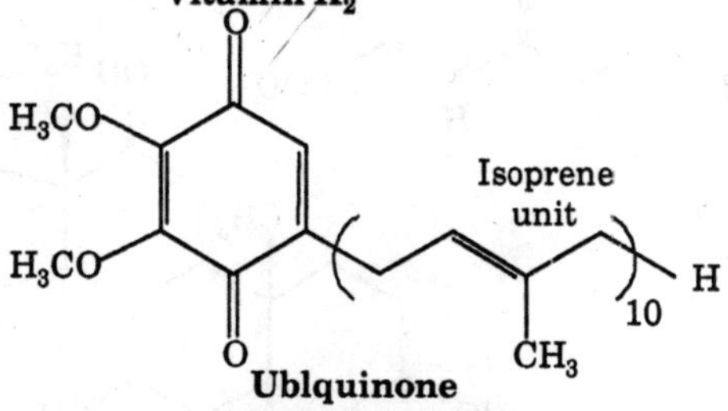

Ublquinone (Coenzyme Q_{10})

We have already encountered several molecules that contain isoprenoid side chains. The C_{30} hydrocarbon *side chain of vitamin K_2*, an important molecule in clotting is built from 6 isoprene (C_5) units. *Coenzyme Q_{10}* in the mitochondria! respiratory chain has a side chain made up of 10 isoprene units. Yet another example is the *phytol side chain of chlorophyll,* which is formed from 4 isoprene units. Many proteins are targeted to membranes by the covalent attachment of a farnesyl (C_{15}) or a geranylgeranyl (C_{20}) unit to the carboxyl-terminal cysteine residue of the protein. *The attachment of isoprenoid side chains confers hydrophobic character.*

Isoprenoids can delight by their color as well as by their fragrance. The color of tomatoes and carrots comes from *carotenoids.* These compounds absorb light because they contain extended networks of single and double bonds and are important pigments in photosynthesis. Their C_{40} carbon skeletons are built by the successive addition of C_5 units to form *geranylgeranyl pyrophosphate, a C_{20} intermediate,* which then condenses tail-to-tail with another molecule of geranylgeranyl pyrophosphate.

$$C_5 \rightarrow C_{10} \rightarrow C_{15} \rightarrow C_{20} \rightarrow C_{40}\ \text{(phytoene)}$$

Phytoene, the C_{40} condensation product, is dehydrogenated to yield lycopene. Cyclization of both ends of lycopene gives δ-carotene, which is the precursor of retinal, the chromophore in all known visual pigments. *These examples illustrate the fundamental role of isopentenyl pyrophosphate in the assembly of extended carbon skeletons of biomolecules.* It is evident that isoprenoids are ubiquitous in nature and have diverse significant roles, including the enhancement of the sensuality of life.

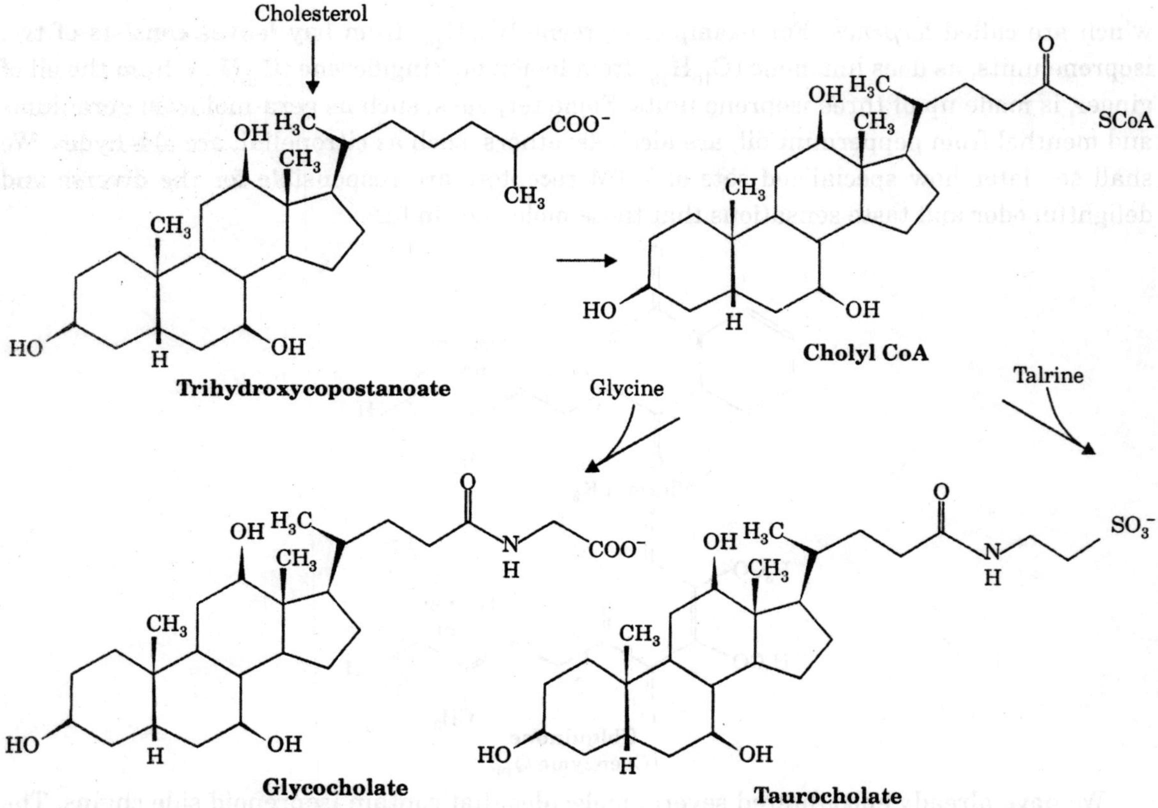

Cholesterol

OH H_3C COO^- CH_3 CH_3 CH_3 HO H OH

Trihydroxycopostanoate

O OH H_3C CH_3 SCoA CH_3 HO H OH

Cholyl CoA

Glycine

Talrine

O OH H_3C CH_3 N H COO^- CH_3 HO H OH

Glycocholate

O OH H_3C CH_3 N H SO_3^- CH_3 HO H OH

Taurocholate

Fig. 7.18. Synthesis of Bile Salts. Pathways for the formation of bilie salts from cholesterol.

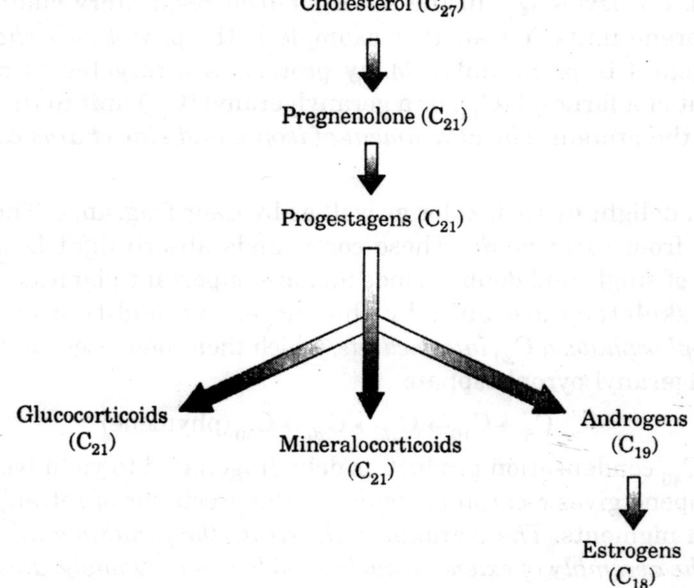

Fig. 7.19. Biosynthetic Relations of Classes of Steroid Hormones and Cholesterol.

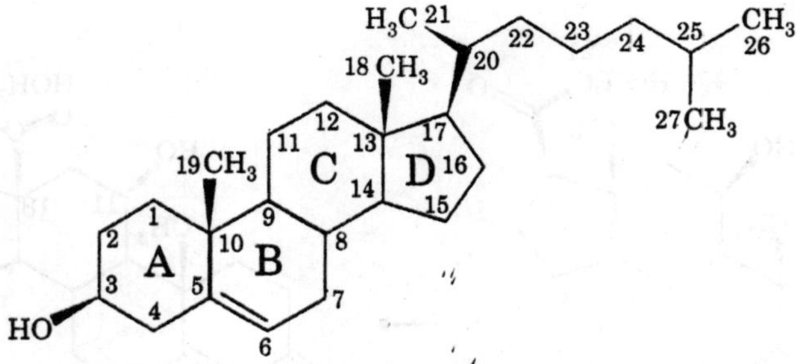

Fig. 7.20. Cholesterol Carbon Numbering. The numbering scheme for the carbon atoms in cholesterol and other steroids.

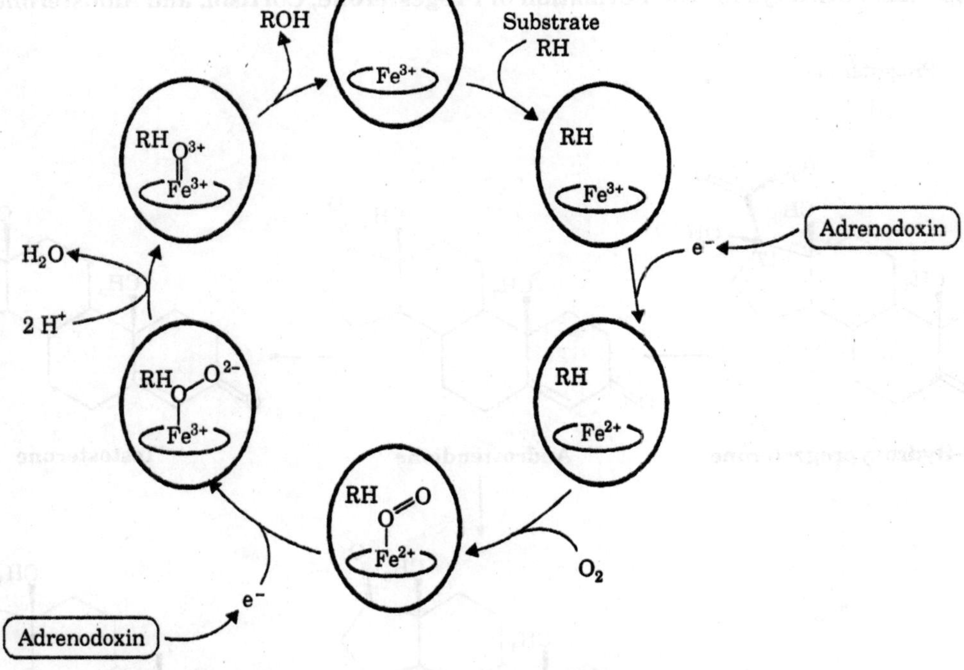

Fig. 7.21. Cytochrome P450 Mechanism. These enzyme-bind O_2 and use one oxygen atom to hydroxylate their substrates.

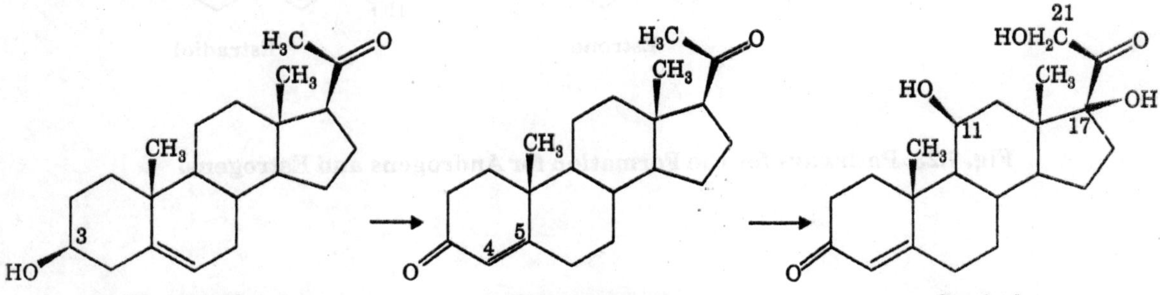

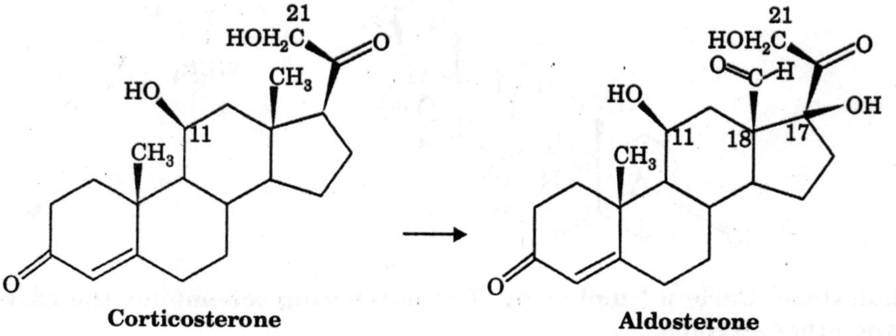

Fig. 7.22. Pathways for the Formation of Progesterone, Cortisol, and Aldosterone.

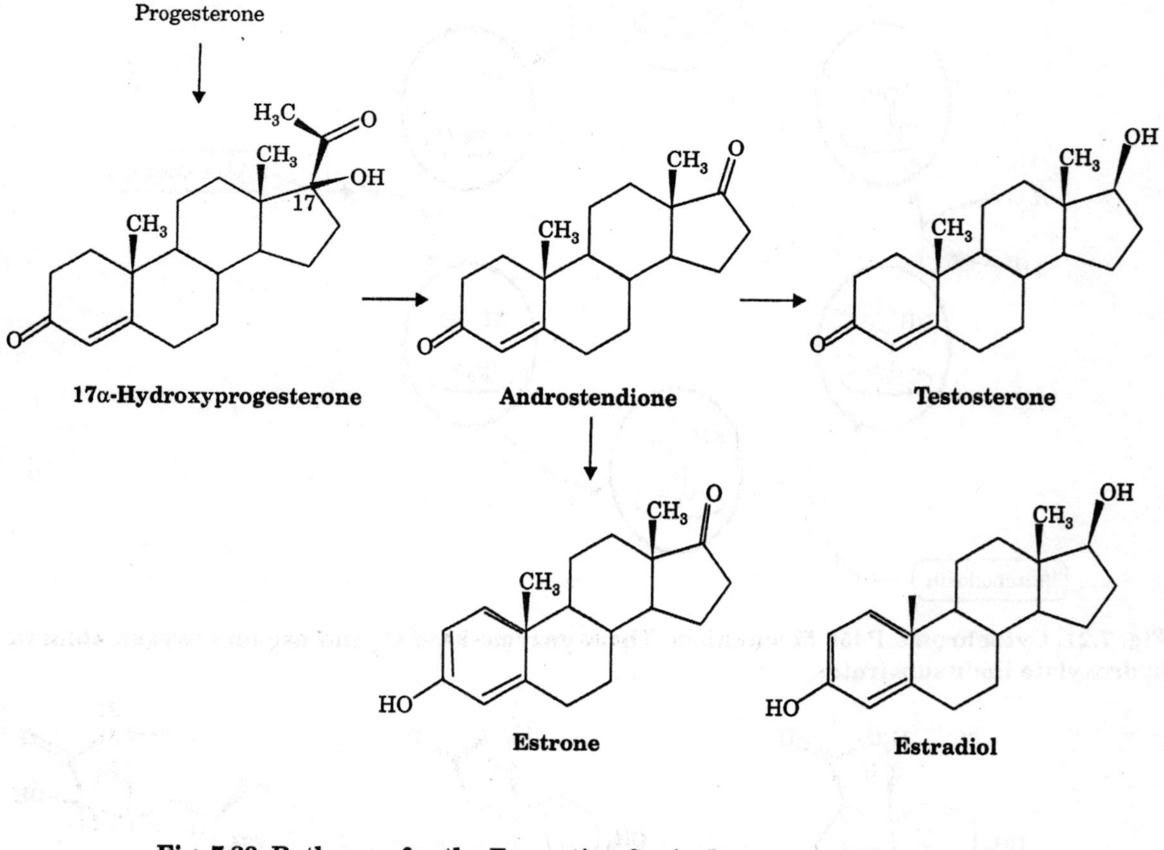

Fig. 7.23. Pathways for the Formation for Androgens and Estrogens.

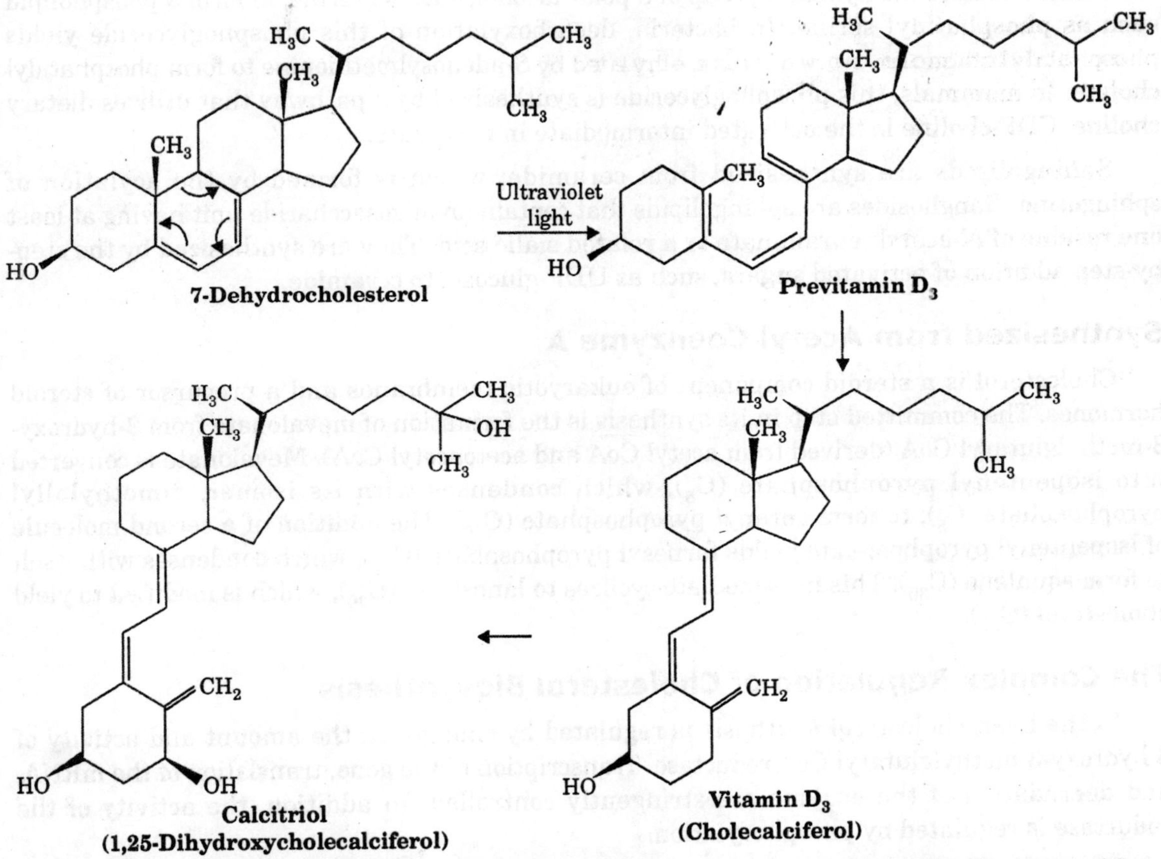

7-Dehydrocholesterol

Ultraviolet light

Previtamin D_3

Calcitriol (1,25-Dihydroxycholecalciferol)

Vitamin D_3 (Cholecalciferol)

Fig. 7.24. Vitamin D Synthesis. The pathway for the conversion of 7-dehydrocholesterol into vitamin D_3 and then into calcitriol, the active hormone.

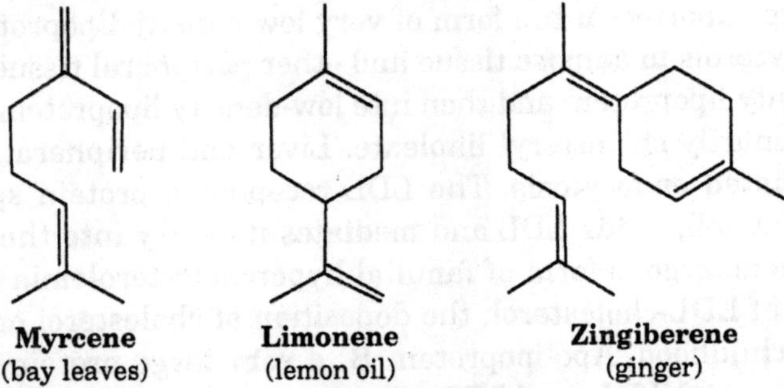

Fig. 7.25. Three Isoprenoids from Familiar Sources.

Synthesis of Phospholipids and Triacylglycerols

Phosphatidate is formed by successive acylations of glycerol 3-phosphate by acyl CoA. Hydrolysis of its phosphoryl group followed by acylation yields a triacylglycerol. CDP-diacylglycerol, the activated intermediate in the de novo synthesis of several phosphoglycerides, is formed from phosphatidate and CTP. The activated phosphatidyl unit is

then transferred to the hydroxyl group of a polar alcohol, such as serine, to form a phospholipid such as phosphatidyl serine. In bacteria, decarboxylation of this phosphoglyceride yields phosphatidylethanolamine, which is methylated by *S*-adenosylmethionine to form phosphatidyl choline. In mammals, this phosphoglyceride is synthesized by a pathway that utilizes dietary choline. CDP-choline is the activated intermediate in this route.

Sphingolipids are synthesized from ceramide, which is formed by the acylation of sphingosine. Gangliosides are sphingolipids that contain an oligosaccharide unit having at least one residue of *N*-acetylneuraminate or a related sialic acid. They are synthesized by the step-by-step addition of activated sugars, such as UDP-glucose, to ceramide.

Synthesized from Acetyl Coenzyme A

Cholesterol is a steroid component of eukaryotic membranes and a precursor of steroid hormones. The committed step in its synthesis is the formation of mevalonate from 3-hydroxy-3-methylglutaryl CoA (derived from acetyl CoA and acetoacetyl CoA). Mevalonate is converted into isopentenyl pyrophosphate (C_5), which condenses with its isomer, dimethylallyl pyrophosphate (C_5), to form geranyl pyrophosphate (C_{10}). The addition of a second molecule of isopentenyl pyrophosphate yields farnesyl pyrophosphate (C_{15}), which condenses with itself to form squalene (C_{30}). This intermediate cyclizes to lanosterol (C_{30}), which is modified to yield cholesterol (C_{27}).

The Complex Regulation of Cholesterol Biosynthesis

In the liver, cholesterol synthesis is regulated by changes in the amount and activity of 3-hydroxy-3-methylglutaryl CoA reductase. Transcription of the gene, translation of the mRNA, and degradation of the enzyme are stringently controlled. In addition, the activity of the reductase is regulated by phosphorylation.

Triacylglycerols exported by the intestine are carried by chylomicrons and then hydrolyzed by lipases lining the capillaries of target tissues. Cholesterol and other lipids in excess of those needed by the liver are exported in the form of very low density lipoprotein. After delivering its content of triacylglycerols to adipose tissue and other peripheral tissue, VLDL is converted into intermediate-density lipoprotein and then into low-density lipoprotein. IDL and LDL carry cholesteryl esters, primarily cholesteryl linoleate. Liver and peripheral tissue cells take up LDL by receptor-mediated endocytosis. The LDL receptor, a protein spanning the plasma membrane of the target cell, binds LDL and mediates its entry into the cell. Absence of the LDL receptor in the homozygous form of familial hypercholesterolemia leads to a markedly elevated plasma level of LDL-cholesterol, the deposition of cholesterol on blood-vessel walls, and heart attacks in childhood. Apolipoprotein B, a very large protein, is a key structural component of chylomicrons, VLDL, and LDL.

Bile Salts and Steroid Hormones

In addition to bile salts, which facilitate the digestion of lipids, five major classes of steroid hormones are derived from cholesterol: progestagens, glucocorticoids, mineralocorticoids, androgens, and estrogens. Hydroxylations by P450 monooxygenases that use NADPH and O_2 play an important role in the synthesis of steroid hormones and bile salts from cholesterol.

P450 enzymes, a large superfamily, also participate in the detoxification of drugs and other foreign substances.

Pregnenolone (C_{21}) is an essential intermediate in the synthesis of steroids. This steroid is formed by scission of the side chain of cholesterol. Progesterone (C_{21}), synthesized from pregnenolone, is the precursor of cortisol and aldosterone. Hydroxylation of progesterone and cleavage of its side chain yields androstenedione, an androgen (C_{19}). Estrogens (C_{18}) are synthesized from androgens by the loss of an angular methyl group and the formation of an aromatic A ring. Vitamin D, which is important in the control of calcium and phosphorus metabolism, is formed from a derivative of cholesterol by the action of light.

In addition to cholesterol and its derivatives, a remarkable array of biomolecules are synthesized from isopentenyl pyrophosphate, the basic five-carbon building block. The hydrocarbon side chains of vitamin K_2, coenzyme Q_{10}, and chlorophyll are extended chains constructed from this activated C_5 unit. Prenyl groups that are derived from this activated intermediate target many proteins to membranes.

CHAPTER 8

Metabolism of Lipids

The basic pathways for both synthesis of fatty acids and for their β-oxidation have been described. However, there are many variations to these pathways, and additional sets of enzymes are needed to synthesize the complex array of lipids present in most organisma. We will consider these details in this chapter. Like most other organisms, human begins are able to synthesize triacylglycerols (triglycerides) phospolipids, and glycolipids needed for cell membranes. Glucose can serve as the starting meterial. However, dietary lipids are alos a major source. For this reason, we will start with a discussion of the digestion and uptake of lipids and of the distribution by way of the bloodstream of ingested lipids and of lipids synthesized in the liver or in other tissues.

DIGESTION, SYNTHESIS, AND DISTRIBUTION OF TRIACYGLYCEROLS IN THE HUMAN BODY

Digestion of triglycerides begins in the stomach with emulsification and partial digestion by gastric lipase. Within the small intestine the ~100-residue protein called **colipase** binds to the surface of the fat droplets and provides an attachment site for the 449-residue **pancreatic lipase.** This Ca^{2+}-dependent serine esterase cleaves each triglyceride to two molecules of fatty acid and one of a 2-monoacylglycerol. These products are emulified by bile salts and are then taken up by the cells of the intestinal lining. The fatty acids are converted to acyl-CoA esters which transfer their acyl groups to the monoacylglycerols to resynthesize the triacylglycerols. The latter are incorporated into the very large lipoprotein particles called **clylomicrons** and enter the bloodstream via the lymphatic system Free fatty acids are also transported as complexes with serum albumin.

Synthesis of lipids from carbohydrates is an efficient process, which occurs largely in the liver and also in intestinal epithelial cell. The newly synthesized ricylglycerol, together with smaller amounts of phospolipids and cholesterol, combine with specific **apolipoproteins,** which are also synthesized in the liver, to from **very low dinsity lipoprotein (VLDL)** particles which are secreted into the blood stream. They transport the newly formed triacylglycerols from the liver to other body cells including the adipocytes, which store excess fat.

1. Plasma Lipoproteins

The small particles of plasma lipoprotein. which carry triacylglycerols, can separated according to their buoyant densities by centrifygation. They have been classified into five groups of increasing density byt smaller size as **chylomicrons,** very low density lipoproteins **(VLDL),** intermediate density lipoproteins **(IDL),** low density **(IDL),** and high densithy lipoproteins **(HDL)**. Each lipoprotein particle contains one or more apolipoproteins, whose sizes vary from the enormous 4536-residye apo B-100 to apoC-II and apoC-III. each of which contains just 79 residues and the 57-residue apoC-I.

The larger lipoproteins are spherical micelles containing a core of tricylglycerols and esters of cholesterol surrounded by a 2-to 3-nm-thick layer consisting of phospholipid, free cholesterol, and the apolipoprotein components. The size of the lipoprotein particles also veries from a 200 to 500-nm diameter for chylomicrons to as little as 5 nm for the smallest HDL paricles. The difference in volume is more impressive. If as has been estimated, a 22-nm diameter LDL particle contains about 2000 cholesterol and chelesteryl ester molecules and 800 phospholipids, a small HDL particle of 7-nm diameter will have room for only about 60 molecules of cholesterol and 90 of phospholipid, while a chylomicron may carry 10 million molecules of triacylglycerol. HDL particles are quite heterogeneous. As is indicated in. They are sometimes dividing into HDL2 and HDL3 density groups. In addition, there is a pre-HDL with lower phospholipid content and discoid forms low in cholesterol. Models of a recontituted lipoprotein disccontain two molecules of apoA-I and ~160 phosphatidylcholines that form a bilayer core.

Each apolipoprotein has one or more distinct functions. The apoB proteins probably stabilize the lipoprotein micelles. in addition, apoB-100 is essential to recognition of LDL by its receptors. The 79-residue apoC-II has specific function of activating the lipoprotein lipase that hydrolyses the triacylglycerols of chylomicrons and VLDL. Lack of either C-II or the lipase results in a very high level of triacylglycerols in the blood.

The large apolipoprontein B-100 is synthesized in the liver and is a principal component of VLDL, IDL, and LDL. It is the sole protein in LDL, accounting for nearly 20% of the mass of LDL particles, Partly because of its insolubility in water, its detailed structure is uncertain. If it were all coiled into an α helix, it would be 680 nm long and could encircle the LDL particle nearly 10 times! while the true structure of apoB-100 is unknown, it is thought to be extended and to span at least a hemisphere of the LDL surface. It consists of at least five domains. Sixteen cysteines are present in the first 25 residues at the N terminus, forming a crosslinked high-cysteine region. There are also 16 *N*-glycosylated sites. Domain IV (residues 3071-4011) is thought to contain the site that binds to its specific receptor, the LDL receptor. Heterogeneity in the amide I band of the infrared absorption spectrum suggests that about 24% is α helix 23% β sheet, and that a large fraction consists of turns, and unordered and extended peptide structures.

In intestinal epithelial cells the same apoB gene that is used to synthesize apoB-100 in the liver is used to make the shorter **apoB-48** (48%) protein. this is accomplished in an unusual way that involves "editing" of the mRNA that is formed. Condon 2153 in the mRNA for the protein is CAA, encoding glutamine. However, tha cytosine of the triplet is acted on by a deaminase, an editing enzyme, to form UAA, a chain termination condon. A third form apoB is found in **lipoprotein (a)** (Lpa). This LDL-like particle contains apoB-100 to which is covalently attached by a single disulfide linkage (probably to Cys 3734 of apoB-100) a second protein,**apo (a).** The latter consists largely of a chain of from 11 to over 50 kringle domains resembling the 78-residue kringle-4 of plasminogen as well as a protease domain. This additional chain may cause tighter binding to LDL receptors and may cause lipoprotein (a) to displace plasminogen from cell surface receptors. The amount of Lp (a) varies over 1000-fold among individuals and is genetically determined. The number of kringle domains also varies. Although the presence of high Lp (a) is associated with a high risk of atherosclerosis and stroke, many healthy 100-year olds also have high serum Lp(a).

Apolipoprotein A-I is the primary protein component of HDL. Most of the 243 residues consist of a nearly continuous amphipathic α Helix with kinks at regularly spaced proline residues. Two disulfide linked ApoA_I molecules may form a blet that encircles the dicoid lipoprotein. ApoA-II is the second major HDL protein, but no clearly specialized function has been indentified, Apo A-I, II, and IV, apoC-I,II and III, and apoE all have multiple repeats of 22 amino acids with sequences that suggest amphipathis helices, The 391-residue ApoA-IV has 13 tandem 22-residue repeats. Proline and glycine are present in intervening hinge regions. This enable these proteing to spread over and penetrate the surfaces of the lipoprotein micelles. Most of these proteins are incoded by a related multigene family.

the 299-residue apolipoprotein E plays a key role in metabolism of both triacylglycerols and cholesterol. Loke apoB-100 it binds to cells surface receptors. Absence of functional apoE leads to elevated plasma triacylglycerol and cholesterol, a problem that is considered in chapter 22, section D. The N-terminal domain, from residues 23 to 164, forms a 6.5 nm-long four-heliz bundle, which binds to te LDL receptors. There are three common isoforms of alipoprotein E (apoE2, apoE3, and apoE4). APo E3 is most common. The presence of apo E4 is associated with an increased risk of Alzheimer diseases chapter 30.

The major lipoproteins of insect hemolymph, the **lpophorins,** transport diacylglycerols. The apolipophorins have molecular masses of~250, 80 and sometime 18 kDa. The three-dimensional structure of a small 166-reisdue lipophorin (apolipophorin-III) is that of a four-heloz bundle. It has been suggested that it may partially unfold into an eztended form, whose amphipathic helices may bind to a phospholipid surface of the lipid micelle of the lipophorin. A similar behaviour may be involved in binding of mammalian apoliporpoteins. Four-heliz lipid-binding proteins have also been isolated from plants. Specialized lipoproteins known as **lipovitellins** store

Table 8.1. Classes of Lopoprotein Particles

Class	*Diameter (nm)*	*Density (G/ml)*	*Composition (weight percent)*				
			Surface cmpmanents			*Core lipids*	
			Protein	*Phospho-lipid*	*Cholesterol*	*Cholesteryl esters*	*Triaculgly-cerol*
Chylomicrons	75-1200	0.93	2	7	2	3	86
VLDL	30-80	0.93-1.006	8	18	7	12	55
IDL	25-35	1.006-1.019	19	19	9	29	23
LDL	18-25	1.018-1.063	22	22	8	42	6
HDL2	9-12	1.063-1.125	40	33	5	17	5
HDL3	5-9	1.125-1.21	45	35	4	13	3
Lp (a); show pre-β	25.30	1.04-1.09					

These are averages and there is considrable variation.

Table 8.2. Properties of Major Plasma Apolipoproteins

Designation	*No. residues*	*Mass(kDa)*	*Source*	*Function*
A-I	243	29		Major HDL protein
A-II	–	17.4	Liver and intestine	
A-IV	376	44.5		
B-100	4536	513	Liver	VLDL formation; ligand for LDL receptor
B-48	2152	241	Intestine	Chylomicron formation ligand for liver chylomicron receptor
C-I	57	6.6		
CII	79	8.9	Liver	Cofactor for lopoprotein lipase
C-III	79	8.8		
D	–	31	Many tissues	A lipocalin
E	299	34	Liver, VLDL	Ligand for LDL receptor
(a)	Variable			Ligand liver chylomicron receptor

phospholipid in eggs whether from nematodes, frogs, or chickens. There is some sequence similarity to that of human apolipoprotein B-100.

Movement of Lipid Materials Between Cells

After the synthesis ansd release of chylomicrons into the lymaphatic circulation, various exchange processes occur by which apolipoproteins, as well as enzymes and other proteins, may be added or removed. These very complex and incompletely understood phenomena are presented in simplified form in Chylomicrons donate apilipoproteins of the A and C families to HDL particles which, in turn, donate apoE and may also return some apoC protein to the chylomicrons.

Both chylomicrons and VLDL particles undergo similar processes in the capillary blood vessels, where their processes in the capillary blood vessels, where their tricylglycerols are hydrolyzed to glycerol and free fatty acids by **lipoprotein lipase.** This enzyme requires for its activity the apoliprprotein C-II which is present in the chylomicrons and VLDL particles. Lipoprotein lipase is also known as the "clearing factor" because it clears the milky chylomicron-containing lymph. It is secreted by adipocytes and other cells and becomes attached to heparan sulfate proteglycans on surfaces of capillary endothelial cells, a major site of its action. Hereditary absence of functional lipoprotein lipase causes **chylomicronemia,** a massive buildup of chylomicrons in plasma. The condition does not cause atherosclerosis but may lead to pancreatitis if not treated. Restriction of dietary fat to 20 g/day or less usually prevents problems. Naturally occurring mutations of lipoprotein lipase involing both the asparate of the catalytic triad and the flexible loop that covers the active site have been discovered.

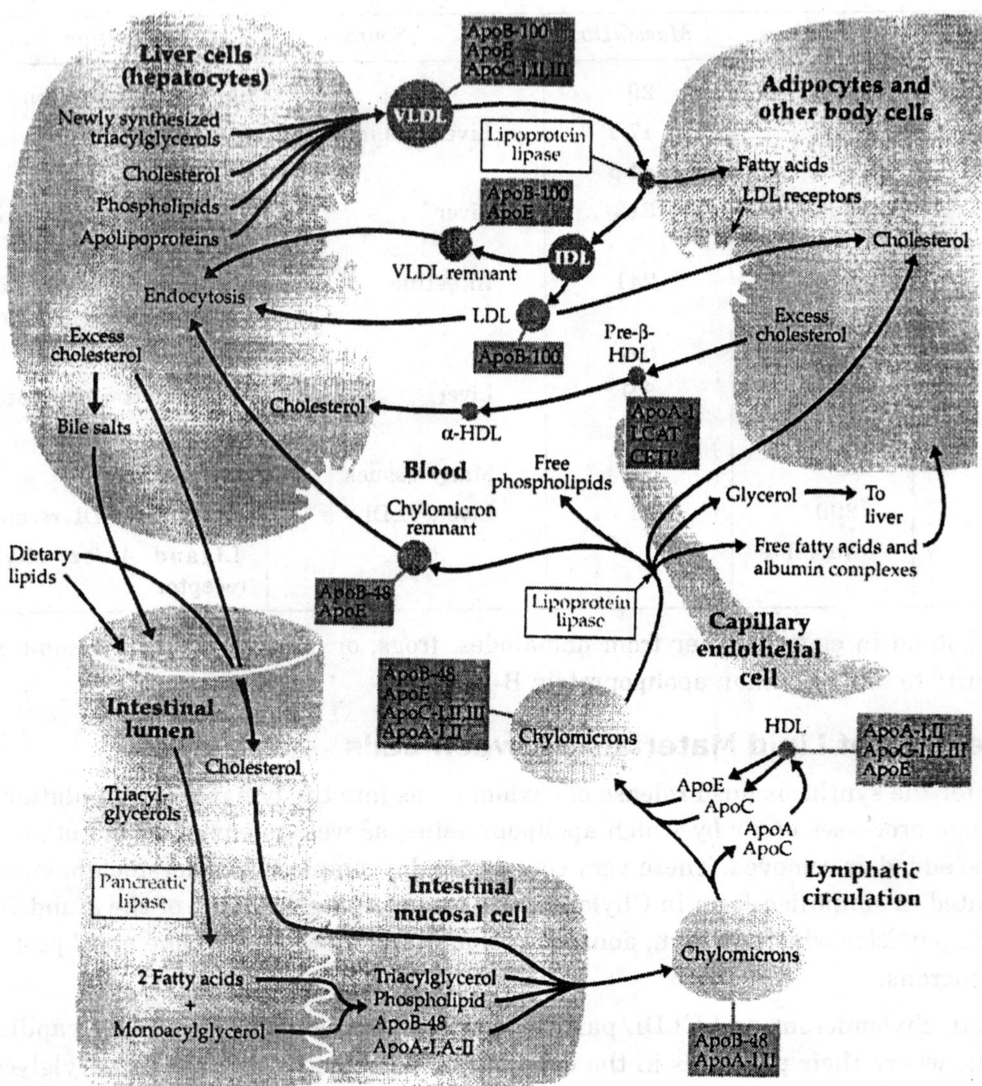

Fig. 8.1. Movement of triacylglycerols from liver and intestine to body cells and lipid carriers of blood. VLDL; very low density lipoprotein which contains triacylglucerols, phospholipids, cholesterol, and apolipoproteins, B, and C. IDL; intermediate density lipoproteins found in human plasma; LDL; low density lipoproteins which have lost most of their triacyglycerols. ApoB-100, etc., are apolipoproteins listed in Table 11.2. LCAT, lecithin: cholesteroal acyltransferse; CETP, cholesterylester transfer protein.

Both lipoprotein lipase and the less well understood **hepatic lipase** are related structurally to pancreatic lipase. In addition to hydrolysis of the triacylglycerols, the uptake of materials from lipoproteins probably involves shedding of intact phospholipids, perhaps as liposome-like particles.

The free fatty acids and glycerol are taken up by mammalian tissue cells leaving the choleseterol and some of the phospholipids of the VLDL particles as LDL. In humans intermediate density lipoproteins (IDL) are formed initially, but some are converted to LDL later. Both LDL particles and the shrunken **chylomicron remnants and VLDL remnants** are taken up by endocytosis in coated pits and are degraded by body cells, principally of the liver. The best known of these receptors is the 839-resideue **LDL receptor,** which has a specific affinity for ApoB-100. The related **VLDL receptor** (apoE receptor) has a higher affinity for apoE and may function in uptake of both VLDL and chylomicron remnants. The **LDL receptor-related protein** functions as a third lipoprotein receptor. In addition, a series of **scavenger receptors,** found in abundance in macrophages, take up oxidized lipoproteins and other materials. Scavenger receptor B1 (SR-B1), which is also found in liver cells, is involved in uptake of cholesterol from HDL particles by hepatocytes. Liver cells, and other cells as well, contain **lipocalins** and **fatty acid binding proteins** that help to carry these relatively insoluble acids to their destinations within the cells. Serum albumin is also a major carrier of free fatty acids. Within the adipocytes the fatty acids are reconverted to triacylglycerols. The low density (LDL) and high density (HDL) lipoproteins are involved primarily in transporting cholesterol to and from cells, a topic.

Fatty acids are carried to tissues for use in synthesis of triacylglycerols, phospholipids, and other membrane lipids. The mobilization of fatty acids from triacylglycerol stores and from cholesterol esters depends upon **hormone-sensitive lipase**. This enzyme is activated by cAMP-dependent phosphorylation and moves from the cytoplasm to the surfaces fo lipid droplets in response to catecholamines and other lipolytic hormones. Fatty acids are a major fuel for aerobic cells. Their conversion to acyl-CoA derivative and oxidation to CO_2 by beta oxidation and other pathways are discussed.

THE BIOSYNTHESIS OF FATTY ACIDS AND THEIR ESTERS

The synthesis of fatty acids two carbon atoms at a time from acetyl-CoA has already been describe elsewhere in this text. In this pathway, which resembles the β oxidation sequence in reverse, the products are saturated fatty acids with an even number of carbon atoms. In this section, we will consider some of the factors that lead to variations in the chain lengths and types of fatty acids.

Fatty Acid Synthases

Both bacteria and plants have separate enzymes that catalyze the individual steps in the biosynthetic sequence. The fatty acyl group grows while attached to the small acyl carrier protein (ACP). Control of the process is provided, in part, by the existence of isoenzyme forms. For example, in *E. coli* there different β-oxoacyl-ACP synthases. They carry out the transfer of any acyl primer from ACP to the enzyme, decarboxylate malonyl-ACP, and carry out the Claisen condensation. One of the isoenzymes is specialized for the initial elongation of acetyl-ACP and also provides feedback regulation. The other two function specifically in synthesis of unsaturated fatty acids.

In a few bacteria and protozoa and in higher animals the fatty acid synthase consists of only one or two multifunctional proteins. That from animal tissues contains six enzymes and an acyl carrier protein (ACP) domain as well. The human enzyme contains 2504 amino acid residues organized as a series of functional domains. Pairs of the 272-kDa chains associate to form 544-kDa dimers. The complex protein may have arisen via an evolutionary process involving fusion of formerly separate genes. The enzyme contains an ACP-like site with a bound 4-phosphopantetheine near the C terminus as well as a cysteine side chain enar the N terminus in the second acylation site. Since the two-SH groups can be crosslinked by dibromopropanone, an antiparallel linear arrangement of the two chains was proposed. Locations of the six enzymatic activities in each chain are indicated on. According to his picture, the ACP domain of one chain would cooperate with the β-oxacyl sunthase (KS) domain of the second chain. However, more recent studies indicate greater flexibility with the ACP, MAS, and KS domains of a single chain also being able to function together. Animal fatty acid synthases produce free fatty acids, principally the C_{16} palmitate. The final step is cleavage of the acyl-CoA by a thioesterase, obe of the six enzymatic activities of the synthase.

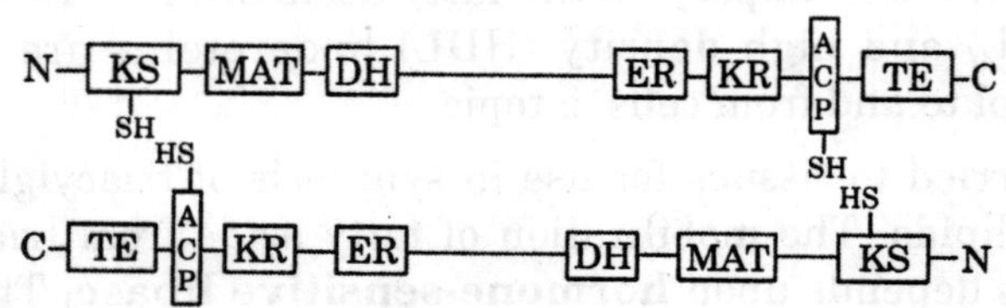

Abbreviation	*Enzymatic activity*	*Residue numbers*
KS	β-Oxacyl (Ketoacyl) synthase	1 — 406
MAT	Malonly and acetyl transferase	428 — 815
DH	Dehydratase	829 — 969
Central region	Structure core (?)	970 — 1629
ER	Enoyl reductase	1630 —1850
KR	β-Oxacyl (ketoacyl reductase)	1870 — 2100
ACP	Acyl carrier protein	2114 — 2190
TE	Thioesterase	2200 — 2505

Organization of eukaryotic fatty acid synthase.

Yeast fatty acid synthase has an α_6 β_6 structure where the 208-kDa α subunit contains the ACP-like site, the active site-SH, and there catalytica activities. The 220-kDa β subunit has five catalytic activities. The yeast enzyme contains tha FMN thought to act as $FMNJ_2$ in the second reduction step. As in bacteria, the products of the complex are molecules of acyl.CoA of chain longths C_{14}, X_{16}, and C_{18}.

Control of Chain Length

The length of fatty acid chains is controlled largely by the enzymatic activity that releases the fatty ancyl-CoA molecules or the free fatty acids from the synthase complex. In the animal enzymes the thioesterase, which is built into the synthase molecule, favors the release

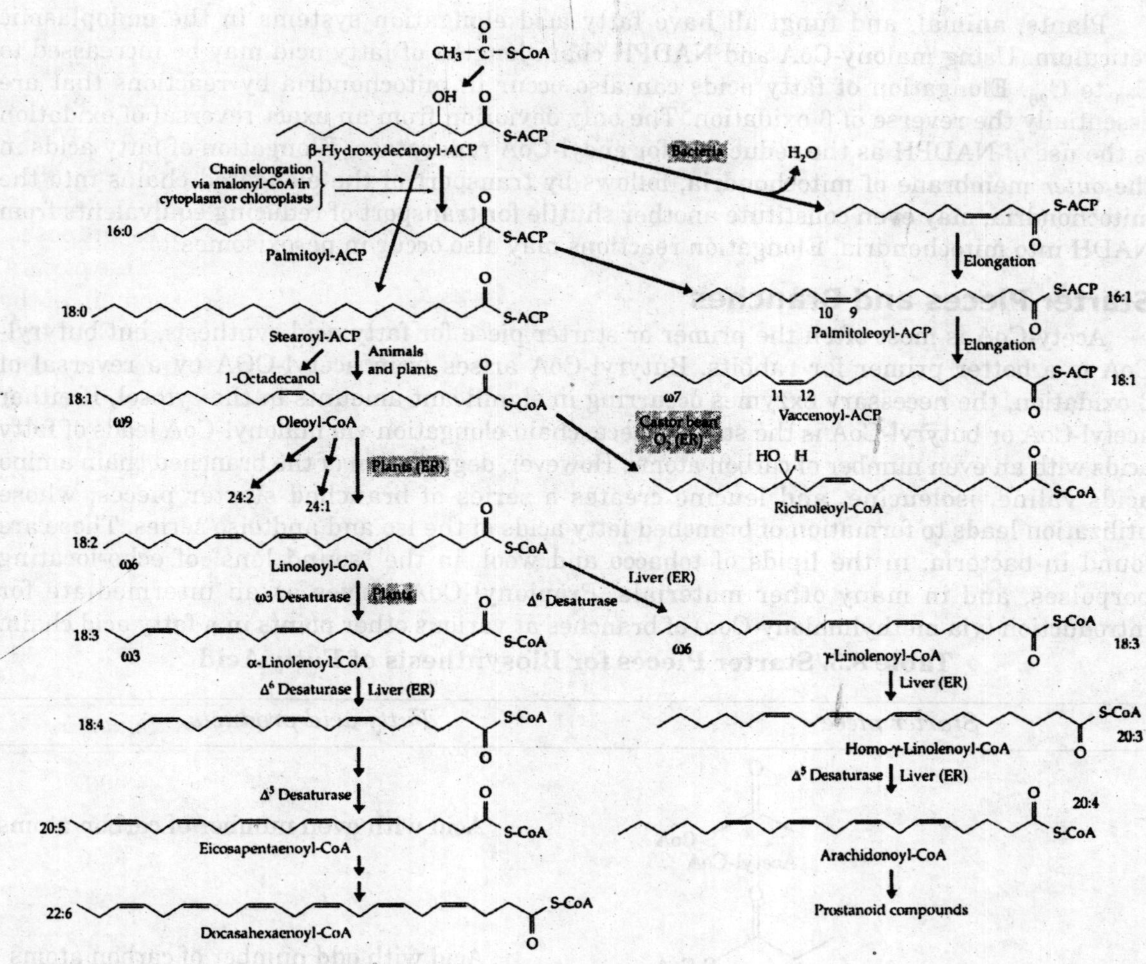

Fig. 8.2. Some biosynthetic reactions of fatty acids. Green arrows indicate transformations carried out by the human body.

primarily of the 16-carbon saturated (16.0) palmitic acid. However, in mammary glands and in the urophgil gland (preen glands) of waterfowl shorter chain fatty acids predominate. These are released from the synthase by reaction with a second thioesterase, a 29-kDA protein that catalyzes the otherwise premature relase of shorter fatty acids Cow's milk contains significant amounts of C_4-C_{14} acid as well as those with longer chains, whereas rabbit's milk contains laegely C_8 to C_{10} fatty acid.

In plants most biosynthesis occurs in the chloroplasts or in the protoplastids of seeds. There are two different synthase systems in chloroplasts, one that forms primarily the 16.0 palmitoyl-ACO and the othere the 18.0 stearoyl-ACP. Hydrolysis of the palmitoly-ACP releases palmitate, one major product of chloroplasts. However, the stearoyl-ACP is desaturated to oleoyl-ACP before hydrolysis to free oleate or conversion to oleoyl-CoA. In many species oleic acid is almost the sole fatty acid exported by the chloroplasts. However, it undergoes a variety of modification reactions in the plans cytosol.

Plants, animal, and fungi all have fatty acid elongation systems in the endoplasmic reticulum. Using malony-CoA and NADPH chain lengths of fatty acid may be increassed to C_{20} to C_{26}. Elongation of fatty acids can also occur in mitochondria by reactions that are essentially the reverse of β oxidation. The only deviation from an exact reversal of oxidation is the use of NADPH as the reductant for enoyl-CoA reducatase. Elongation of fatty acids in the *outer* membrane of mitochondria, follows by transport of the elongated chains into the mitochondria, may even constitute another shuttle for transport of reducing equivalents from NADH into mitochondria. Elongation reactions may also occur in peroxisomes.

Starter Pieces and Branches

Acety-CoA is most often the primer or starter piece for fatty acid synthesis, but butyryl-CoA is a better primer for rabbits. Butyryl-CoA arises from acetyl-COA by a reversal of β oxidation, the necessary exzymes occurring in significant amounts in the cytosol. If either acetyl-CoA or butyryl-CoA is the starter piece, chain elongation via malonyl-CoA leads of fatty acids with an even number of carbon atoms. However, degradation of the branched chain amino acids valine, isoleucine, and leucine creates a series of branched starter pieces, whose utilization leads to formation of branched fetty acids of the iso and andteiso series. These are found in bacteria, in the lipids of tobacco and wool, in the "sound lens" of echo-locating porpoises, and in many other materials, Propionyl-CoA serves as an intermediate for introduction (via methylmalony-Coa) of branches at various other points in a fatty acid chain.

Table 8.3. Starter Pieces for Biosynthesis of Fatty Acid

Starter piece	*Fatty acid products*
O, S-CoA Acetyl-CoA	Acid with even number of carbon atoms
O, S-CoA Propionyl-CoA	Acid with odd number of carbon atoms
Valine → O, S-CoA Isobutyryl-CoA	Iso series (even)
Leucine → O, S-CoA Isovaleryl-CoA	Iso series (odd)
Isoleucine → O, S-CoA	Anteiso series (odd)

For example, 2R-and 4R-Methylthezanoic acid, 2,4,6,8-tetramethyldecanoic acid, and a variety of other braches chain acids are estarified with long-chain alcohols (mainly

1-octadecanol) to form the waxes of the preen glands fo ducks and geese. the C_{32} **mycocerosic acid** of *Mychobacterium tuberculosis* is also formed using both malonyl-CoA and methylmalonly CoA for chain cleongation. This acids is present in mycobacteria cell walls eserified with long-chain diols.

Synthesis by the Oxoacid Chain Elongation Process

The carbon skeleton of leucine is dderived from that of valine by elongation of the corresponding oxoacid by a single carbon atom that is derived from acetyl CoA. AT least in plants some branched chain fatty acid of medium length are formed via the saem oxoacid elongation process, which extends tha chain one carbon atom at a time using 2-oxobutyrate as a starting compound. The same process can be used to form medium length straight-chain fatty acids (up to ~C_{12}) with either an odd or an even number of carbon atoms.

2-Oxoglutarate
Acetyl-CoA, NAD^+ → CO_2
Acetyl-CoA, NAD^+ → CO_2
Acetyl-CoA, NAD^+ → CO_2
Oxoacid chain elongation process. Used three times
2-Oxosuberate → Biotin
NAD^+ Oxidative decarboxylation → CO_2
Sulfur
7-Mercaptoheptanoate
Threonine, ATP
7-Mercaptoheptanoylthreonine
ATP
7-Mercaptoheptanoylthreonine phosphate (HS-HTP)

2-Oxoglutarate can also serve as a starger piece for elongation by the oxoacid pathway. Extension by three carbon atoms yield 2-**oxosuberate**. This dicarboxylate is converted by reactions into biotin and in archaebacteria into the coenzyme 7-mercaptoheptanoyl-threonine phosphate (HTP). Lopoic acid is also synthesized from a fatty acid, the eight-carbon octanoate. A fatty acid synthase system that utilizes a mitochondrial ACP may have as its primary function the synthesis of octanoate for lipoic acid formation. The mechanisam of insertion of the two sulfur atoms to form lipoate is uncertain. It reqhires an iron Sufur protein and is probably similar to the corresponding process in the sunthesis of biotin and in formation of HTP. One component of the archaebacterial cofactor **methanofuran** is a tetrracarboxylic acid that is formed from 2-oxoglutarate by successive condensation with two malonic acid units as in fatty acid synthesis.

Unsaturated Fatty Acids

Fatty acids containing one or more double bonds provide necessary fluidity to cell membranes and serve as precursors to other components of cell. Significant differences in the methods of introduction of double bonds into fatty acids are observed among various organisms. Bacteria such as *E. coli* that can live anaerobically often from **vaccenic acid** as the principal unsaturated fatty acid. It is formed by chain elongation after introduction of a cis double bond at the C_{10} stage of synthesis. The bacteria possess a **β-hydroxydecanoyl thioester dehydratase,** which catalyzes elimination of a β-hydroxyl group to yield primarily the cis-β, γ rather than the *trans*-α, β-undsturated product. The mechanism may resemble that of enoyl hydratase. The indicated *trans*- α, β– unsaturated intermediate (enzyme bound) being isomerized to the cis-β, γ–unsaturated product through an allulic rearrangement. The product can then be elongated to the C_{16} **palmitoleyl-ACP** and C_{18} **vaccenoly-ACP** derivatives. However, dephdration of β-hydroxydecanoyl-ACP lies at a branch point in the biosynthetic sequences. The *trans*-α, β- unsaturated fatty acyl compounded lies on the usual foute of chain elongation to palmitoyl-COA.

H3C S – ACP (Acyl carrier protein) HO H O

β-Hydroxydecanoyl-ACP

H_2O

H3C S – ACP O

trans-α,β-Unsaturated product

H3C S – ACP O Chain elongation

cis-β,γ-Unsaturated product

Longer saturated fatty acids

Chain elongation

Longer unsaturated fatty acids

In higher plants, animal, protozoa, and fungi, saturated fatty acids are acted upon by **desaturases** to introduce double bonds, ususlly of the cis (Z) configuration. The substrates may be fatty acyl-ACP, fatty acyl-CoA molecules, membrane phospholipids, or glycolipids. The Δ^9 desaturase, isolated from liver of from yeast, converts stearoyl-CoA to oleoyl-CoA. This membrane-associated enzyme system utilizes NADH as a reductane, passing clectrons via cytochrome b_5 reductase and cytochrome b_5 itself to the desaturase. The pro-*R* hydrogens are removed at both C-9 and C-10.

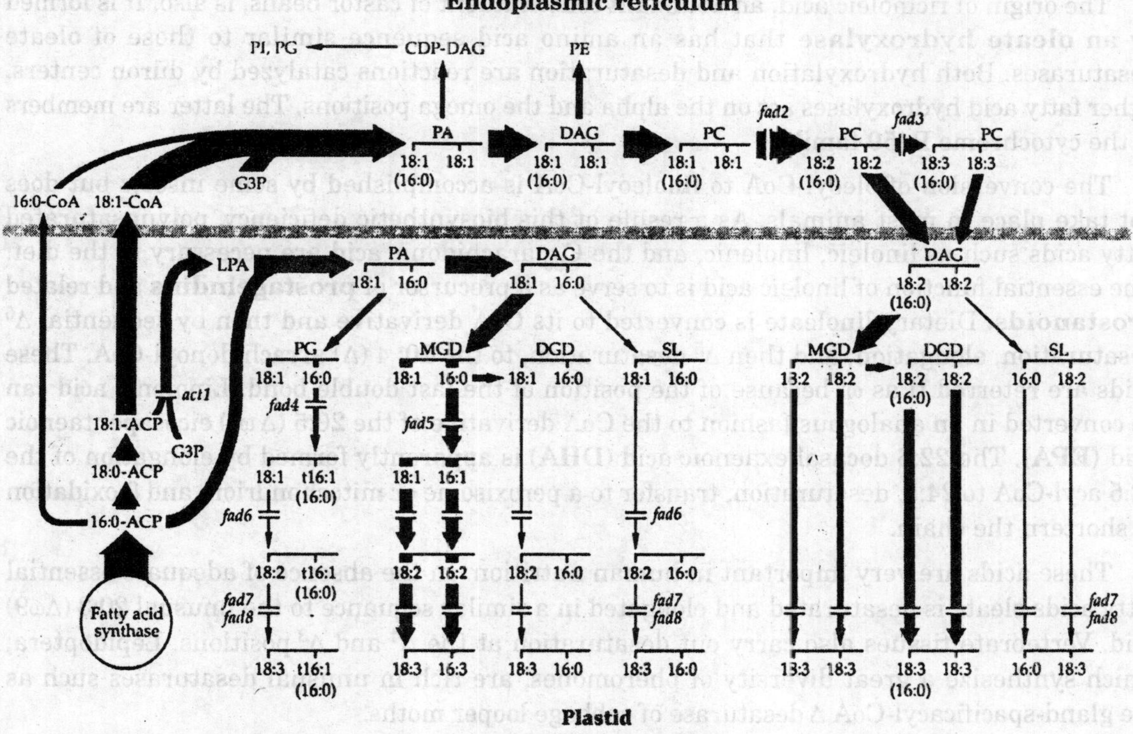

Fig. 8.4. Major pathways of synthesis of fatty acids and glycerolipids in the green plant *Arbidopsis*. The major site of fatty acid synthesis is chloroplasts. Most is exported to the cytosol as oleic acid After conversions to its conezyme A derivative it is converted to phosphatidic acid (PA), diacylglycerol (DAG), and the phospholipids: phosphatidyclcholine (PC) phosphatidylinositol (PI)] phosphadylglycerol (PG), and phosphatidylethanolamine (PE). Desaturation also occurs, and some linoleic and linolenic acids are reutrned to the chloroplasts. Form Sommerville and Brwse. Other abbreviations: monogalacosyldiacyglcerol (MGD), digalactosyldiacyglycerol (DGD), sulfolipid (SL, glycerol 3-phosphate (G3P), lysophosphatidic acid (LPA), Acyl carrier protein (ACP), cytidine diphosphate-DAG (CDP-DAG).

$$O_2 + 2H^+ + 2e^- + \text{stearoyl-CoA} \rightarrow \text{oleoyl-CoA} + 2\ H_2O$$

In plants a similar enzyme catalyzes formation of the first double bond in a fatty acyl group converting stearoyl-ACP into oleyl-ACP in the choloroploasts. The soluble basteraroyl-ACP desaturase has a diiron-oxo active site. Electrons are donated from light-generated reduced ferredoxin. In addition to the desaturase both plants and cyanobacteria usually desaturate

C_{18} acids also at the Δ^{12} and Δ^{15} positions and C_{16} acids at the $\Delta^{7,}$ Δ^{10} and $\Delta^{13}(\omega 3)$ positions. Desaturation of oleate occurs primarily in the ER after convarsion of the free acid to its coenzyme A derivative or to phosphatidycholine. Consecutive introduction of two double bonds forms linoleoyl-CoA (18:2) and linolenoyl-CoA. All double bonds are cis. The membrane lipids of chloroplasts contain both linoleic and linolenic acids, which have apparently been rerurned to the chloroplasts from the cytosol as indicated. Plants grown at colder temperatures have a higher content of these trienoic acids than those grown at higher temperature.

The origin of ricinoleic acid, an abundant constitutuent of castor beans, is also. It is formed by an **oleate hydroxylase** that has an amino acid sequence similar to those of oleate desaturases. Both hydroxylation and desaturation are reactions catalyzed by diiron centers. Other fatty acid hydroxylases act on the alpha and the omega positions. The latter are members of the cytochrome P450 family.

The conversion of oleoyl-CoA to linoleoyl-CoA is accomplished by some insects but does not take place in most animals. As a resule ot this biosynthetic deficiency, polyunsaturated fatty acids such as linoleic, linolenic, and the C_{20} arachidonic acid are necessary in the diet. One essential function of linoleic acid is to serve as a precursor of **prostagelndins** and related **prostanoids.** Dietary linoleate is converted to its CoA derivative and then by sequential Δ^6 desaturation, elongation, and then Δ^5 desaturation, to the 20: 4 (Δ) arrachidonoyl-CoA. These acids are reterred ti as ω^6 because of the position of the last double bond. Linolenic acid can be converted in an analogous fashion to the CoA derivative of the 20:5 ($\Delta\ \omega^6$) eicosapentaenoic acid **(EPA)**. The 22:6 docasahexaenoic acid (**DHA**) is apparently formed by elongation of the 22.5 acyl-CoA to 24:5, desaturation, transfer to a peroxisome or mitochondrion, and β oxidation to shortern the chain.

These acids are very important in human nutrition. In the absence of adequate essential fatty acids oleate is desaturated and elongated in a similar sequance to the unusual 20:3 ($\Delta\omega 9$) acid. Vertebrate tissues also carry out desaturation at the Δ^4 and Δ^3 positions. Lepidoptera, which synthesize a great diversity of pheromones, are rich in unusual desaturases such as the gland-spacificacyl-CoA Δ desaturase of cabbage looper moths.

Polyunsaturated fatty acids, cintaining 4,5, or 6 double bounds and chain lengths of up to C_{36}, are found in phosphatidylcholine of vertebrate retains. Although the double bonds are rarely in conjugated positions in food fats and in animal bodies, some plants convert oleic or linoleic acid into fatty acids with as many as three or four conjugated double bound. **Conjugated linoleic acid** (9c,11t 18.2) can be formed from 11-*trans*-ovtadecenoate in the human body. This compound is also found in meat and dairy products. It has been reported to have anticancer properties and may be another beneficial dietary consitituent. An isomerase isolated from red algae converts polyunsaturated acids into forms with conjegated double bonds. For example, arrachidonic acid (5Z, 8Z, 11Z, 14Z) is convertae to (5Z, 7Z, 9Z, 14Z)-eicosatetraenoic acid.

Cyclopropane Fatty Acids and Mycolic Acids

Fatty acids contaning one or more cyclopropane rings are present in many bacteria. The extra carbon of the cyclopropane ring is added from *S*-adenosylmethionine (AdoMet) at the site of a cis double bond in a fatty acyl group of a phosphatidyle-thanolamine molecule in a membrane. The same type of intermediate carbocation can yield either a cyslopropane fatty acid or a metheny fatty acid. the latter can be reduced to a branched fatte acid. This is an alternative way of introducing methyl branches that is used by some bacteria.

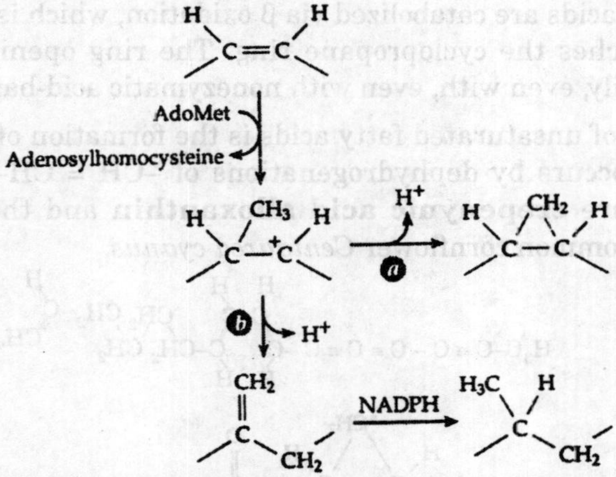

Mycobacteria are rich in cyclopropane-containing fatty acids. These **mycolic acids** are major components of the cell walls and may account for 30% of the dry weight of the cells. The most abundant mucolic acid of *M. tuberculosis* consists of C_{52} fatty acid containing two cyclopropane rings joined via a Claisen-type condensation with a C_{26} carboxylate fatty acid. A similar mycolic acid formed by *M. smegmatis* jas double bouds instead of cyclopropane rings as indicated below. There are other variations. In place of a double boud or cyclopropane group there may be –OH,–OCH_3, C = O, epoxide or Ch_3.

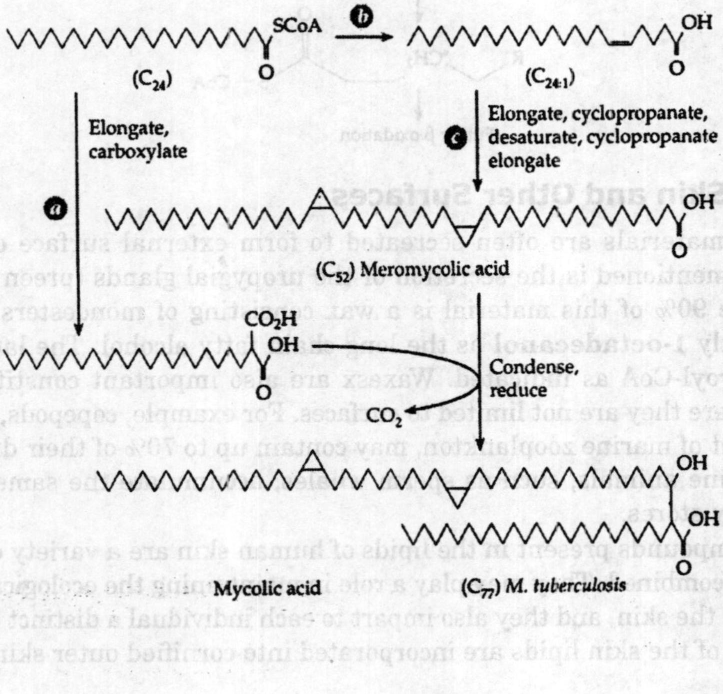

Mycolic acid (C_{79}) *M. smegmatis*

Cyclopropane fatty acids are catabolized via β oxidation, which is modified as in when the chain degradation reaches the cyclopropane ring. The ring opening of the cyclopropanol derivatives occurs readily, even with, even with nonezymatic acid-base catalysis.

Another alteration of unsaturated fatty acids is the formation of acetylenic groups (–C ≡ C–). This apparently occurs by dephydrogenations of –CH = CH–. Example of naturally occurring acetylenes are **crepenynic acid alloxanthin** and the following remarkable hydrocarbon from the common cornflower *Centaurea cyanus*.

First two steps of β oxidation

Futher β oxidation

The Lipids of Skin and Other Surfaces

Special fatty materials are often secreated to form external surface of organisms. An example already mentioned is the secretion of the uropygial glands (preen glands) of water fowl. In the goose 90% of this material is a wax consisting of monoesters of various acids with predominantly **1-octadecanol** as the long-chain fatty alcohol. The latter is formed by reduction of stearoyl-CoA as indicated. Waxesx are also important constituents of marine environments, where they are not limited to surfaces. For example, copepods, which constitute a major component of marine zooplankton, may contain up to 70% of their dry weight as wax esters. Some marine animals, such as sperm whales, accumulate the same esters in major amounts as energy stores.

Among the compounds present in the lipids of human skin are a variety of branched fatty acid both free and combined. They may play a role in maintaining the ecological balance among microorganisms as the skin, and they also impart to each individual a distinct odor or "chemical fingerprint" Some of the skin lipids are incorporated into cornified outer skin surface.

Surface lipids of plants : The thick cuticle that covers the outer surfaces of green plants consists largely of waxes and other lipids but also contains a complex polymeric of **cutin** (stems and leaves) or **suberin** (roots and wound surfaces). Plant waxes commonly have C_{10}–C_{30} chains in both acid and alcohol components. Methyl branches arfe grequently present. A major function of the waxes is to inhibit evaporation of water and to protect the outer cell layer. In addition, the methyl branched component may inhibit enzymatic breakdown by microbes. Free fatty acids, free alcohols, aldehydes, ketones, β-dicketones, and alkanes are also present in plant surface waxes. Chain lengths are usually C_{20}–C_{35}. Hydrocarbon formation can occur in other parts of a plant as well as in the cuticle. Thus, normal **heptane** constitutes up to 98% of the voltatile portion of the turpentine of *pinus jeffreyi.*

Cutin is largely a polyester with a high content of ω-hydroxypalmitic acid and related fatty acids, which are also hydroxylated at a second position:

$$HO—CH_2—(CH_2)_x—\underset{\displaystyle OH}{\underset{|}{CH}}—(CH_2)_y—COOH$$

Cutin monomers. C_{16} acids in which $y = 8,7,6$, or 5 and $x + y = 13$

This allows branching of the polymer. Monomers of other chain lengths as well as aromatic components related to lignin are also present and polymerized into a high molecular mass branched structure. Suberin is a more complex ligninlike polymer with a high content of pheolic constituents such as vanillin.

Formation of hydrocarbons : Alkanes and alkenes occur in plants, in preen gland secretions, and in insects. The alkanes of plant cuticle are thought to be formed by elongation of a C_{16} acid followed by loss of the carboxyl group. The mechanisms are not obvious. However, these hydrocarbons are often two carbon atoms shorter than the starting fatty acid. The pathway between them might begin by α-oxidation to form an α-peroxy acid which would decarboxylate to form an aldehyde, a reaction similar to that of. Alternatively, a long-chain acyl=CoA may be reduced directly to an aldehyde. In fact, when suitable inhibitors are present aldehydes do accumulate in tissues that are forming hydrocarbons. Conversion of an aldehyde intermediate to an alkane may occur by **decarbonylation** (loss of CO). This has been demonstrated in pea (*Pisum sativum*) leaves, in uropygial glands, in flies, and in a colonial green alga, *Botyro-coccus brunii.* In the last case 32% of the dry weight of the cells is C_{27}, C_{29}, and C_{31} hydrocarbons. They appear to be formed by action of a decarbonylase that apparently contains a cobalt prophyrin. Plants require cobalt for growth, but an enzymatic function has not previously been established.

In contrast, the sex pheromone of the female housefly is (Z)-9-tricosene, a hydrocarbon apparently formed by an oxidative decarboxylative process from a precursor aldehyde by an enzyme that require NADPH and O_2 and is apparently a cytochrome. P_{450}Oxidative deformylation by a cytochrome P450 converts aldehydes to alkenes Presumablyvia a peroxo intermediate. Formation of an alkene by decarboxylation has also been proposed, but a mechanism is not obvious.

Insect waxes, hydrocarbons, and pheromones : Surface lipids of many insects contain esters of long-chain (as long as 66 carbon atoms) alcohols and long-chain acids. On the other hand, waxex of the tobacco hornworm consist largely of 11-and 12-oxoderivatives of a C_{28} alcohol, which may be esterified to short-chain acids. A major hydrocarbon of cockroaches is 6,9-heptacosadiene.

Insects communicate through the use of a great variety of volatile pheromones. As mentioned already, some mothjs utilize acetate esters of various isomers of Δ^7 and Δ^{11} unsaturated C_{14} fatty acids as sex pheromones. Some other moths convert the *trans*-11-

tetradecenyl acetate into the corresponding C_{14} aldehyde or alcohol, while others use similar compounds of shorter ($C_{11} - C_{12}$) chain length. Some ants use ketones, such as 4-methyl-3-heptanone, as well as various isoprenoid compounds and phyrazines as volatile signaling compounds. Other insects also utilize isoprenoids, alkaloids, and aromatic substances as pheroones.

SYNTHESIS OF TRIACYLGLYCEROLS, GLYCOLIPIDS, AND PHOSPHOLIPIDS

Reduction of dihydroxyacetone phosphate yields *sn*-glycerol 3-phosphate, the starting compound for formation of the glycerol-containing lipids. Transfer of two acyl group from ACP or CoA to the hydroxyl groups of this compound (steps *b* and *c*) yields 1,2-diacylglycerol 3-phosphate (phosphatidic acid.) Two different acyltansferases are required. Unsaturated fatty acids are incorporated preferentially into the 2-position. The intermediate1-acyl-*sn*-glycero-3-phosphate, often called **lysophosphatidic acid** (LPA), is formed in excess in activated platelets and has a veriety of signaling activities. LPA for signaling is derived by turnover of existing phospholipids. An alternative route of LPA formation in liver is the transfer of one acyl group onto dihydroxyacetone phosphate and reduction prior to addition of the second acyl group.

CH_2OH
O=C
CH_2-O-P

Fatty acyl-CoA (ACP)
CoA-SH

$CH_2-O-C(=O)-R$
O=C
CH_2-O-P

NADH

$CH_2-O-C(=O)-R$
H
HO C
CH_2-O-P

LPA

Phosphatidic acid lies at a metabolic branch point. On the one hand, the phospho group can be removed by a specific phosphatase (step *d*) and another acylgroup (most often an unsaturated acyl group) map be transferred onto the resulting diacylglycerol (DAG, diglyceride, step *e*) to form a **triacylglycerol** (triglyceride). Alternatively, the phosphatidic acid may be converted to a **CDP-diacylglycerol** (step *g*), a key intermediate in phospholipid synthesis both in eukaryotes and in bacteria. Not only can phosphatidic acid be hydrolyzed to 1,2-diacylglycerols, but the reverse process can occur by action of a kinese. This presamably permits recycling of the diacylglycerol formed by turnover of membrane phospholipids.

Diacylglycerols can also be converted to a variety of glycolipids such as the **galactolipids** of chloroplasts. These are the major lipids of photosynthetic membranes. Some bacteria, e.g., the mycoplasma *Acholeplasma laidlawii,* contain both monoglucosyl-and diglucosyl-DAG. Changes in the ratio of these two membrane components may regulate the phase equilibrium between bilayer and nonbilayer forms. 1,2-Diacylglycerol can also react with UDP-sulfoquinovose to form the characteristic sulfolipid of choloroplasts.

1,2-Diacylglycerol

UDP-Gal

Monogalactosyldiacylglycerol

UDP-Gal

Digalactosyldiacylglycerol

In animals a principal regulatory point for lipid synthesis is in the activation of acetyl-CoA carboxylase by citrate Beyond that, a complex hormonal control is exerted on both biosynthesis and the catabolism of triglycerides stored in liver and adipose tissues. For example, adrenaline and glucagon, by stimulating production of cAMP, stimulate acetyl-CoA carboxylase, activate lipases that cleave triacylglycerols, and mobilize depit fath. Insulin, on the other hand, promotes lipid storage. It increases the activity of the exzymes of lipogenesis from the ATP-dependent citrate cleavage enzyme and inhibits cAMP production, thus blocking lipolysis within cells. At the transcriptional level sterols bind to activator proteins (**sterol regulatory element binding proteins,** SREBPs) and activate genes for acetyl CoA carboxylase and for steroyl-CoA desaturases. Fatty acid synthases, which play a central role in lipid formation, are controlled by both hormonal and nutritional factors at the transcriptional and translational levels. Environmental factors also have indirect effects. For example, the Δ^9 fatty acid desaturase activity of poikilothermic (cold-blooded) animals is increased at low temperatures. The resulting increased synthesis of unsaturated fats leads to increased fluidity of the membrane bilayer. As metioned on the same is true for green plants.

Phospholipids

Bacterial and also some eukayotic phospholipids are formed following conversion of phosphatidic acids to CDP-diaclglycerols, which are able to react with a variety of nucleophiles with displacement of CMP. Reaction with L-serine (step *h*) leads to **phosphatidylserine,** and reaction with glycerol 3-phosphate (step *i*), which enters cells via a special transporter, produces **phosphatidyglcerol 3-P.** The enzyme catalyzing the formation of phosphatidylserine appears to occur naturally as in integral membrane protein of the ER. Some is also bound to ribosomes and to mitochondria. In contrast, most of the other enzymes of phospholipid formation are closely associated with embedded in the cytoplasmic membrane. One of these, a pyruvoyl group dependent enzyme, Catalyzes decarboxylation of phosphatidylserine to **phosphatidylethnolamine**. This reaction had been thought unimportant in animals, but result with cultured cells show decarboxylation of phosphatidylserine is often the major route of formation of phosphatidylethanolamine in mammalian cells. This phospholipid also accounts for 75% of total phospholipid of the *E. coli* cell envelope. It is synthesized on the cytosolic side of the inner membrane, but it is also translocated to the outer membrane, where it is a major consituent of the inner bilayer leaflet. PE is essential for viability of *E. coli* cells. It provides dipolar ionic head groups and apparently serves as a chaperone for folding of some membrane proteins.

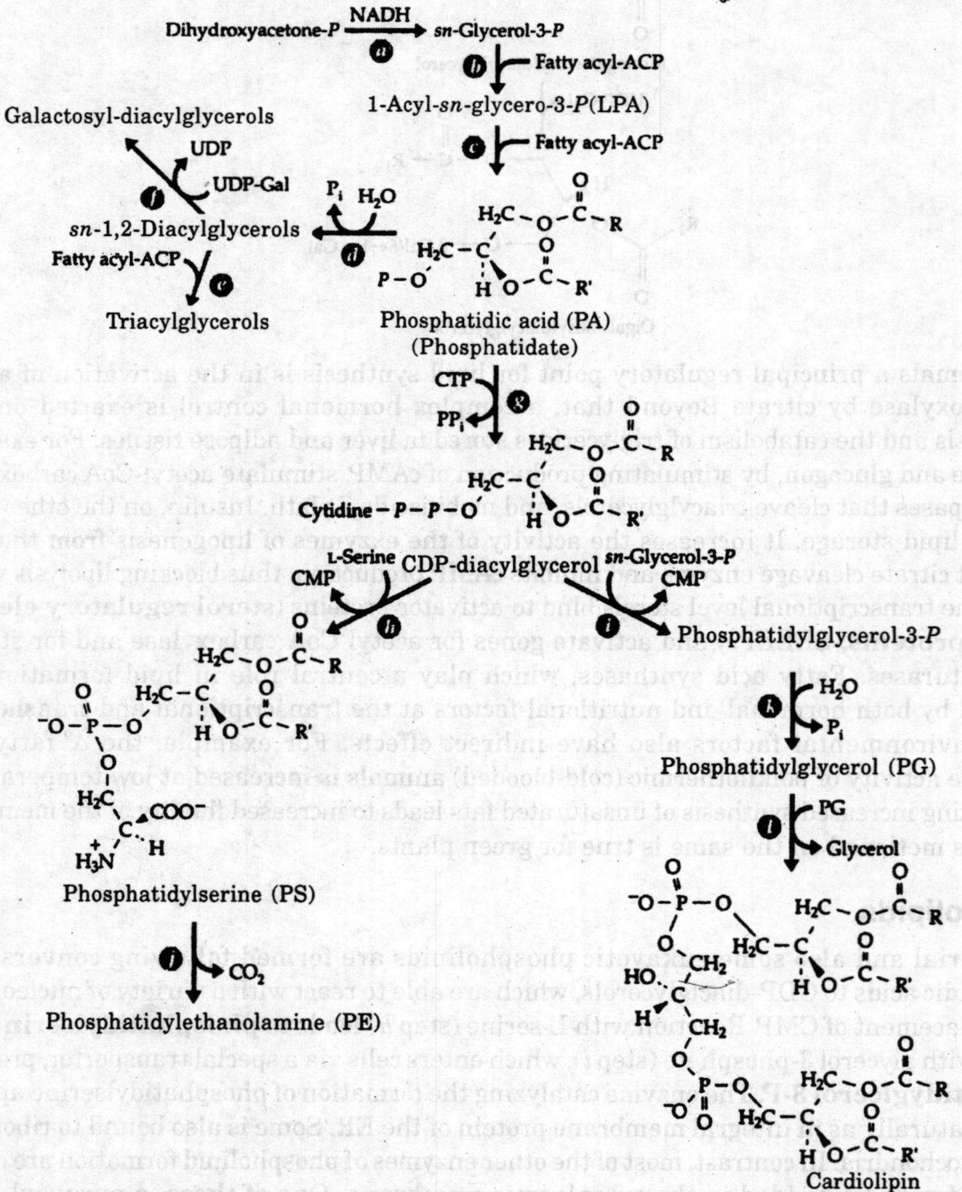

Fig. 8.5. Biosynthesis of triacylglycerols, glycolipids and major phospholipids that are formed both in prokaryotes and eukaryotes. More complete schemes of phospholipid synthesis. Green arrow: pathway occurring only in eukaryotes.

After removal of a phosphate from phospharidylglycerol *3-P*, the resulting phosphatidylglycerol can be converted to **diphoshatidylglycerol** (known as **cardiolipin**). One manner in which this is accomplished in bacteria is indicated by step *l*. One molecule of glycerol is displaced as two molecules of phosphatidylglycerol are coupled. The alternarive pathway is followed in eukaryotic mitochondria and perhaps in some bacteria. The entire phosphatidic acid group is transferred from. CDP-diaylglycerol to phosphatidylglycerol with displacement of CMP. Gram-negative bacteria also synthesize a second set of membrane phospholipids, compounds such as **lipid A** that are based on acylated glucosamine.

Phosphatidylcholine, which is rarely present in bacteria, is formed in eukaryots from phosphatidyl-ethanolamine by three consecutive steps of methylation by *S*-adenosylmethionine. This pathway is of major importance in eukaryotic cells. However, alternative pathways (the Kannedy pathways), which are represented by black line on the left side of are also used for formation of both phosphatidylcholine and phosphartidylethanolamine. In both cases, the free base, choline, or ethanolamine is phosphorylated with ATP. Choline phosphate formed in this manner is then converted by reaction with CTP to CDP-choline. Phosphatidylcholine is formed from this intermediate while CDP-ethanolamine is used to form phosphatidylethanolamine. These synthetic reactions occur within cells unclei as well as on surfaces of cytoplasmic membranes.

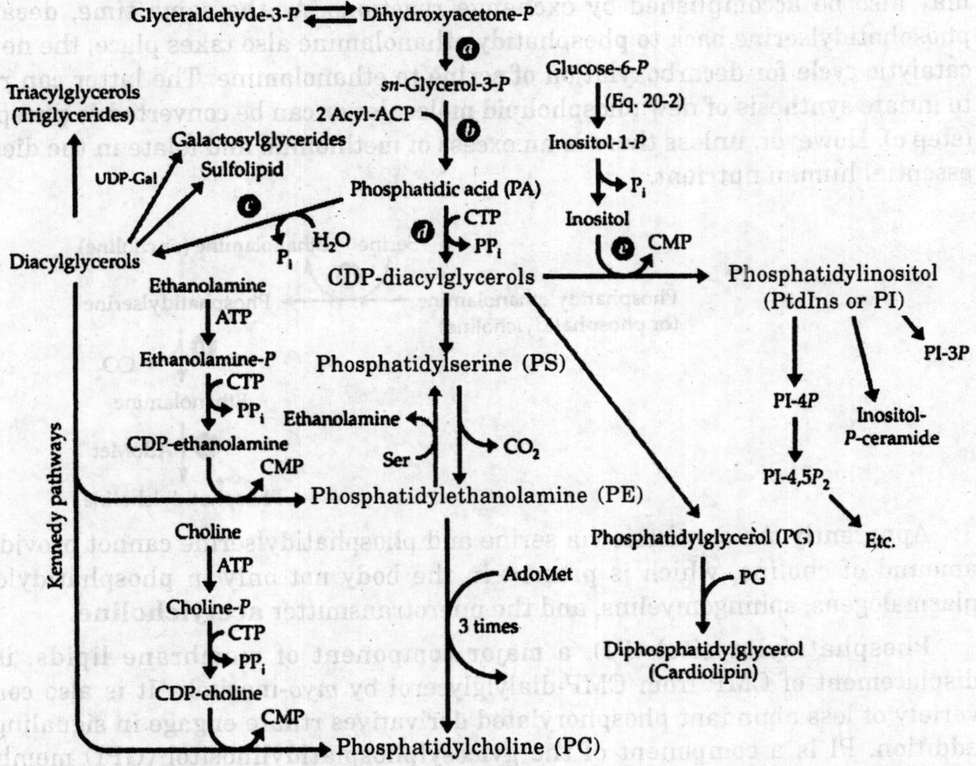

Fig. 8.6. A more complete outiline of the biosynthesis of triacylgcerols, glycolipids, and phospholipids including characterstic eukaryotic pathways. Green lines indicate pathways utilized by both bacteria and eukaryotes. Structures of some of the compounds are the gray arrow show the formation of phosphatidylserine by exchange with ethnolimine.

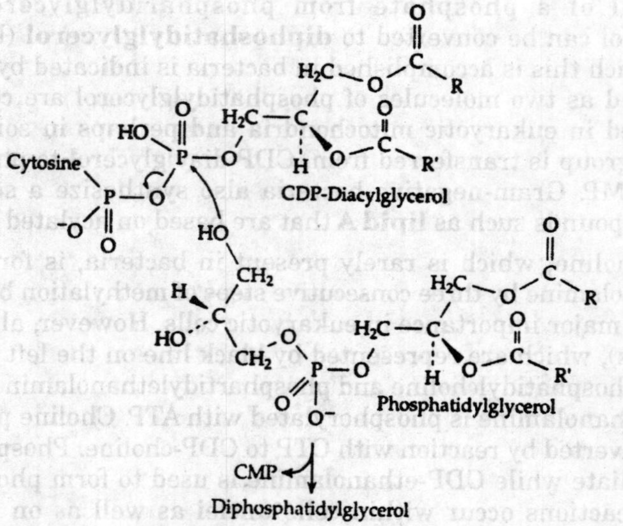

Cytosine
CDP-Diacylglycerol
Phosphatidylglycerol
CMP
Diphosphatidylglycerol

The formation of phosphatidylserine and possibly other phospholipids in animal tissues may also be accomplished by exchange reactions. At the same time, decarboxylation of phosphatidylserine back to phosphatidylethanolamine also takes place, the net effect being a catalytic cycle for decarboxylation of serine to ethanolamine. The latter can react with CTP to intiate synthesis of new phospholipid molecules or can be converted to phosphatidylcholine (step *c*). However, unless there is an excess of methionine and folate in the diet, choline is an essential human nutrient.

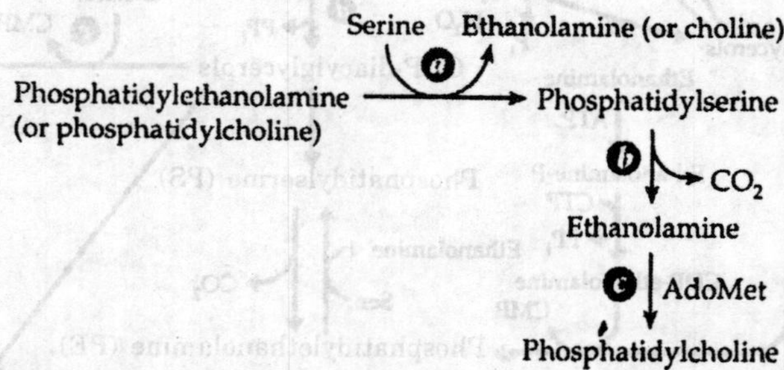

Apparently the sunthesis via serine and phosphatidylserine cannot provide an adequzte amound of choline, which is present in the body not only in phosphatidylcholine but in plasmalogens, sphingomyelins, and the nuerotransmitter **acetylcholine.**

Phosphatidylinositol (PI), a major component of membrane lipids, is formed buy displacement of CMP from CMP-dialylglycerol by *myo*-inositol . It is also converted into a veriety of less abundant phosphorylated derivatives rthate engage in signaling activities. In addition, Pl is a component of the gylcosylphosplatidylinositol (GPI) membrane anchors for surface proteins. Free. GPI anchors, lacking bound proteins, are also present in membranes.

They are especially abundant in many parasitic protozoa and may carry additional glycosyl groups.

Regulation of phospholipid synthesis, which has been studied in detail in yesast, is complex but highly coordinated. The committed in the synthesis of PE and PC is the hydrolysis of phosphatidate (PA) by a phosphatase to generate diacylglycerols. Reaction of PA with CTP (Step *d*) also affects synthesis of the other major phospholipids. Much of the coordinate regulation arises at the transcriptional level. For example, genes for the synthesis of PC or PI are repressed by inositol alone and in combination with choline. Regulation of CTP synthetase controls the formation of CDP-diacylglycerols. In mammalian cells PC synthesis appears to occur only during the S-phase of the cell cycle. The CTP: phospholine cytidyltransferase that catalyzes CDP-choline formation is controlled by storage in a resevoir in the nucleus from which it is transferred to ER membranes.

The Ether-Linked Lipids

Closely related to both the triacylglycerols and phospholipids, the ether-linked lipids contain in place of one ester group an *alkoxy* (–OR) or *alkenyl* (–O–CH = CH–R) group. Phospholipids containing the alk-1-enyl group, the **plasmalogens,** were first recognized in 1924 by Feulgen and Voit, who were developing histological staining precedures. They observed that treatment of tissue slices with acid resulted in the liberation of aldehydes, which were later shown to be formed by breakdown of the alkenyl lipids. Over 10% of the lipid in the human central nervous sysrem is plasmalogen and about 1% is alkoxy lipid. Among the latter is the **platelet activating factor**. In many mammalian cells the ethanolamine plasmalogen **plasmenylethanolemine** represents the major storage depot of arachidonic acid.

$$\mathrm{Y{-}O{-}\overset{H}{C}{=}\underset{H}{C}{-}R} \xrightarrow[\;Y{-}OH\;]{H_2O,\ H^+} \mathrm{O{=}\overset{H}{C}{-}CH_2{-}R}$$

Ether-linker lipids constitute up to 35% of the total phospholipid in molluscs. Although they are usually regarded as animal constituents, small amounts of ether-linked lipids have been identified in plants. The major phospholipids of archaebacterial membranes are ether-linked derivatives of the polyprenyl phytanyl group and of the demeric biphytanyl group.

Biosynthesis of ether lipids begins with formation of fatty acyl derivatives of dihdroxyacetone phosphate. The acyl group is then displaced, along with the oxygen atom to which it is attached, by an alkoxy group of a long-chain fatty alcohol, which is formed by reduction of the corresponding acyl-CoA. The oxygen of the alcohol (designated by an asterisk) is retained in the product. The reaction differs significantly from displacements discussed. The pro-*R* hydrogen atom (marked by the dagger, +) at C-1 exchanges with the mediuum during the reaction suggesting that enolization of the diydroxyacetone phosphate takes place. A possible mechanism would be to add the incoming R–O⁻, generated as in serine proteases, to the double bond of the enol at C-1. This would generate a transient carbanion on C-2. It could then eliminate the carboxylate containing the original C-1 acyl group, and the enol could then ketonize.

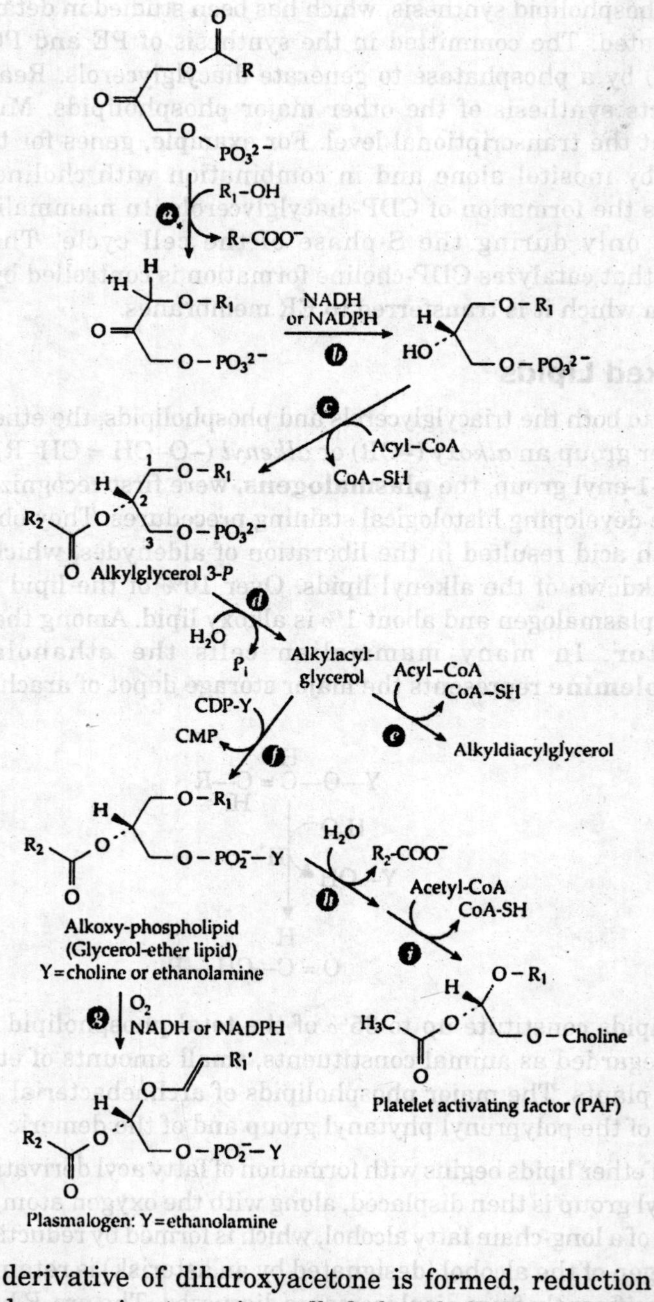

R–OH
R–COO⁻
a
NADH or NADPH
b
c
Acyl–CoA
CoA–SH
Alkylglycerol 3-P
d
H_2O
P_i
Alkylacyl-glycerol
Acyl–CoA
CoA–SH
e
Alkyldiacylglycerol
CDP-Y
CMP
f
H_2O
R_2-COO⁻
h
Acetyl-CoA
CoA-SH
i
Alkoxy-phospholipid (Glycerol-ether lipid)
Y = choline or ethanolamine
g
O_2
NADH or NADPH
Platelet activating factor (PAF)
Plasmalogen: Y = ethanolamine

Once an alkoxy derivative of dihdroxyacetone is formed, reduction to the 2-OH form, further acylation, and conversion to various alkyl phospholipids and neutral lipids can occur. The pathways are closely akin. The conversion of alkoxy lipids to plasmalogens occurs by oxidative desturation. The intial steps in the synthesis of ether-linked lipids take place principally in the peroxisomes. Enzymes catalyzing both the acylation of diydroxyacetor phosphate and the synthesis of alkyl-diydroxyacetone-*P* are found in high amounts in ani peroxisomes. In the rate autosomal recessive disorder known as the **Zellweger syndr** peroxisomes are completely lacking. Both the synthesis of ether linked lipids and the β oxidation of very-long-chain fatty acid are depressed. These acids, principally C 26:0 and C26:1, accumulate in tissues of patients with this severe disease, which is usually fatal during the first four months of life.

The platelet activating factor is formed in noutiophils marcoplages from alkylacyl-*sn*-glycero-3-phosphocholine by the action of phospholipase A_2. This exzyme removes the C_2 acyl group , which is then replaced by an *acetyl group* transferred from acetyl-CoA to form PAF. Alternatively, a phosphachaline group may be transferred onto 1-alkyl-2cayl-*sn*-glycerol from CDP-choline as in the formation of phosphatidylcholine. PAF can undergo hydrolytic removal of its acetyl group in tissues but can also transfer it to such acceptors as lysoplasmalogens or sphingosine. Hydrolytic loss of the acetyl group from PAF destroys biological activities including induction of allergic and inflammatory responses. The various signaling activities of PAF arise from binding to G-protein linked receptors in many cells and tissues.

Sphingolipids

Sphinoglipids are phospholipids and glycolipids derived from **sphingosine** and other "long-chain bases." At least 60 bases of this type have been identified. They very in chain length from C_{14} to C_{26} an include members of the iso and anteiso series. Up to two double bonds may be present. The C_{18} compound, usually called sphingosine. is derived from condensation of palmitoyl-CoA with serine. CArbon bioxide is lost from the serine during the condensation reaction, and the resulting ketone is reduced with NADPH (step *b*) to form **sphinganine,** a common component of animal sphingolipids. It may hydroxylated to phytosphingosine in plants and fungi (step *c*). Sphinganine is converted to long-chain amides by acyl transfer from acyl-CoA (step *d*) and then undergoes desturation (step *e*) to form **ceramides,** the precursors to more than 100 gangliosides, to the phospholipid **sphingomyelins**, and also to free sphingosine (step *h*). This last reaction is degradative and on the pathway of breakdown of gengliosides. Further catabolism of sphingosine is thought to take place by a PLP-mediated chain cleavage to palmitadehyde.

The essential functions of sphingolipids, including the complex ganglisides, are only now being clarified. The latter are abundant in brain and are thought to function in cell-cell recognition. On blood cell surfaces they carry blood group antigens. They play an essential role in spermatogenesis and may function in various signaling processes. In the outer cornified layers of skin, ceramides with very long chain (C_{28} –C_{36}) fatty acyl groups undergo ω hydroxylation and become esterified to glutamate side chains of specific skin proteins called **involucrins**. The long hydrocarbon chains are thought to pass entirely through the lipid bilayer to form rigid lamellae of a water-impermeable outer skin carrier. An important hypothesis is that sphingolids associate with cholesterol to form "lipid rafts," which float in a sea of glyhcerolipids serve as bases for various signaling processes. The long hydrophobic acyl chains of the sphingolipids pack well with cholesterol to form a rigid lipid structure of high meltinig temperature.

Complex Lipids in Signaling

While pathways of synthesis of complex lipids have been described, we are far from understanding the dynamics of the synthesis and turnover of the membranous structures built

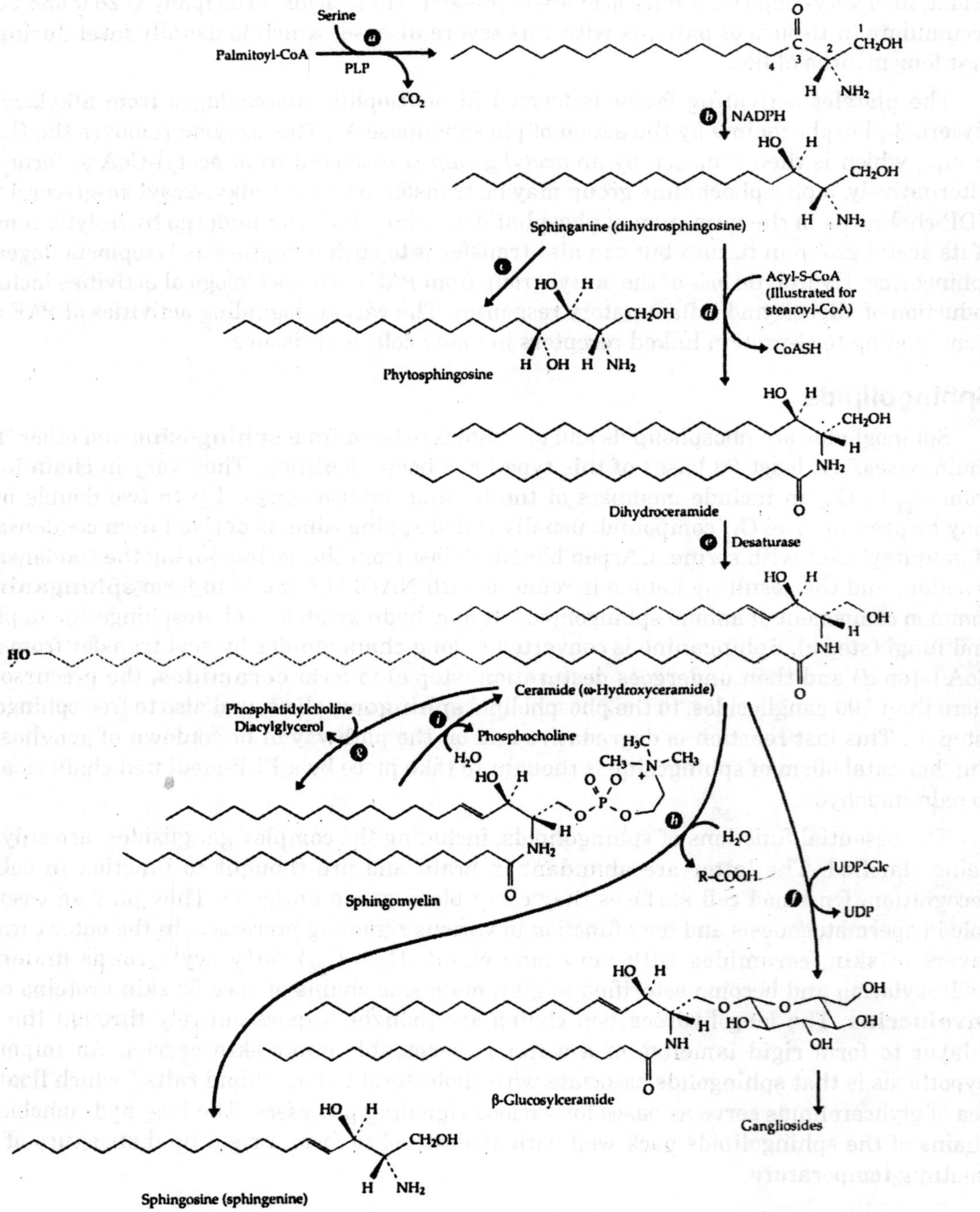

Serine
Palmitoyl-CoA
PLP
CO_2
NADPH
Sphinganine (dihydrosphingosine)
Phytosphingosine
Acyl-S-CoA (Illustrated for stearoyl-CoA)
CoASH
Dihydroceramide
Desaturase
Ceramide (ω-Hydroxyceramide)
Phosphatidylcholine
Diacylglycerol
Phosphocholine
H_2O
Sphingomyelin
H_2O
R-COOH
UDP-Glc
UDP
β-Glucosylceramide
Gangliosides
Sphingosine (sphingenine)

Fig. 8.7. Pathways of synthesis and metabolism of sphingolipids. Gray arrows indicate catabolic pathways. The green extension on the ceramide structure is that of long-chain ω-hydroxyceramide that is covalently bound to protein in human skin.

from them. The fact that the lipid bilayer of a cell membrane is so thin means that any sudden changes in composition at a particular location will cause changes in physical properties and a wave of diffusion that will travel along the membrane. The membrane seems to be ideally structured to receive and propagate messages from outer surface or internal receptors, or massages sent along the bilayer.

One of the most studied examples of signaling with membrane lipids is provided by the **phosphoinositide cascade,** which is pictured. Six or more phosphate esters of phosphatidylinositol (PI) are generated by the action of kinases. More than 100 extracellular singaling molecules activate specific isozyme forms of **phospholipase C,** releasing 20 or more different inositol phosphates from these phosphoinositide esters. The relased inositol phosphates, which act as water-soluble messengers, are further modified by the action of several phosphatases. At the same time, **diacylycerols** are left in the membrane. With loss of the negative charges of the PL phosphates there will be immediate electrostatic effects in the membrane, which may alter the ionic environment, open ion channels, etc. The diacylglycerols, which diffuse within the membrane, may lose arachidonic acid from the *sn*-2 position to supply substrate for the arachidonate cascades described in Section D. Diacylglycerols also activate the 11 isozyme forms of **PROTEIN KINASE**. Some of these exzymes not only are activated by diacylglycerols, but also require **phosphatidylserine** for activity.

Other lipid-based signaling cascades arise from reactions that modify phosphatidyclcholine molecules. Unsaturated fatty acids in the *sn*-2 position readily oxidized by free radicals with cleavage of the hydrocarbon chains to form alcohols, aldehydes, and carboxylic acid. These mimic PAF in their biological activities. Phosphatidate molecules with saturated or monounsaturated fatty acids in the *sn*-2 position arise from breakdown of phosphatidylcholine catalyzed by **pheroholipase D**. Phosphatidate may also be formed by a family of lipid **diacylglycerol kinases**. Phosphatidates containing saturated or monounsaturated fatty acids also have a variety of signaling activities. An arachidonoyl-diacylglycerol kinase is thought to function in many process. An example is a PI-mediated cycle in invertebrate vision. Sphingomyelin breakdown release diffusible ceramides that have been implicated as singnaling molecules in cell proliferation, differentiation, growth arrest, and apoptosis. Sphingosine and sphingosine 1-*p* also have signaling functions.

Phospholipids have been shown to exchange between different membranes, *e.g.*, of mitochondria and the ER, Exchanges of phosphatidylcholine, phosphatidylinositol, and sphingomyelin are catalyzed by specific **exchange proteins**. These protein may also participate in signaling, but their major function may be to transport the phospholipids from their sites of synthesis to the various membranes of the cell.

Peroxidation of Lipids Rancidity

Storage of fats and oils leads to **rancidity**, a largely oxidative deterioration causes development of unpleasant tastes, odors, and toxic compounds. Similar chemical changes account for the "drying" of oil-based paints and varnishes. These reactions occur most readily with polyunsaturated fatty acids, wheater free or in ester linkage within triacylglycerols. The reactions are initiated by free redicals, which may be generated by oxidative enzymes within or outside of cells, or by nonenzymatic reactions catalyzed by traces of transition metals or by environmental pollutants. Characteristic of rancidity is an autocatalytic chain reaction.

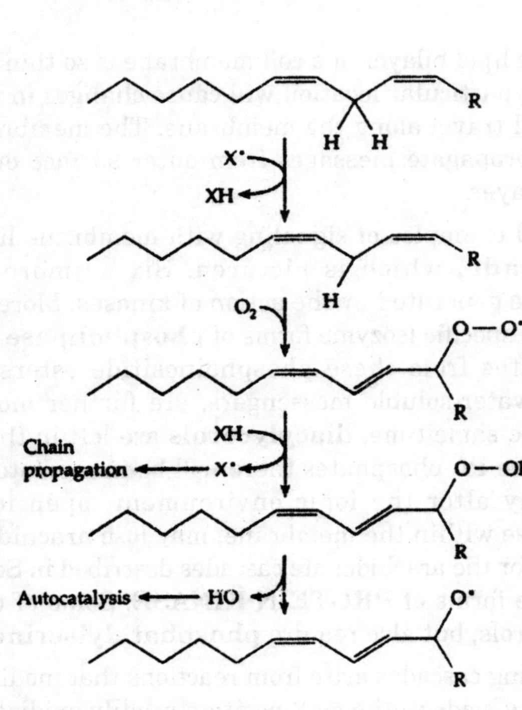

R
H H
X•
XH
R
H
O_2
O—O•
R
XH
Chain propagation ← X•
O—OH
R
Autocatalysis ← HO
O•
R

Radical X*, which initiates the reaction, is regenerated in a chain propagation sequence that, at the same time, produces an organic peroxide. The latter can be cleaved to form two additional radical, which can also react with the unsaturated fatty acids to set up the autocatalytic process. Isomerization, chain cleavages, and radical coupling reactions also occur, especially with polyunsaturated fatty acids. For example, reactive unsaturated aldehydes can be formed

$$\text{R–CH=CH–CH(}^{\bullet}\text{O)–R}' \longrightarrow \text{R–CH=CH–CHO} + {}^{\bullet}\text{R}'$$

An intermediate in may be converted to **4-hydroxy-2-nonenal,** a prominent product of the peroxidation of arachidonic or linoleic acids. However, other biosynthetic pathways to this compound are possible. 4-Hydroxy-2-nonenal can react with side chains of lysine, cysteine, and histidine to form fluorescent products such as the follwoing cyclic compound generated by an oxidative reaction.

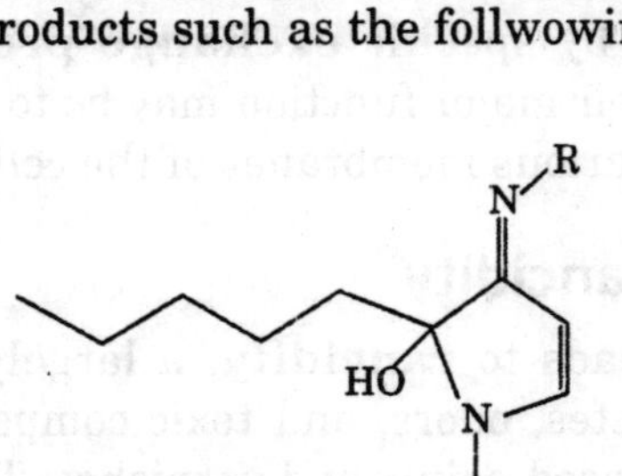

N–R
HO
N
R′

Polyenoic acids also give rise to malondialdehyde, a reactive untagenic compound can be reduced and dehydrated to acrolein, a toxic compound which also reacts with both lysine and serine to produce products excreted in the urine. More dangerous are similar reactions of

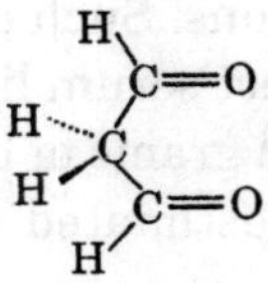

Malondialdehyde

$CH_2 = CH–CHO$

Acrolein

these aldehydes with proteins of the body in conditions such as diabetes or renal insufficiency. The bifunctional malondialdehyde forms Schiff bases with protein amino groups and acts as a crosslinking agent. **Age pigments** (also called **lipofuscin**), which tend to accumulate within neurons and other cells, are thought to represent precipitated lipid-protein complexes resulting from such reactions. The reactions are similar to those of proteins with the products of sugar breakdown. Organisms have developed multiple enzymatic mechanisms for detoxification of products of both glycation and oxidative degradation.

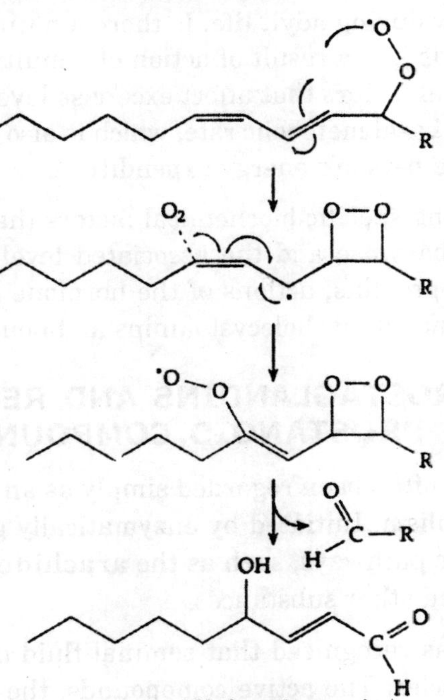

4-Hydroxy-2-nonenal

The oxidative degradation represented by the foregoing reactions is referred to as peroxidation. Peroxidation can lead to repid development of rancidity in fats and oils. However, the presence of a small amount of tocopherol inhibits this decomposition, presumably by trapping the intermediate radicals in the form of the moire stable tocopherol radicals, which may dimerize or react with other redicals to terminate the chain.

Catalytic hydrogenation of vegetable oils is widely used to form harder fats and to decrease the content of polyunsaturated fatty acyl groups. The products have a greatly increased resistance to rancidity. However, they also contain fats with trans double bonds as well as isomers with double bonds in unusual positions. Such compounds may interfere with normal fatty acid metabolism and also appear to affect serum lipoprotein levels adversely. Trans fatty acids are present in some foods. One hundred grams of butter contain 4–8 g, but hydrogenated fats often contain much more. It has been estimated that in the United States trans fatty acids account for 6–8% of total dietary fat.

Some Nutritional Questions

While many of the poorer people on earth starve to death the problems of atherosclerosis and obesity affect many in wealthier societies. The fat content of food is often blamed, and, the quality of fatty acids in the diet is very important. However, like fatty acids, carbohydrates are also metabolized via acetyl-CoA and can readily be converted to both fatty acids and cholesterol. Obesity is largely a problem of excessive total caloric intake.

Why do some pepole stay slim while others be come obese? What are the regulatory mechanisms that affect appetite and body composition? The human body weight tends to be stable or to increase slowly during adylt life. Is there a natural set point for each individual? No, an apparent set point is just a result of action of a multitude of factors including genetic variations and psychological factors that affect excercise levels, eating habits, etc. It is worthwhile to recognize that the basal metabolic rate, which is also affected by many factors, account for a very high fraction of a person's energy expenditure.

The following are among specific biochemical factors that act on the energy balance: the activity of acetyl-CoA carboxylase and the associated level of malonyl-CoA; the activity of mitochondrial uncoupling proteins; actions of the hormone leptin (which have been hard to interpret), and hormones including cholecystokinins and neuropeptide.

PROSTAGLANDINS AND RELATED PROSTANOID COMPOUNDS

Lipid peroxidation has often been regarded simply as an undesirable side reaction, but it is a normal part of metabolism. Initiated by enzymatically generated radicals, peroxidation occurs as specific metabolic pathways, such as the **arachidonate cascade,** which leads to a varity of local hormones and other substances.

As early as 1930, it was recognized that seminal fluid contains materials that promote contraction of uterine muscles. The active comopounds, the **prostaglandins,** were isolated and crystallized in 1960 and were identified shortly therefter. As many as 14 closely related compounds found in human seminal fluid, one of the richest known sources. Prostaglandins are present in seminal fluid at a total concentration of ~1 imM, but their action on smooth muscles has been observed at a concentration low as 10^{-9} M. The structures and biosynthetic pathways of several of the prostaglandins are indicated. Prostaglandins are ussally abbreviated PG with an additional letter and numerical subscript addes to indicate the type. The E type are β-hydroxyketones, the type 1,3-diols, and the A type α,β-unsaturated ketones. Series 2 prostaglandins arise from arachidonic acid, while series 1 and 3 arise from fatty acids containing one fewer or one more double bond, respectively. Additional forms are known.

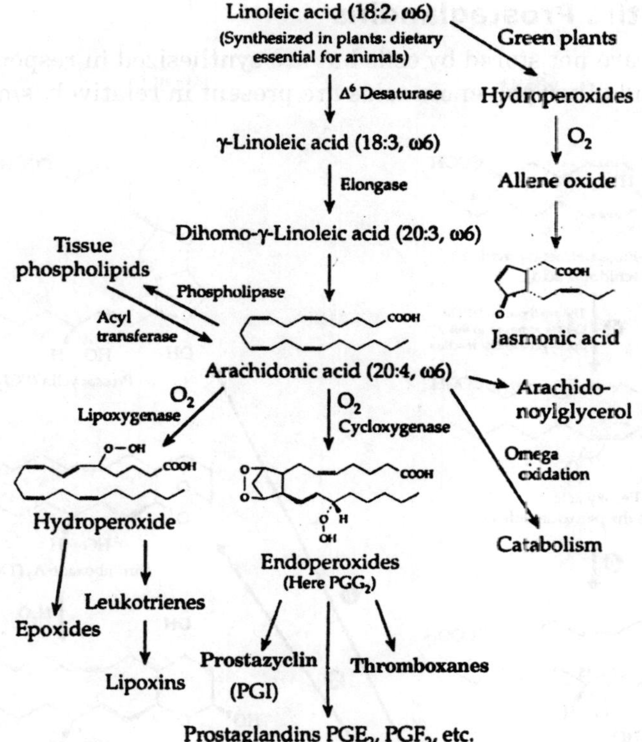

15 12 8 5

COOH

Arachidonic acid

C20 : 4 : ω6

13

COOH

α-Linolenic acid

C18 : 3 : ω3

13

COOH

H_S H_R

Linoleic acid

C18 : 2 : ω6

$E-Fe^{3+}$

H^+ $E-Fe^{2+}$

O_2 COOH H

H^+

$^{\cdot}O-O$ COOH

$E-Fe^{3+}$

HO–O H H^+ $E-Fe^{2+}$

COOH

(13S)-9,11 (Z,E)-Octadecadienoic

Metabolism of the Prostaglandins

Prostaglandins are not stored by cells but are synthesized in response to external stimuli. Arachidonic acid and other polyenoic acids are present in relatively small amounts (*e.g.*, ~1%

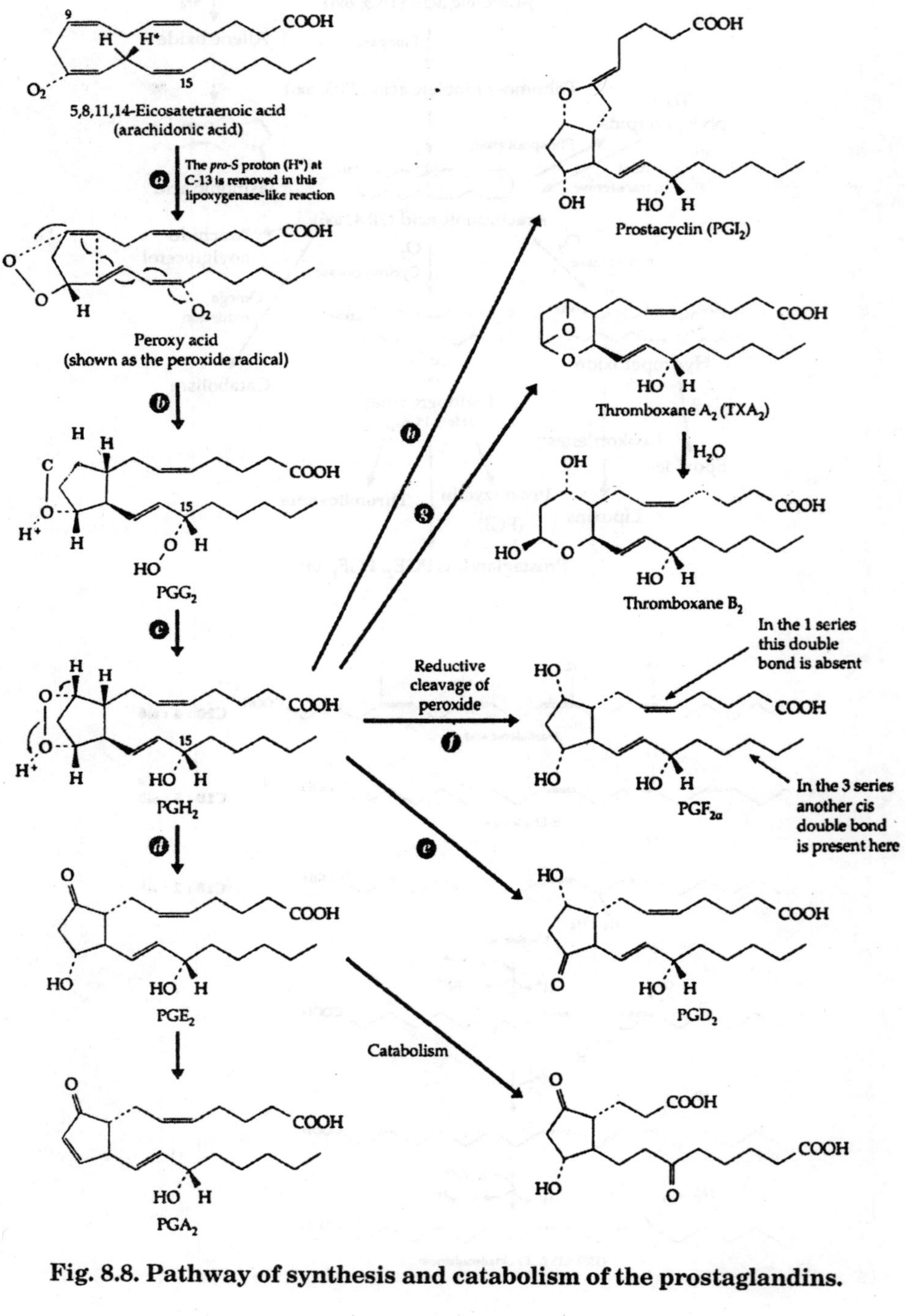

Fig. 8.8. Pathway of synthesis and catabolism of the prostaglandins.

of total plasma fatty acids), but they are concentrated in the 2-position of phospholipids. This is in part a result of phospholipid "remodeling". Acyl groups are hydrolyzed from the *sn*-2 position by action of phospholipase A_2. An acyltransferase with a preference for arachidonoyl groups then transfers esterified arachidonic acid from phosphatidylcholine and other phospholipids to the lyso forms of the phospholipids, which lack a 2-acyl group. The enzyme has a strong preference for the lyso-ethanolamine plasmalogens. As a consequence, in the plasmenylethanolamine of platelets arachidonoyl group account for 66% of the acyl residues at the 2-position. An arachidonate-specific acyl-CoA synthetase repidly reconverts any free arachidonate that is not used for prostanoid synthesis back into phospholipids.

The synthesis of prostaglandins, which was elucidated by Samuelsson, begins with the release of arachidonate and other polyenoic acid precursors from phospholipids through the action of phospholipase A_2. The released arachidonate is then acted upon by **prostagelndin H synthases,** which catalyze two consecutive reactions at adjacent but distinct sites in a single protein. The first, **cyclooxygenase** or prostaglandin endoperoxide synthease reaction, forms PGG_2 from arachidonate and the second, reaction, generates PGH_2. There are two major mammaliam isozymes of prostaglandin synthase, which are often called cyclooxygenase-1 (COX-1) and cyclooxygenase-2 (COX-2). From studies of stereospecifically synthesized H-containing fatty acid precursors, it was establisjed that the first step in cyclooxygenase action involves removal of the *pro*-S proton at C-13 of the fatty acid. The O_2-requiring cyclooxygenase resembles lipoxygenase. The product is a peroxy acid, possibly in the form of the peroxide redical. This redical (or peroxide anion) undergoes cyclization with sychoronous attack by a separate O_2 molecule at C-15 to give the endoperoxide PGG. Reduction of the latter to an OH group by the NADPH-dependent peroxidase (step *c*) yields PGH. The entire sequence is catalyzed by the single 70-kDa PGH synthase, which contains a single heme prosthetic group. During the cyclooxygenase reaction the enxyme appears to accept electrons from cytochrome b_5. During the peroxidase step the heme group undergoes formation of the characteristic peroxidase intermediate compounds I and II. It has been suggested that a tyrosyl radical is generated in the peroxidase active site (on Y 385 of COX-1) and is used to form an arachidonate radical that reacts with O_2 in the cyclooxygenase reaction. Alternatively, a carbocation mechanism is also possoble.

PGH can break down in three ways tio give the E and F series of prostaglandins. In one the proton at C-9 is eliminated (step *d*) as indicated by the small arrows by the PGH_2 structure. An alternative isomerization (step *e*) gives. The F prostaglandins are formed by reductive cleavage of the endoperoxide (step *f*). The A series and other prostaglandin arise by secondary reactions.

A biochemical characteristic of the prostaglandins is repid catabolism. The product in (lower right) arises by oxidation of the 15-OH to a carbonyl group, permitting reduction of the adjacent trans double bond. Two steps of β oxdations as well as ω oxidation are also required to produce the dicarboxylic acid product shown. However, a series of products appears, and the distribution varies among species. Catabolism of prostaglandins is especially active in the lungs, and any prostaglandins entering the bloodstream are removed by a single pass throuth the lungs. This observation has led to the conclusion that prostaglandins are not hormones in the classical sense but ac on a more local basis.

Thromoboxanes and Prostacyclins

In blood platelets and in some other tissues PGG is also transformed to another series of compounds, the **thromboxanes,** which were identified in 1975. Labile Hemicacetals, the thromboxanes A (TXA, are derived by rerrangement of PGH (step *g*). Thromboxane synthase, which catalyzes the reaction, has characteristics of a cytochrome. Cytochromes are known to react with peroxides as well as with O_2, and the endoperoxide of PGH may be opened by the synthase prrior to rearrangement to TXA. Thromboxane A_2 is so unstable that its half life at 37°C in water ~36s. It is spontaneously converted to TXB_2, which contain an –OH group at C-15. The thromboxanes B are much more stable than TXA but are not very active physiologically.

By 1976, Vane and associates hads indentified another prostanoid compound, **prostacyclin** (or PGI_2). This compound also arises from PHG_2 by action of a cytochrome P450-like prostacyclin synthase. It is thought to be an important vasoprotective molecule. As with the thromboxanes, prostacyclin undergoes rapid inactivation by hydrolysis to the physiologically inactive 6-ox-$PGF_1\alpha$.

Lipoxygenases

Lipoxygenases, of which the enzyme from soybeans has been studied the most, also catalyze oxiadation of polyunsaturated fatty acid in lipids as indicated. Formation of the hydroperoxide product is accompanied by a shift of the double bond and conversion from cis to trans configuration. Soybean lipoxygyenase is a member of a family of related lipoxygenases that are found in all eukaryotes. All appear to have similar iron-or manganes-containing active sites and to act by similar mechanisms. The major substrate in animals is arachidonic acid (probably as the arachidonate ion). As marked on the structures above, there are 6-, 8-, 12-, and 15-lipoxygenases, which catalyze reaction with dioxygen at the indicated places. Linoleic and linolenic acids are the primary substrates in plants. Soybeen lipoxygenase acts on the 13-position of linoleic acid. However, this enzyme is often referred to as a 15-*lipoxygenase* because it acts on arachidonate at C15. The 100-kD enzyme from soybeans contains one atom of Fe(II), which is bound by a cluster of His and Tyr side chains. It must be oxidized to the Fe(III) state before becoming active. The initial reaction with O_2 may occur via an intermediate radical.

Leucktrienes, Lipoxins, and Related Compounds

Yet another series of products results from the action on arachidonate of tissue lipoxygenases, which compete with prostaglandin-forming cyclooxygenases. The 5-lipoxygenase produces the unstable peroxide **5-hydroperoxy 6, 8, 11, 14-eicosatetraenoic acid** (usually abbreviated **5-HPETE**). This enzyme require ATP and Ca^{2+} and appears to be regulated by a series of other metabolites. The 12-and 15-lipoxygenases, whose distribution varies among different mammalian organs and tissues, form the corresponding 12-and 15-HPETES as well as 5,15-, 8,15- 14, 15-diHPETEs. Some of these peroxide have physiological effects of their own, but they are largely transformed peroxodases to more stable compounds such as the corresponding alcohol (hydroxy-icosatetraenoic acids or HETEs.

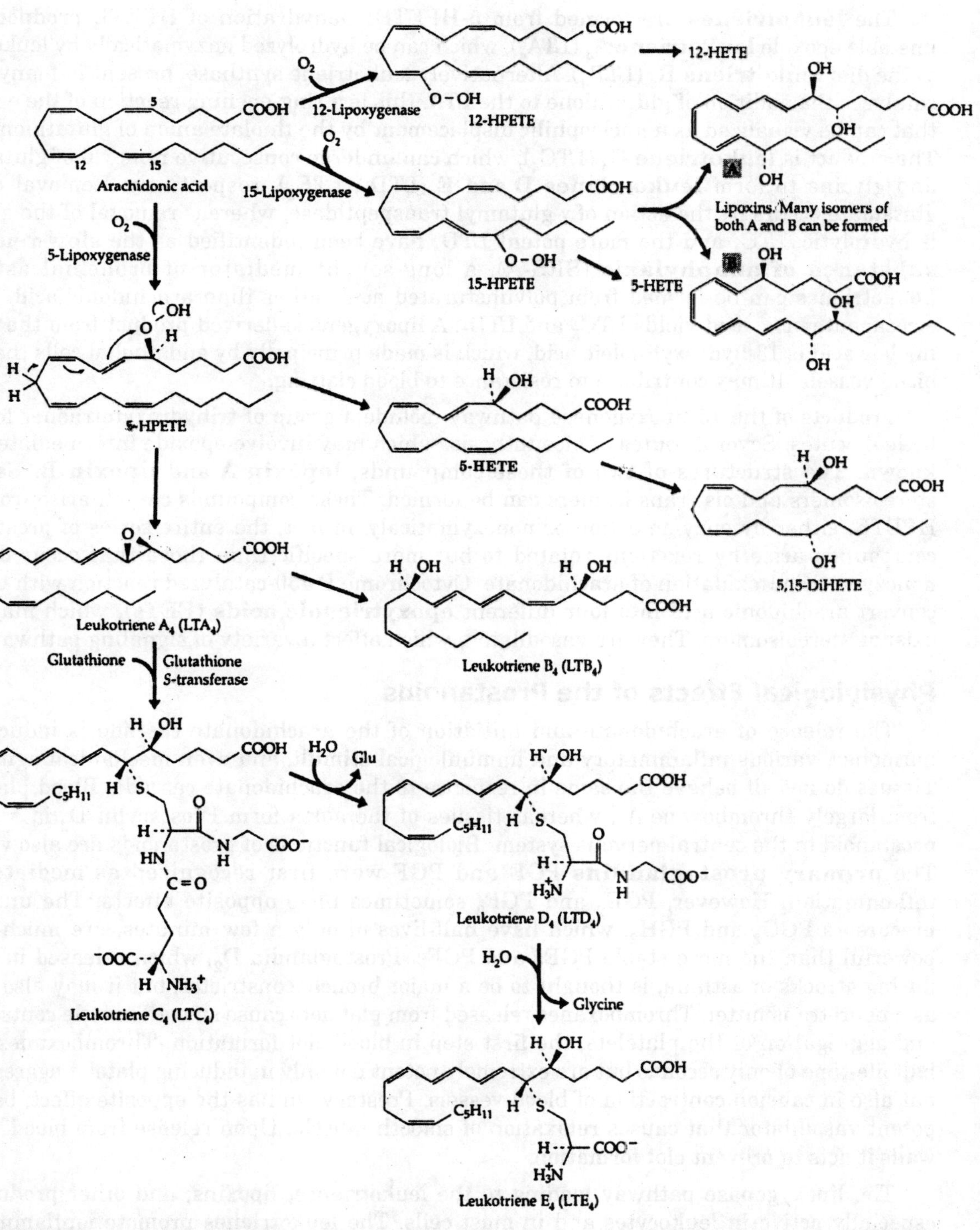

12-Lipoxygenase
12-HPETE
12-HETE
Arachidonic acid
15-Lipoxygenase
15-HPETE
5-HETE
Lipoxins. Many isomers of both A and B can be formed
5-Lipoxygenase
5-HPETE
5-HETE
5,15-DiHETE
Leukotriene A_4 (LTA_4)
Leukotriene B_4 (LTB_4)
Glutathione
Glutathione S-transferase
H_2O
Glu
Leukotriene C_4 (LTC_4)
Leukotriene D_4 (LTD_4)
H_2O
Glycine
Leukotriene E_4 (LTE_4)

Fig. 8.9. The 5-lipoxygenase pathway for biosynthesis of the hydroperoxy-and hydroxyeicosanoic acids, leukotrienes, and lipoxins.

The **leukotrienes** are formed from 5-HPETE. Dehydration of HPETE produces the unstable epoxide **leukotriene A_4** (LTA_4), which can be hydrolyzed enzymatically by leukocytes to the diol **leukotriene B_4** (LTB_4). Alternatively leukotriene synthase, present in many cells, catalyzes the addition of glutathione to the LTA. this is a ring-opening reaction of the epoxide that can be visualized as a nucleophilic displacement by the thiolate anion of glutathione C-6. The product is **leukotriene C_4** (LTC_4), which can undergo consecutive removal of glutamate and glycine to form **leukotrienes D** and **E** (LTD_4, LTE_4), respectively. Removal of the glutamate occurs by the action of γ-glutamyl transpeptidase, whereas removal of the glycine is hydrolytic. LTC_4 and the more potent LTD_4 have been indentified as the **slow-reacting substance of anaphylaxis** (SRS-A), a long-sought mediator of bronchial asthma. Leukotrienes can be formed from polyunsturated acids other than arachidonic acid. Thus, eicosapentaenoic acid yields LTC_5 and LTD_5. A lipoxygenase-derived product from the C18:2 linoleic acid is 13-hydroxylinoleic acid, which is made principally by endothelial cells that line blood vessels. It may contribute to resistance to blood clotting.

Products of the 15-lipoxygenase pathway include a group of trihydroxytetraenes formed by leukocutes. Several routes of biosunthesis, which may involve epoxide intermediates, are known. The structures of two of these compounds, **lopoxin A** and **lipoxin B.** Several stereoisomers and cis-trans isomers can be formed. These compounds can all arise from 15-HPETE, either by enzyme action or nonezymaticaly. in fact, the entire series of prostanoid compounds arise by reactions related to but more specific than those that occur during nonezymatic autoxidation of arachidonate. Cytochrome P 450-catalyzed reaction with O_2 can convert arachidonic acid into four different **epoxytrienoic acids (EETs)**, which may also exist as stereoisomers. They are vasodilators which affect a variety of signaling pathways.

Physiological Effects of the Prostanoids

The release of arachidonate and initiation of the arachidonate cascade is induced by hormones, various inflammatory and immunlogical stimuli, and even mechanical agitation. Tissues do not all behave the same in response to the arachidonate cascade. Blood platelets from largely thromboxane A_2, whereas tissues of the aorta form Prostacylin D_2 ia a major prostanoid in the central nervous system. Biological functions of prostanoids are also varied. The **primary prostaglandins** PGE and PGF were first recognized as mediators of inflammation. However, PGE_2 and PGF_2 sometimes have opposite effects. The unstable precursors PGG_2 and PGH_2, which have half-lives of only a few minutes, are much more powerful than the more stable PGEs and PGFs. Prostaglandin D_2, which released in lungs during attacks of asthma, is thought to be a major bronchoconstrictor, but it may also serve as a neurotransmitter. Thromboxanes released from platelets cause smooth muscle contraction and aggregation of the platelets, the first step in blood clot formation. Thromboxanes have half life-time of only seconds but are extremely potent not only in inducing platelet aggregation but also in causion contraction of blood vessels. Prostacyclin has the opposite effect, being a potent vasodilator that causes relaxation of smooth muscle. Upon release from blood vessel walls it acts to prevent clot formation.

The lipoxygenase pathway leading to the leukotrience, lipoxins, and other products is especially active in leukocytes and in mast cells. The leukotrienes promote ionflammation but lipoxins A_4 and B_4 are antiinflammatory. The release of leukotrience LTC_4, LTD_4, and LTE_4 in lung tissue is correlated with the long-lasting contractions of smooth muscle of the bronchi that are characteristic of asthma. Leukotrience LTC_4 is ~1000 times more powerful than histamine in inducing such contraction. Leckotriencee have also been found in the central nervous system.

Some effects of prostaglandins are mediated throuth cell surface G-protein coupled receptors. Some other prostanoids bind to and activate nuclear peroxisome proliferator-activated receptors. PGL_2 may inhibit fatty acid sunthesis and fat deposition in adipose tissue through these receptors. Some of the prostanoid derivatives enter membranes and may become incorporated into phospholipids and exert their effects there.

A number of medical uses of prostaglandins have been discoverad and more will probably be developed. While prostaglandins may be required for conception, small amounts of PGE_2 or $PGF_{2\alpha}$ induce abortion, $PGF_{2\alpha}$ is also used to induce labor. Prostaglandins are widely employed to control breeding of farm animals, to synchronize their estrus cycles, and to improve the efficiency of artificial insemination.

Inflammation

Special interest in the prostaglandins has focused on pain of inflammation and alleric respinses. The medical significance is easy to see. Five million Americans have **rheumatoid arthritis,** an inflammatory disease. Branchial asthma an dother allergic diseases are equally important. Our most common medicine is **aspirin**, an anti-inflammatory drug. Both the inflammatory response and the immune response are normal parts of the defense mechanisms of the body, but both are potentially harmful, and it is their regulation that is probably fault in rheumatoid arthritis and asthma. Overproduction of prostaglandins may be a cause of menstrual cramps.

Prostaglanding have been implicated both in the induction of inflammation and in its relief. In inflammation small blood vessels become dilated, and fluid and proteins leak into the interstitial spaces to produce the characteristic swelling (edema), many polumorphonuclear leukocytes attracted by chemotactic factor that include LTB_4 migrate into the inflamed area, engulfing dead tissue and bacteria. In this process lysosomes of the leukocytes release phospholipase A, which hydrolyzes phospholipids and initiates the arachidonate cascade. The leuktrience that formed promote the inflammatory response. However, cAMP can suppress inflammation, and PGE_2 has a similar effect. Indeed, E prostaglandins, when inhaled in small amounts, relieve asthma.

The synthesis of prostaglandins is inhibited by aspirin and many other analgesic drugs. Aspirin is an acetylating reagent, and the inhibition has been traced to acetylation of the side chain –OH group of a single serine residue, Ser 530 of COX-1 or Ser 516 of COX-2 in the arachidonate binding channel. Other nonsteroidal antiinflammatory drugs (NSAIDs), *e.g.*, ibuprofen, are competitive inhibitors of the cyclooxygenases. COX-3, a variant form of COX-1, may be the target for acetaminohen. However, the same drugs also inhibit activition of neutrophils and may thus exercise their anti-inflammatory action in more than one way. Since PGE_1 is a potent **pyrogen** (fever-inducing agent), a relationship to the ability of aspirin to reduce fever is also suggested. Unfortunately, all of these drugs inhibit COX-1 of platelets. Small regular doses of aspirin may be useful in preventing blood clots in persons with arterial disease, but can be disastrous. Thousands of people die annually of hemorrhage caused by aspirin.

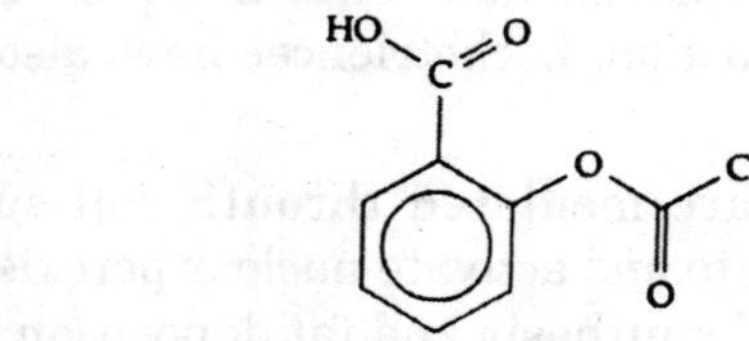

HO C O O CH3 O

Aspirin (acetylsalicylic acid)

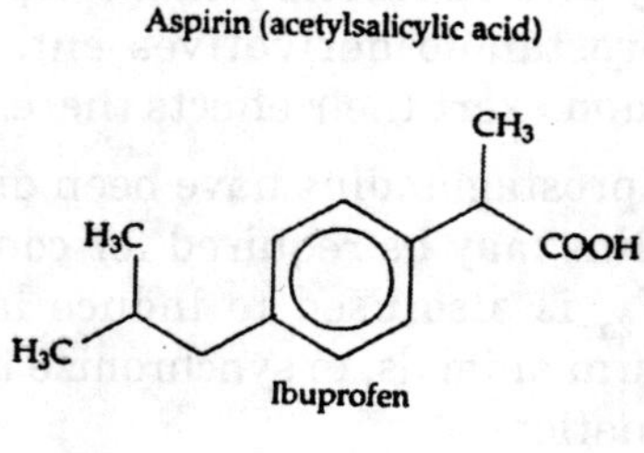

CH3 H3C COOH H3C

Ibuprofen

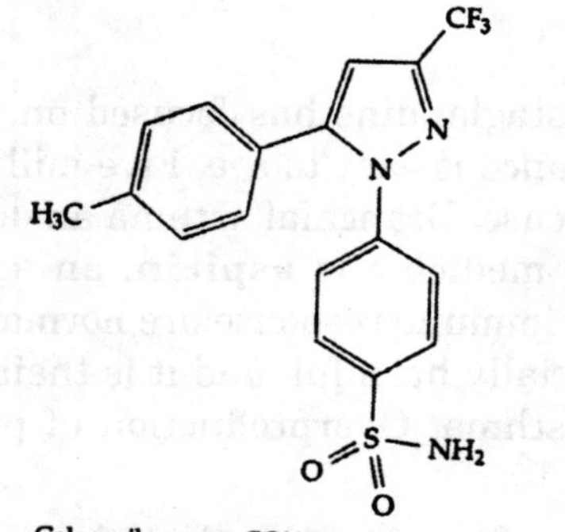

CF3 N N H3C S NH2 O O

Celecoxib, a new COX-2 inhibitor

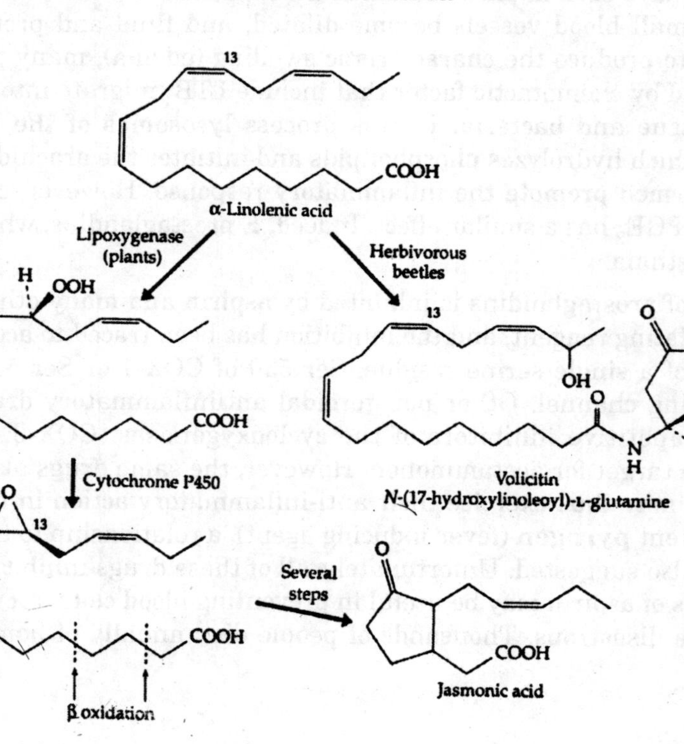

13 COOH

α-Linolenic acid

Lipoxygenase (plants)

Herbivorous beetles

H OOH COOH

13 O NH2 OH O COO⁻ N H H

Volicitin

N-(17-hydroxylinoleoyl)-L-glutamine

Cytochrome P450

O 13 COOH

β oxidation

Several steps

O COOH

Jasmonic acid

Recently it was recognized that COX-1 provides eicosinoid for homeostatic purposes, while it is COX2 that is inducible and generates prostoglandins for production of leukotrienes and induction of the inflammatory response. Now there is a major effort, with the first drugs already in use, to develop specific inhibirors for COX-2, which do not inhibit COX-1. It is hoped hat these will be safer than aspirin. However, these drugs can also cause dangerous side effects. COX-2 of marcophages is also inhibited by γ-tocopherol, a major form of vitamin E.

Plant Lipoxygenases and Jasmonic Acid

An octadecenoid signaling pathway, which resembles the arachidonate cascade in some respects, plays an important role in green plants. Alpha-linolenic acid is acted upon by a lipoxgenase in plastids to form a 13-hydroperoxy derivative. This is dehydrated and cyclized by allene oxide synthase. Although this doesn's appear to be an oxidation reduction process, the enzyme seems to be a cytochrome P450 and to initiate the cyclization to the unstable epoxide **allene oxide** by homolytic cleavage of the peroxy group. Allene oxide synthase and the cyclase that acts in the next step may be cytosolic, while the β oxidation that shortens the chain occurs in plant exclusively in peroxisomes of glyoxysomes.

Jasmonic acid is a plant growth regulator that affects many aspects of plant development as well as responses to environmental signals. A very important function is mobilization of plant defenses in response to damage by herbivores, by bacterial or fungal pathogens, or by ultraviolet light. The sunthesis of protease inhibitors as well as phytoalexins is induced. A curious variant of the jasmonate pathway is the acquisition of α-linoleic acid from plants by chewing caterpillars. The linoleic acid is hydroxylated by the insect, and conjugated with glutamine to form **volicitin** (*N*-(17-hydroylinoleoyl)L-glutamine;. Some of this compound reenters the plant from material regurgitated into wounds by the caterpillars. Volicitin induces the plant to release volatile terpenes and other compounds. The value to the caterpillars is not clear, but not only does volicitin induce defensive reactions in plant but also the released volatile compounds attrat wasps that parasitize the caterpillars. Plants also form a group of **isoprostanes** E_1 from α-linolenic acid.

Allene oxides are unusual biological products. However, they are formed from arachidonic acid by some corals and are evidently precursors to prostaglandin esters, which may be present in high concentrations. Allene oxides are also present in starfish oocytes.

THE POLYKETIDES

In 1907, Collie proposed that polumers of ketene ($CH_2 = C = O$) might be precursors of such compounds as **orsellinic** acid, a common constituent of lichens. The hypothesis was modernized in 1953 by Birch and Donovan, who proposed that several molecules of acetyl-CoA are condensed but *without the two reduction steps required in biosynthesis of fatty acids.* As we now know they were correct in assuming that the condensation occurs via malonyl-CoA and an acyl carrier group of an enxyme. The resulting **β-polyketone** can recent in various ways to give the large group of compounds known as polyketides.

$$nH_3C-C(=O)-S-CoA \longrightarrow H\left(CH_2-C(=O)\right)_n S-CoA$$

Acetyl-CoA β-Polyketone

β-Polyketones can be stabilized by ring formation through ester or aldol condensations. Remaining carbonyl groups can be reduced (prior to or after cyclization) to hydroxyl groups, and the latter can be eliminated as water to form benzene or other aromatic rings illustrates two ways in which cyclization can occur. One involves a Claisen ester condensation during which the enzyme and its SH group are eliminated. Enolization of the product gives a trihydroxy-acetophenone. The second cyclization reaction is the aldol condensation. Following the condensation water is eliminated, and the prodyct is hydrolyzed and enolized to form orsellinic acid. Another product of fungal metabolism is **6-methlsalicylic acid,** which lacks one OH group of orsellinic acid. This synthesis can be explained by assuming that the carbonyl group at C-5 of the original β-polyketone was reduced to an OH group at some point during the biosynthesis. Elimination of two molecules of water together with enolization of the remaining ring carbonyl gives the product.

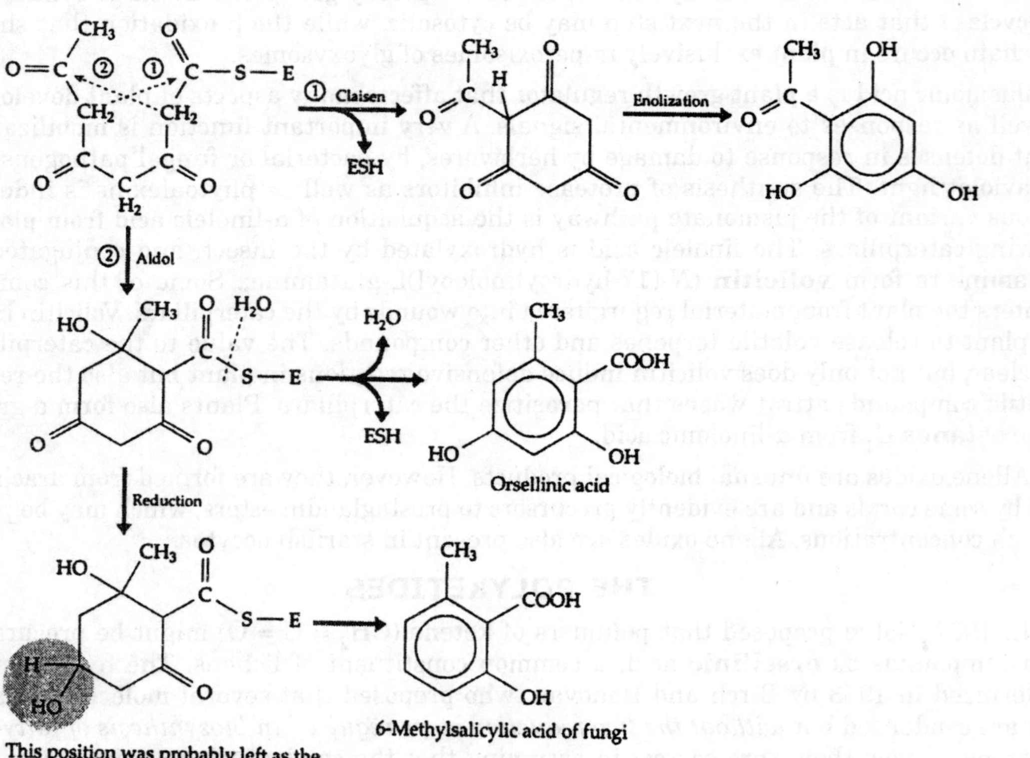

① Claisen
ESH
Enolization
② Aldol
H_2O
ESH
Orsellinic acid
Reduction
6-Methylsalicylic acid of fungi
This position was probably left as the alcohol during the original chain synthesis

Fig. 8.10. Postulated origin of orsellinic acid and other polyketides.

By allowing a few variations in the basic polyketone structure, the biosynthesis of a large number of unsual compounds can be explained Extra oxygen atom can be added by hydroxylation and methyl groups may be transferred from *S*-adenosymethionine to form methoxyl groups. Occasionally a methyl group may be transferred directly to the carbon chain. Glycosyl groups may also be attached. Many starter pieces other than acetyl-CoA may intiate polyketide synthesis. These include the branched-chain acids of nicotinic and benzoic acids,

4-coumaroyl-CoA, and a 14:1 Δ^9-ACP. The last of these starter pieces is formed by desaturation of the corresponding 14:0-ACP and is converted via poloyketide synthesis to one of a family of **anacardic acids,** which provide pest resistance to a variety of dicotyledenous plants. The CoA derivative of malonic acid amide is the starter piece for synthesis of the antibiotic **tetracycline** as indicate. Polyketide origins of some other antibiotics are also indicated in this figure.

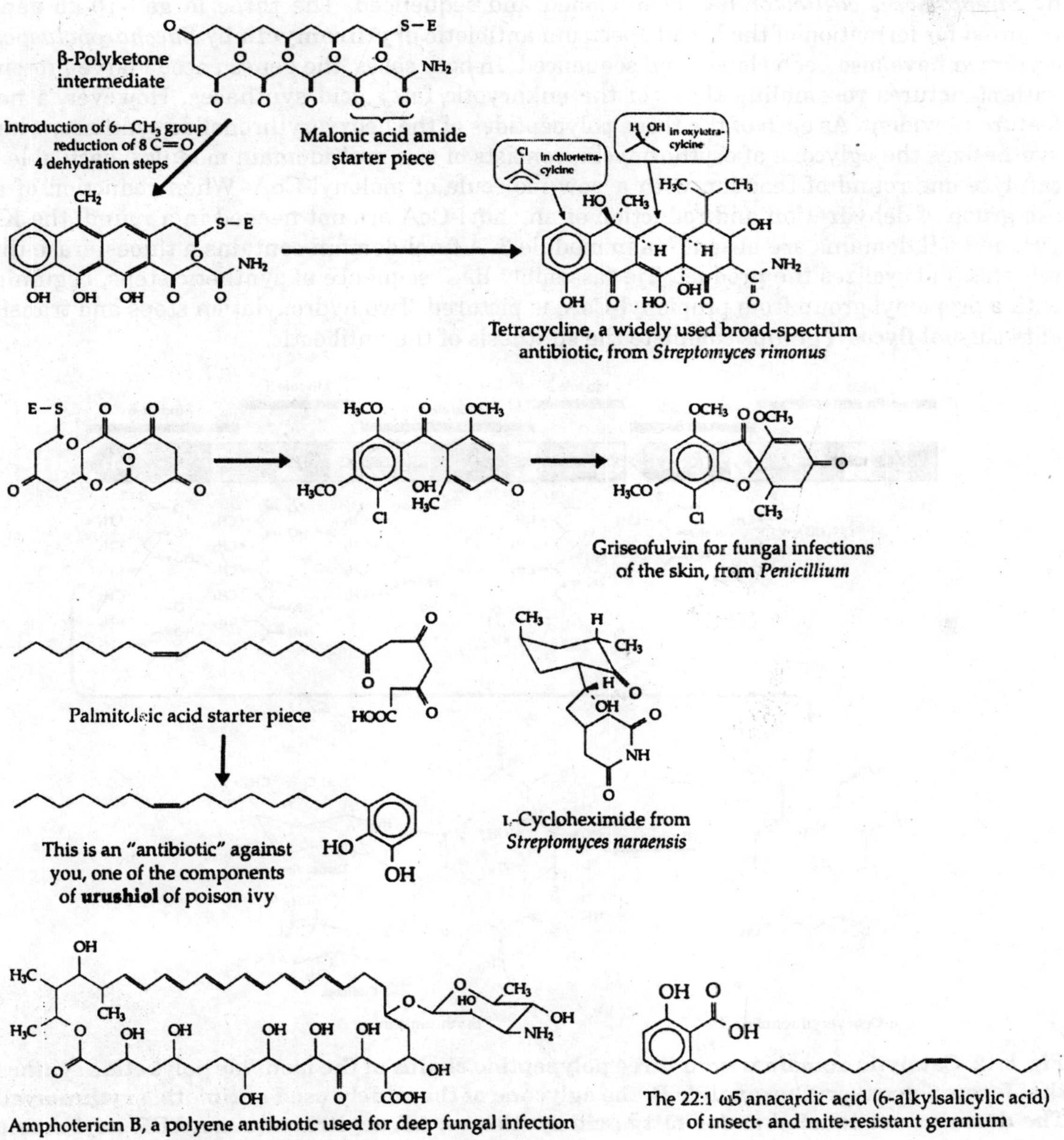

Fig. 8.11. Some important polyketide antibotics and plant defensive compounds.

The cloning and sequencing of genes for enzymes involved in synthesis of polyketides of fungi and actinomycetes has shown that these enxymes are closely related to fatty acid sunthases and, like the latter, have a multidomain structure. The possibilities of engineering these genes, together with the urgent need for new antibiotics, has led to an explosion of information about polyketide synthases.

A 26-kb gene cluster encoding enzymes for synthesis of the blue antibiotic **actinorhodin** by *Streptomyces coelicolor* has been cloned and sequenced. The three large ~10-kb genes required for formation of the broad-spectrum antibiotic **erythromycin** by *Saccharopoluspora erythraea* have also been cloned and sequenced. In botb cases, the genes encode large proteins with structures resembling those of the eukaryotic fatty acid synthases. However, a new feature is evident. As each of the three polypeptides of the deoxyerythronolide synthase, which synthesizes the aglycone of erythormycin, consists of two multidomain modules, each able to catalyze one round of reaction with a new molecule of molonyl-CoA. When reduction of an oxo gruop of dehydration and reduction of an enoyl-CoA are not needed in a round, the KR, DH, and ER domains are absent (as in module 3. A final domain contains a thioesterase that releases and cyclizes the product. Tje "assembly line" sequence of synthetic steps, beginning with a propionyl group from propionyl-CoA, is pictured. Two hydroxylation steps and transfer of two usual flycosyl groups complete the sunthesis of the antibootic.

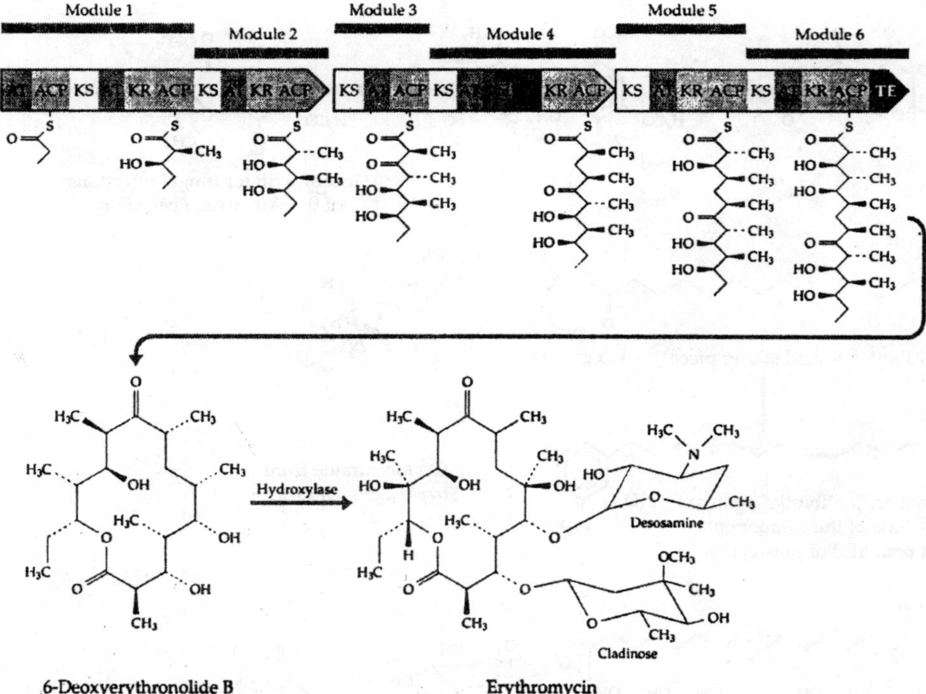

Fig. 8.12. Catalytic domain within three polypeptide chains of the modular polyketide synthase that forms 6-deoxyerythoronolide B, the aglycone of the widely used antibiotic erythromycin. The domains are labeled as for fatty acid synthases; AT, acyltransferase; ACP, acyl carrier protein; KS, β-ketoacyl-ACP synthase; KR, ketoreductase; DH, dehydrase; ER enoylreductase. TE, thioesterase.

Other medically important polydetides include the antibootics **doxorubicin** (14-hydroxydaunomycin; rifamycin, and the antifungl **pimaricin, griesofulvin,** and amphotericin, the HMG-CoA reductase inhibitor **lovastatin,** the 2-butanyl-4-methylthreonine of cyclosporin A, and other immunosuppressants such as **repamycin.** Many characteristic plant products, including **stilbenes** and **chalcones**, are polyketides. A veriety of different polyketides serve as phytoalexins. Some such as **aflatoxin** are dangerous toxins. Ants ans ladybird beetles make toxic polyamine alkaloids using a polyketide pathway.

Avermectin, a widely usxed antibiotic against canine heartworms, is formed by a polyketide synthase with an unusually broad specificity for starter units. More than 40 alternative carboxulic acids are accepted. By grafting the first multidomain module of the erythromycin-forming synthase onto the wide-specificity loading module of the avermectin forming sunthase, a whole new series of antibiotics have been createe. This only one of many steps being taken to create new aliphatic and aromatic, liner and macrocuclic polyketides by genetic engineering. Combinatorial biosynthesis is also being developed and has even been discovered in nature.

LIPOCALINS, FATTY ACID-BINDING PROTEINS, AND LIPOPHORINS

Small hydrophobic molecules, which might easily bind in biologically undesirable ways, are chaperoned in animals, plants, and bacteria by binding proteins that provide hydrophobic cavities or crevices appropriate for holding these molecules in readily releasable forms. The **lipocalins,** most of which are extracellular proteins, have a conserved structural motif consisting of an 8-stranded β barrel arranged as two stacked orthogonal sheets with a C-terminal α helix that blocks one end. The other end is able to open and allow a small hydrophobic molecule to bind in the internal cavity. Only three short amino acid sequences are conserved within a large family of lipocalins which includes **plasma retinol-binding protein**, mammalian **odorant-binding proteins, α-lactalbumin, apolipoprotein D**, and the blue biliverdinbinding protein **insecticyanin** of insect hemolymph. Most lipocalins are soluble, but some such as the plasma **α1-microglobulin**, which plays a role in the immune system have additional functions that require them to bind to other proteins or to cell surfaces. The **gelatinase-associated lipocalin** of human neutrophils binds bacterially derived *N*-formylpeptides that act as chemotactic agents and induces release of materials from intracellular granules. A few lipocalins have enzymatic activity. For example, **prostaglandin D synthase** is both an enzyme and a carrier of bile pigments and thyroid hormones. Most lipocalins have been found in higher animals, but at least a few bacterial proteins belong to the family. One is the 77-residue *E. coli* outer membrane lipoprotein.

A related family of proteins are represented by **fatty acid-binding proteins** and by the intracellur **retinol**-and **retiinoic acid-binding proteins.** These are 10-stranded antiparallel β-barrels with two helices blocking an end a third group of lipid-binding proteins have a four-helix bundle structure. They include the insect **lipophorins**, which transport diacylglycerols in the hemolymph (see main nonspecific lipid carriers of green plans. An 87-residue four-helix protein with a more open structure binds acylcoenzyme a molecules in liver.

A small 98-residue sterol-binding protein from the fungus *Phystophthora,* and agriculturally important plant plathogen, has a very different folding pattern. The sterol binds into a cavity formed by six helices and two loops. the protein is not only a sterol carrier but an **elictin,** which induces a defensive response in the invaded plant. The function of the protein for the invader may be to acquire sterols for the fungus, which is unable to synthesize them.

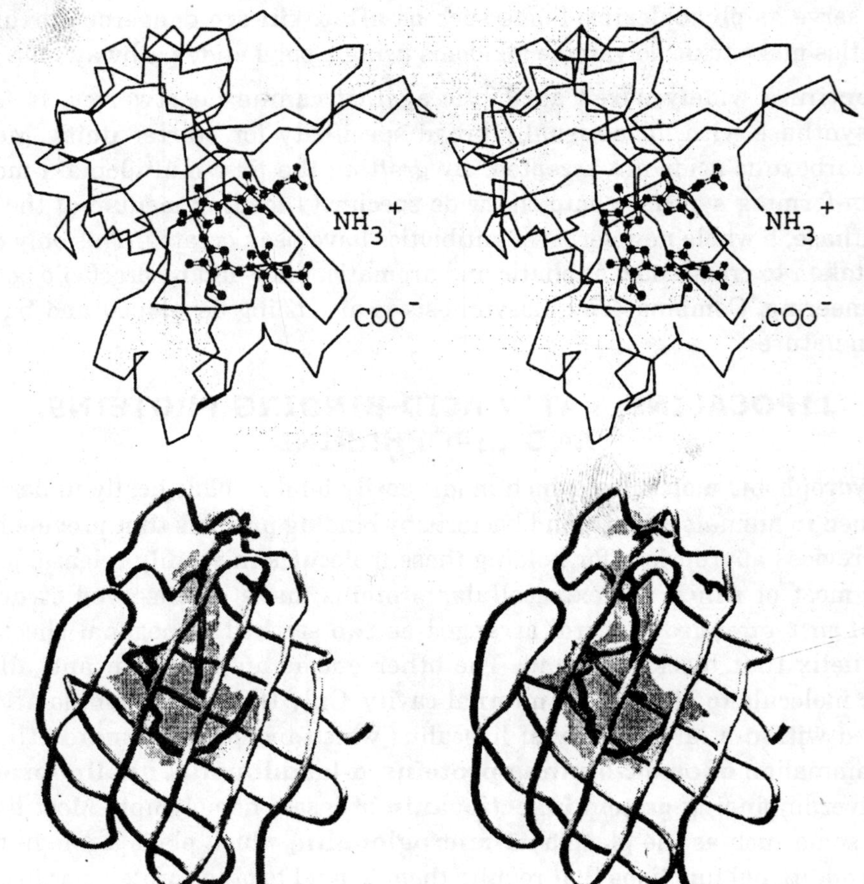

Many larger lipid carrier proteins are known. the 476-residue plasma cholesteryl ester transfer protein is discusses briefly in. Plasm phospholipid transfer proteins are of similar size. A 456-residue humman phospholipid-binding protein interacts with the lipopolyaccharide of the surfaces of gram-negative bacteria and participates in the immune response to the bacteria. It has an elongated boomerang shape with two cavities, both of which bind a molecule of phosphytidylcholine. Other plasma lipid transfer proteins may have similar structures.

THE ESSENTIAL FATTY ACIDS

In 1930, Geroge and Mildred Burr reported that the C18.2 **linoleic acid,** a fatty acid of exculsively plant orgains, cured a disease a dissease condition observed in rats raised on a highly pyrified fat-free diet. These animals grew poorly, developed a scaly dermatitis, and suffered kidney damage and impaired ertility. The sumptoms could be prevented if 1% of the dietary energy was provided by linoleic acid. This C18.2 fatty acid can be converted in animals into a series of other fatty acids by chain elongation and desaturation. All of this series have a double bonds is carbon atoms from the $-CH_3$terminus and form an ω6 (or *n*-6) family.

$$\omega 6 : C18{:}2 \rightarrow C18{:}3 \rightarrow C20{:}3 \rightarrow C20{:}4 \rightarrow C22{:}4 \rightarrow C22{:}5$$

Arachidonate

$$\begin{array}{cc} \downarrow & \downarrow \\ PGE_1, & PGE_2, \\ PGE_{1\alpha} & PGE_{2\alpha} \end{array}$$

The Major known essential founction of linoleic acid is cinversion to the C20%4 **arrchidonic acid,** the major precursor to prostaglandins and other prostaboid compounds (Section D). This conversion occurs in infants as well adults, but the reate may not always be adeqate, and archidonic acid is usually classified as an essential fatty acid. It is not clear wheather linolenic acid has aby essential role of its own. However, while archidonic acid can be converted into the prostaglandins designated PGE_2 and $PGF_{2\alpha}$, linoic acid can also give rise,. via the C20:30 **dihmolinolenic acid,** to PGE_1 and $PGF_1\alpha$ the C18:3 **α-lonolenic acid,** another plant acid, can partially replace linolenic acid and can be converted into PGE_3 and $PGF_{3\alpha}$.

$$\omega 6 : C18{:}3 \rightarrow C18{:}4 \rightarrow C20{:}4 \rightarrow C20{:}5 \rightarrow C22{:}5 \rightarrow C22{:}6$$

$$\begin{array}{ccc} \downarrow & \downarrow & \downarrow \\ PGE_1, & PG_4 & PG_5 \\ PGE_{1\alpha} & & \end{array}$$

Thus, it is not surprising that three acids are not completely equivalent.

Recent interest has focused on the C20.5 **eicosapentaenoic acid** (EPA) and the C22.6 **docosahexenoic acid** (DHA). These ω3 (or *n*-3) polyunsaturated acids are formed fromlinolenic acid by marine algae and are found in fish olis. The C22.5 and C22.6 acids can be converted to prostaglandins of the PG_4 and PG_5 series. DHA together with the ω6 C 22.4 acid consititutes over 30% of the fatty acids in brain phospholipids. In the retina DHA accounts for over 60%

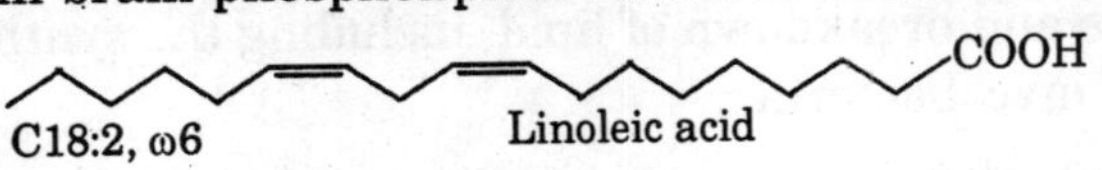

COOH

C18:2, ω6 Linoleic acid

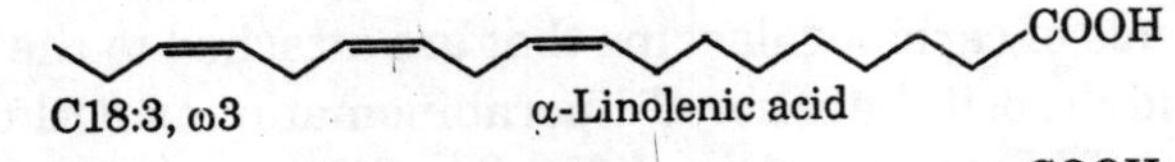

COOH

C18:3, ω3 α-Linolenic acid

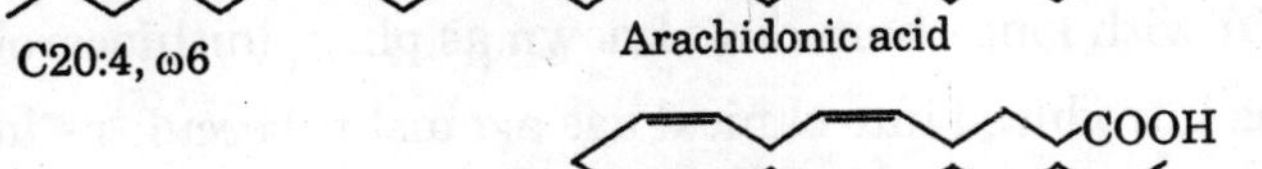

COOH

C20:4, ω6 Arachidonic acid

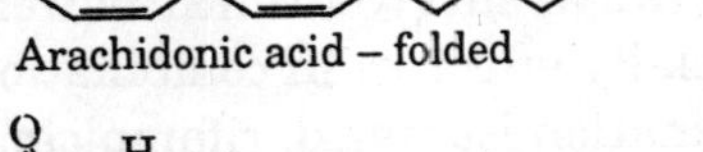

COOH

Arachidonic acid – folded

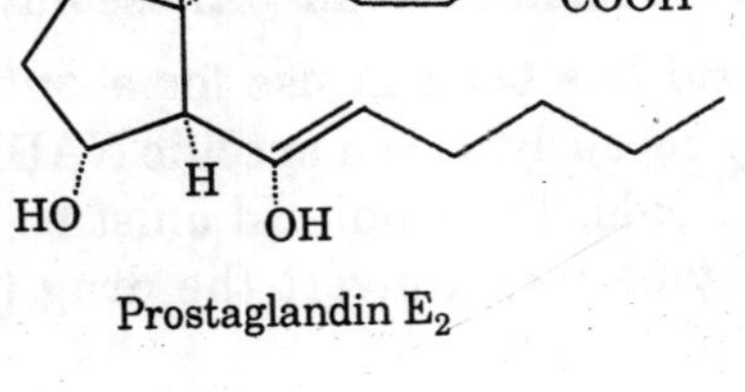

O H COOH H HO OH

Prostaglandin E_2

of the total in the rod outer segments DHA may be formed in the human body from α-linolenic acid obtained from plant sources. However, the rate of synthesis may be inadequate especially in infants and in persons of old age. Deficiency of either EPA or DHA may lead to poor brain development during prenatal and infant life. Formation of neq sunapses between neurons as well as growth of new neurons in some areas of the brain is associated with thinking and with memory formation. Therefore, a lack of the essential ω3 and ω6 fatty acids may contribute to mental deterioration in older adults. Eskimo populations, which consume large amounts of fish, have a very low incidence of coronary heart disease. An inverse relation between fish fonsumption and heart disease has also been demonstrated in other populations. Inclusion of fish oils to 20-30% of total caloric intake in the diet causes a marked dercease in plasma triaculglycerols and very low density lipoproteins (VLDL). This effect has been attributed to the altered composition of the prostanoid compounds known as thromboxanes and prostacyclins (PGI). For similar reasons tha ω3 acids may have an antinflammatory effect. Administration of fish oil to patents with kidney disease has proved beneficial and may decrease risk of some cancers. However, a diet high in ω3 fatty acids has also been reported to increase cancer risk. Long-chain ω3 fatty acids may protect against sudden death from heart disease. They may promote lateral phase separation within membranes to form regions low in cholesterol.

TUBERCULOSIS

As many as one-third of the inhabitants of the earth are infected by *Mycobacterium tuberrculosis.* For most of the infection is dormant, But in some the slow growing bacteria cause a progressive and deadly destruction of the lungs. There are still about three million death annually, and *M. tuberculosis* now, as in the past, kills more people then any other pathogen. The development of drug-resistant strains of the bacteriun and the threat of a world-wide resurrence of tuberculosis has spurred new efforts to understand the unusual metabolism of mycobacteria and to develop new drugs. The complete 4.4 million base-pair sequence of the circular genome is known. An unusually large fraction of the genes encode enzymes involved in synthesis and breakdown of lipid, including the synthesis of **mycolic acids** which are characteristic of mycobacteria.

The mycobacterial cell wall, diacussed in Chapter 8, contains mycolic acids bound covalently at the nonreducing ends of carbinogalactans that are attached to the inner peptidoglycan layer, as well as phosphatidylinositol-anchored liporabinomannans. Other unusual lipids that are also present and accound for some of the difficulty of treatment with antibiotics include esters of **mycoceronic acid** with long-chain diols known as phenolhithiocerols and pithiocerols.

Streptomycin was introduced into clinical use against tuberculosis in about 1943. However, resistant mutants always survived until newer drugs were developed. Isonicotinylhyazide (**isoniazid**) is eepecially effective in combinations with suitable entibiotics and other drugs. The four-drug combination isoniazid, rifampicin, pyrazinamide, and ethambutol is ofter used. Nevertheless, bacteria resistant to all of these have developed.

Although isoniazid has been in use for about 45 years, the enzyme that it inhibits has been recognized only recently. It is a specific NADH-dependent **enoyl reductase** involved in synthesis of mycolic acids. The isoniazid must be activated by action of a bacrerial catalase-peroxidase. This enzyme may convert the drug to a reactive redical that combines with a

NADH-derived redical to form an adducr in the active site of the enzymes. One possible reaction sequence follows. However, the mechanisms are not clear.

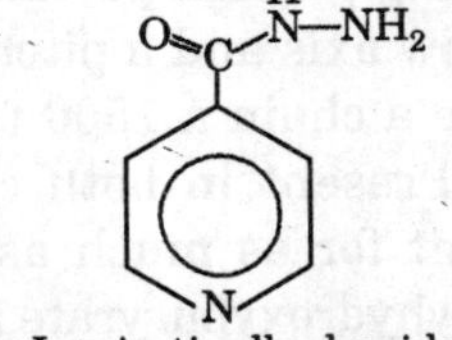

Isonicotinylhydrazide (Isoniazid)

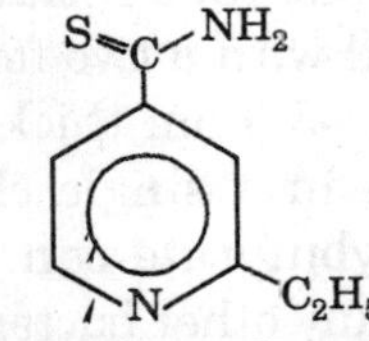

Ethionamide

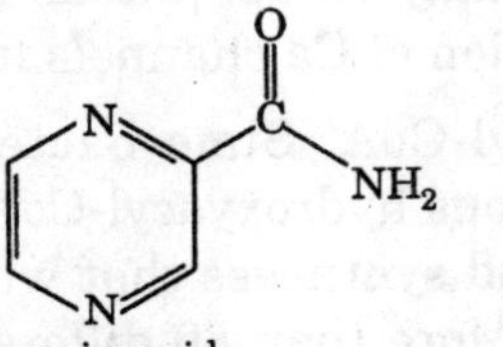

Pyrazinamide

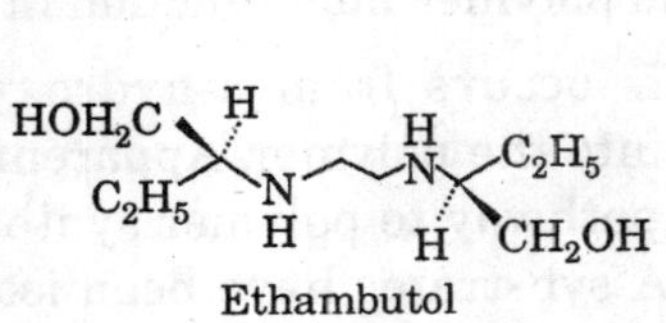

Ethambutol

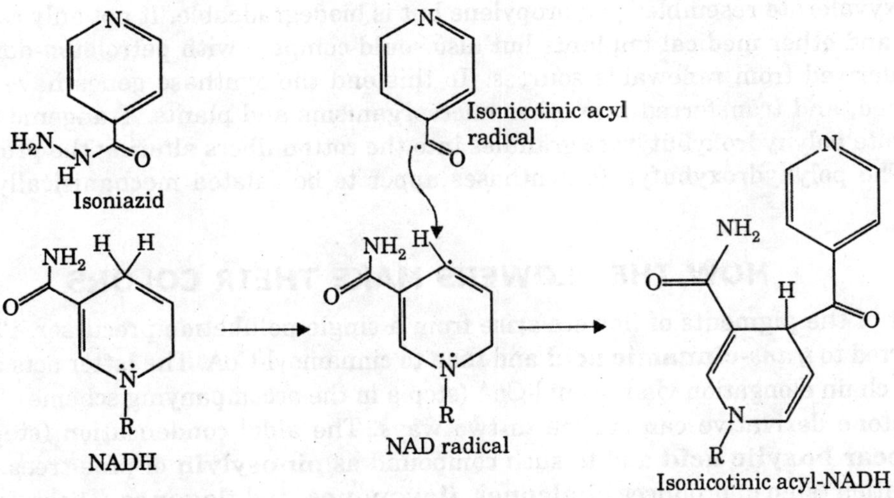

Isoniazid

Isonicotinic acyl radical

NADH

NAD radical

Isonicotinic acyl-NADH

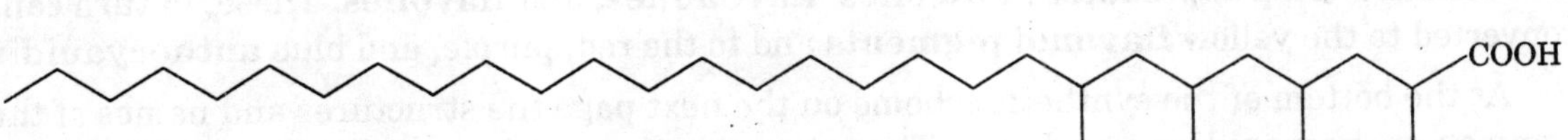

2R,4R,6R,8R-Tetramethylocatacosanoic acid (mycocerosic acid)

POLY-β-HYRDOXYBUTYRATE AND BIODEGRADABLE PLASTICS

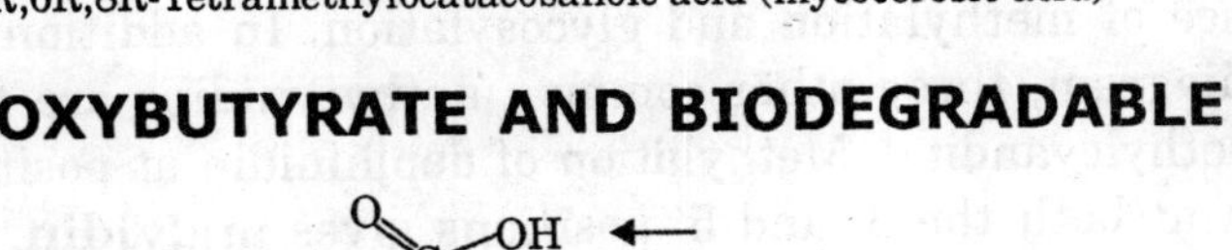

or longer alkyl groups

Linking groups in polymer

D-β-Hydroxybutyric acid

The important bacteria storage material polyhydroxybutyric is related metabolically and structurally to the lipids. This highly reduced polymer is made up of D-β-hydroxybutyric acid units in ester linkage, about 1500 residues being present perchain. The structure is that of a compact right handed coil with a two fold screw axis and a pitch of 0.60 nm. Within it oftern occurs in thisn limallage ~5.0 nm thick. Since a chain a 1500 residues stretches to 440 nm, there must be ~88 folds in a single chain. Present in both cytoplasmic granules and in membranes, polyhydroxybutyrate can account for as much as 50% of the carbon of some bacteria. In *E. coli* and many other bacteria polyhydroxyburyrate is present is a loeer molecular mass form bound to calcium polyphosphates, proteins, or other macromolecules. It has also been extracted from bovine serum albumin and may be ubiquitous in both eukaryotes and prokaryotes. The polymer may function in formation of Ca channels in membranes.

Biosynthesis occurs from 3-hydroxybutyryl-CoA. Sime bacteria incorporate other β-hydrozyacids into the polymer. Apparently various hydroxyacyl-CoAs can be diverted from the β oxidation pathway to polymer synthesis, and synthases that will accept a variety of β-hydroxyacyl-CoA sybstrates have been isolated. More than 80 different hydroxyacyl groups can be incorporated into the polymer. A bacterially produced of β-hydroxybutyrate and β-hydroxyvalerate resembles polypropylene but is biodegradeable. It not only can be used for sutures and other medical implants but also sould compete with petroleum-derived plastics and be derived from renewable sources. To this end the synthase genes have been cloned, engineered, and transferred into other microorganisms and plants. Trangenic cotton plants incorporate polyhydroxybutyrate granules into the cotton fibers altering the properties of the fibers. The polyhydroxybutyrate synthases apper to be ralated mechanisically to bacterial lipases.

HOW THE FLOWERS MAKE THEIR COLORS

Most of the pigments of flowers arise from a single poluketide precursor. Phenylalanine is converted to *trans*-**cinnamic acid** and then to cinnamoyl-CoA. The latter acts as the starter piece for chain elongation via malonyl-CoA (step a in the accompanying scheme). The resulting β-polyketone derivative can cuclize in two ways. The aldol condensation (step *b*) leads to **stilbenecar boxylic acid** and to such compound as **pinosylvin** of pine trees. The Claisen condensation (step *c*) produces **chalcones, flavonones,** and **flavones.** These, in turn can be converted to the yellow **flavonol pigments** and to the red, purple, and blue **anthocyanidins.**

At the bottom of the synthetic scheme on the next page the structures and names of three common anthocyanidins are shown. The names are derived from those of flower from which they have been isolated. The colors depend upon the number of hydroxyl group and on the presence of absence of methylation and glycosylation. In addition to the three pigments indicated in the diagram, three other common anthycynidins are formed by methylation. **Preonidin** is 3-methylcyandine. Methylation of deplhinidin at position 3′ yiels **petunidin,** while methylation at both the 3′ and 5′ positions gives **malyidin**. There are many other anthocyanidins of more limited distribution. Anthocuanidins are nearly insoluble, but they exist in plants principally as glycosides known as **anthocyanins**. The number of different glycosides among the many species of flowing plants is large. Both the 3 and 5 –OH groups may be glycoslated with GLc, Gal, Rha, Ara, and by a large variety of oligosaccharides. The colors of the anthocyanins very from red to violet and blue and are pH dependent. For example,

cyanin (diglucosyl cyanidin) is red in acid solution and becomes violet upon dossociation of the 4′- hydroxyl group:

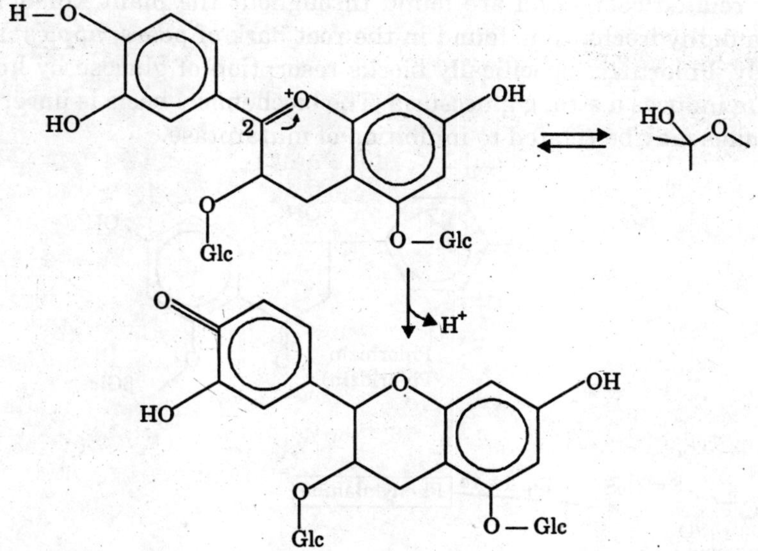

H—O
HO
2 +O
OH
HO O
O—Glc
Glc
H+
O
HO
O
OH
Glc
O—Glc

Dissociation of the 7–OH generates an anion with an extended conjugated π electron system, which will favor absorption of long-wavelength light and a blue colour. Notice that a large number of resonance structures can be drawn for both the anthocyanin and the dissociated forms. Formation of complexes of Mg^{2+} or other metal ions with the 4′–O and adjacent OH group may also stabilize blue colored forms.

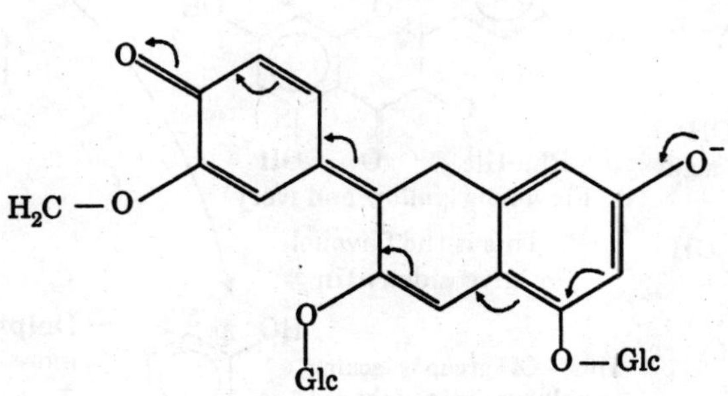

O
H_2C —O
O^-
O
Glc
O—Glc

Most blue flower pigments are based on delphinidin, but "heavenly blue" of the morning glory is a peonidin with a complex caffeolylglucose-containing glycosyl group on the 3-position. Its blue colour has been attributed to the relatively high pH of ~7.7 vacuples. The aromatic within the glocosyl group of this and other complex anthocyanins may fold over the primary chromophore and establize the coloured forms. A competing reaction, which is indicated in green on the forst structure in this box, it is addition of a hydroxy ion at C-2 to give a nearly colourless adduct.

The yellow pigments of flowers are usually flavonols. The most common of all is **rutin,** The 3α rhamnosyl derivative of **quercitin**. An extraordinary number of other foavonols, flavones, and related compound are found throughout the plant kingdom. One of these is **phlorhizin,** a didhydrochalcone found in the root bark of pears, apples, and other plants of the rose family, Phlorhizin specifically blocks resorption of glucose by kidney tubyles. As a result, the drug induces a strong glucosuria. The biochemical basis is uncertain, but the action on kidney tubules may be related to inhibition of mutarotase.

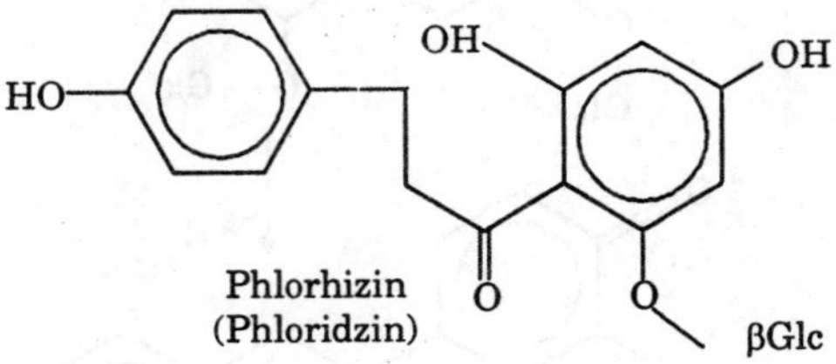

Phlorhizin
(Phloridzin)

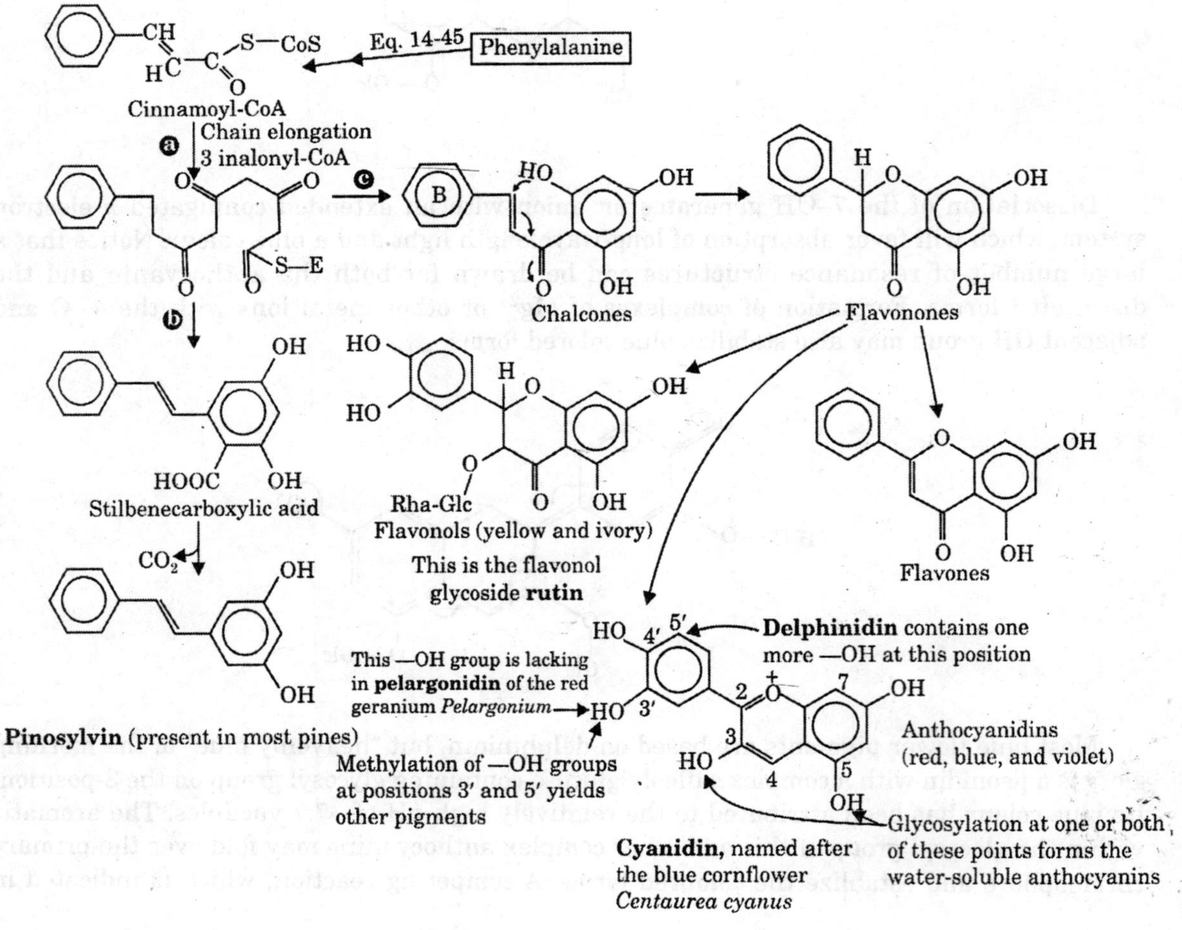

Cinnamoyl-CoA
Eq. 14-45 Phenylalanine
Chain elongation
3 inalonyl-CoA
Chalcones
Flavonones
Stilbenecarboxylic acid
Flavonols (yellow and ivory)
This is the flavonol glycoside **rutin**
Flavones
Pinosylvin (present in most pines)
This —OH group is lacking in **pelargonidin** of the red geranium *Pelargonium*
Methylation of —OH groups at positions 3′ and 5′ yields other pigments
Delphinidin contains one more —OH at this position
Anthocyanidins (red, blue, and violet)
Glycosylation at one or both of these points forms the water-soluble anthocyanins
Cyanidin, named after the blue cornflower *Centaurea cyanus*

The flavone glycoside **hesperidin** makes up 80% of the dry weight orange peels. It has been claimed (but not proved) that this compound, also known as **vitamin P** and **citrus bioflavonoid,** is essential to good health. Another flavone, **maysin,** is a resistance factor for the corm earworm and is present in silks of resistant strains of *Zea mays*.

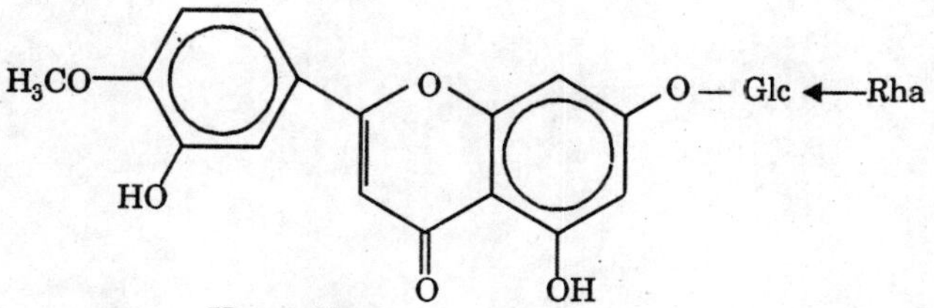

Hesperidin

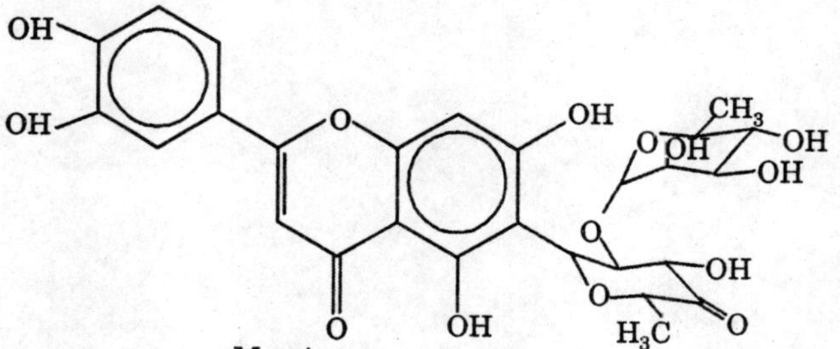

Maysin

Index